Synaptic Plasticity in Neurodegenerative Disorders

This book explores the pivotal role of synaptic plasticity in the pathogenesis, progression, and potential treatment of neurodegenerative disorders. The initial chapter provides an in-depth understanding of the complexity and impact of neurodegenerative conditions. It discusses the association of mitochondrial dysfunction, epigenetic influences, and neuroinflammation with synaptic plasticity in neurodegenerative diseases. The following chapters review the dynamic changes that occur at the cellular and synaptic levels in Parkinson's disease, Alzheimer's disease, and Huntington's disease, paving the way for innovative therapeutic strategies. Furthermore, the book presents various computational tools and methodologies essential for enhancing our understanding of synaptic plasticity. It examines the transformative role of artificial intelligence tools in addressing synaptic impairment across various neurodegenerative diseases.

- Discusses the role of synaptic plasticity in neurodegenerative diseases, shedding light on how dynamic changes occur at the cellular and synaptic levels
- Explores the transformative role of artificial intelligence tools in addressing synaptic impairment across various neurodegenerative diseases
- Provides a comprehensive overview of neurodegenerative disorders, including pathogenesis, etiology, and treatment strategies
- Presents tools and techniques used to simulate the complex system biology of synaptic plasticity
- Examines the role of computational neuroscience in understanding and potentially treating conditions such as multiple sclerosis and amyotrophic lateral sclerosis

Toward the end, the book explores the role of synaptic impairment and computational neuroscience in understanding and potentially treating conditions such as multiple sclerosis and amyotrophic lateral sclerosis. With its multifaceted approach, this book serves as a useful resource for researchers, clinicians, and students in the fields of neuroscience, computational biology, and neurology.

Synaptic Plasticity in Neurodegenerative Disorders

Edited by
Sangeeta Singh, Sachchida Nand Rai,
and Santosh Kumar Singh

CRC Press
Taylor & Francis Group
Boca Raton London New York

CRC Press is an imprint of the
Taylor & Francis Group, an **informa** business

Designed cover image: Shutterstock

First edition published 2025
by CRC Press
2385 NW Executive Center Drive, Suite 320, Boca Raton FL 33431

and by CRC Press
4 Park Square, Milton Park, Abingdon, Oxon, OX14 4RN

CRC Press is an imprint of Taylor & Francis Group, LLC

ISBN: 978-1-032-73528-3 (hbk)
ISBN: 978-1-032-73530-6 (pbk)
ISBN: 978-1-003-46464-8 (ebk)

DOI: 10.1201/9781003464648

Typeset in Times
by KnowledgeWorks Global Ltd.

Contents

Preface

In the vast landscape of neuroscience, understanding the intricate workings of the human brain remains one of the most challenging and compelling pursuits. Amidst this pursuit lies a particularly pressing area of inquiry: neurodegenerative disorders. These conditions, characterized by the progressive degeneration of nerve cells, strike at the core of human cognition, memory, and movement, posing profound challenges to individuals and societies worldwide.

This book, *Synaptic Plasticity in Neurodegenerative Disorders*, delves into the nexus of synaptic plasticity and computational neuroscience within the realm of neurodegenerative diseases. Synaptic plasticity, the ability of synapses to strengthen or weaken over time in response to activity, serves as a cornerstone in understanding the mechanisms underlying learning, memory, and cognition. In the context of neurodegeneration, exploring synaptic plasticity unveils crucial insights into disease progression, pathophysiology, and potential therapeutic interventions.

The chapters within this volume traverse a diverse array of topics, each offering a unique perspective on the intricate interplay between synaptic plasticity, computational neuroscience, and neurodegenerative disorders. From foundational overviews of neurodegenerative diseases to in-depth explorations of specific conditions such as Alzheimer's disease, Parkinson's disease, Huntington's disease, and amyotrophic lateral sclerosis (ALS), this book illuminates the multifaceted nature of synaptic dysfunction in these disorders.

Chapter 1 serves as an introduction to neurodegenerative diseases, providing a comprehensive overview of their etiology, epidemiology, and clinical manifestations. Subsequent chapters delve into the role of synaptic plasticity in various neurodegenerative conditions, highlighting its significance in Alzheimer's disease (Chapter 3), Parkinson's disease (Chapter 7), Huntington's disease (Chapter 8), ALS (Chapter 11), and more.

Moreover, this volume explores the intricate connections between synaptic plasticity and other pathological mechanisms implicated in neurodegenerative disorders. Chapters such as Chapter 5 shed light on the relevance of synaptic plasticity in conjunction with mitochondrial dysfunction, while Chapter 10 delves into the therapeutic potential of targeting synaptic plasticity to mitigate neuroinflammation.

Furthermore, this book delves into the methodological approaches, challenges, and opportunities in studying synaptic plasticity and mental health (Chapter 6), as well as the functional and molecular underpinnings of neural plasticity in conditions such as multiple sclerosis (Chapter 13).

As editors of this volume, we are deeply grateful to the contributors who have shared their expertise and insights, enriching the discourse on synaptic plasticity and computational neuroscience in neurodegenerative disorders. It is our sincere hope that this book will serve as a valuable resource for researchers, clinicians, students, and anyone seeking a deeper understanding of the intricate interplay between synaptic plasticity and neurodegeneration.

Acknowledgments

In embarking on the journey of compiling this book, we are humbled by the support, guidance, and contributions from a multitude of individuals whose collective efforts have made this endeavor possible.

First and foremost, we express our deepest gratitude to the Almighty for granting us the wisdom, strength, and perseverance to undertake this scholarly pursuit. His divine guidance has been our guiding light throughout this endeavor.

To our mothers, the epitome of unconditional love and unwavering support, we owe a debt of gratitude that words cannot adequately express. Their encouragement, sacrifices, and boundless affection have been the cornerstone of our lives and academic endeavors.

We extend heartfelt appreciation to our fathers, whose wisdom, guidance, and encouragement have been instrumental in shaping our intellectual pursuits and moral compass. Their unwavering belief in our potential has been a source of inspiration and motivation.

We are immensely grateful to all the contributors from India and abroad whose scholarly insights, expertise, and dedication have enriched the pages of this book. Your invaluable contributions have expanded the horizons of knowledge in the fields of synaptic plasticity, computational neuroscience, and neurodegenerative disorders.

Finally, we would like to thank our families, friends, and colleagues for their unwavering support, encouragement, and understanding throughout the duration of this project. Your belief in us has been a source of strength and inspiration.

In conclusion, we acknowledge with gratitude all those who have contributed to the creation of this book, whether through their expertise, support, or encouragement. May this collaborative effort serve as a beacon of knowledge and inspiration for future generations of researchers and scholars in the fields of neuroscience and neurodegenerative disorders.

The Editors

Dr. Sangeeta Singh (MNASc) serves as an Assistant Professor at the Department of Applied Science, Indian Institute of Informational Technology, Allahabad (IIIT-A), located in Prayagraj, Uttar Pradesh (UP), India. Prior to her tenure at IIIT Allahabad, she was employed at ICAR-Indian Veterinary Research Institute, Izzatnagar, Bareilly (UP), India as Post-Doctoral Fellow. She obtained her Ph.D. (Life Science) in 2010. Dr. Singh's research interests encompass neurodegenerative illness and regenerative medicine aimed at facilitating neuronal repair and regeneration. She has obtained multiple research funding from the Science and Engineering Research Board (SERB) from the Department of Science and Technology (DST), India. She has authored more than 60 publications in high-impact journals, in addition to book chapters and conference proceedings. Dr. Singh acts as a reviewer for several esteemed journals. including the Journal of Neuroscience Research, Molecular Neurobiology, Cellular and Molecular Neurobiology, etc. Dr Singh awarded with the prestigious, The National Academy of Sciences, India (NASI)-Membership 2023. Dr. Singh is a Life Member of several societies, including the Indian Academy of Neurosciences, The Association of Microbiologists of India, and Society for Biomaterials and Artificial Organs India, among others.

Dr. Sachchida Nand Rai is currently serving as a Research Associate (ICMR-RA) at the Centre of Experimental Medicine and Surgery, Institute of Medical Sciences, Banaras Hindu University (life member BHU), Varanasi, a position he has held since January 27, 2023. Dr. Rai earned his Ph.D. from the Biochemistry Department at BHU in 2018. His research focuses on exploring the anti-Parkinsonian properties of various herbal plants such as *Mucuna pruriens, Tinospora cordifolia,* and *Withania somnifera*, as well as their bioactive components like ursolic acid, betunilic acid, and chlorogenic acid, particularly in toxin-induced Parkinsonian mouse models. Banaras Hindu University consistently ranks among India's top universities for life sciences research output and quality. Prior to his current role, Dr. Rai served as a Post-Doctoral Fellow at the Centre of Biotechnology, University of Allahabad, Prayagraj, India, from January 27, 2020, to January 26, 2023, accumulating three years of experience. Dr. Rai has made significant contributions to the field, with impressive publications in reputable journals. He serves as an Academic Editor for *BMC Research Notes* and Associate Editor for *3-Biotech* (Springer Nature), *Molecular Neurobiology, Oxidative Medicine and Cellular Longevity, Journal of Inflammation Research*, and *Degenerative Neurological and Neuromuscular Disease* (Dove Press). Additionally, he sits on the advisory board of Heliyon Neuroscience. Dr. Rai has edited over 100 manuscripts and reviewed more than 1050 manuscripts for high-impact journals. He has authored two books with renowned publisher Bentham. Dr. Rai's Google Scholar citations stand at 3063, with an h-index of 30. He is a member of the Indian Academy of Neurosciences. His current research involves investigating the neuroprotective properties of commonly used Indian medicinal plants in toxin-induced Parkinsonian mouse models.

Dr. Santosh Kumar Singh is a distinguished figure in the scientific community, serving as a Senior Scientist and Administrative Incharge at the Centre of Experimental Medicine and Surgery, Institute of Medical Sciences, Banaras Hindu University, Varanasi. He obtained his Ph.D. in Botany from the same institution in 1998 and further enriched his expertise as a Visiting Scientist at the University of South Bohemia, Czech Republic. With an extensive career spanning over 23 years in academia and research, complemented by over 14 years of leadership in university administration, Dr. Singh's primary area of expertise revolves around the discovery and development of plant-derived drugs.

Dr. Singh's contributions are profound. He has supervised 12 Ph.D. and 10 MD/MS theses, authored numerous publications in esteemed journals, and edited an international book. He has also held editorial roles in various journals, showcasing his commitment to advancing scientific discourse. His leadership is evident through successfully spearheading multiple research projects funded by prestigious organizations such as DST, DBT, and SERB, India.

Acknowledged for his remarkable achievements, Dr. Singh has received accolades including the National Award of Science and Technology Communication from DST and the Excellence in Science Communication Award from the Scientific & Education Research Society. His contributions extend beyond academia, as he actively engages in scientific awareness initiatives targeting diverse audiences such as school students, farmers, and rural communities.

Notably, Dr. Singh's dedication to uplifting economically disadvantaged regions through science and technology interventions is noteworthy. Serving as a technical advisor for FAARD-Foundation, he mentors over 60 farmer producer organizations, positively impacting the lives of more than 100,000 farmers.

Contributors

Mohammad Asif
Department of Information Technology and Centre for Cognitive Computing
Indian Institute of Information Technology Allahabad
Devghat, Jhalwa, Prayagraj, Uttar Pradesh, India

Meenakshisundaram Balasubramaniam
Department of Geriatrics
Reynolds Institute on Aging
University of Arkansas for Medical Sciences
Little Rock, AR, USA

Princy Choudhary
National Centre for Cell Science
Ganeshkhind, Pune, Maharashtra, India
and
Department of Psychiatry
National Institute of Mental Health and Neurosciences
Bengaluru, Karnataka, India

Surekha Dabla
Department of Neurology
Post Graduate Institute of Medical Sciences PT.B.D. Sharma University of Health Sciences (PGIMS)
Rohtak, Haryana, India

Anushka Ajit Dandawate
Department of Information Technology
Indian Institute of Information Technology Allahabad
Devghat, Jhalwa, Prayagraj, Uttar Pradesh, India

Deepika
Department of Biochemistry
Central University of Haryana
Mahendergarh, Haryana, India

Shrey Dwivedi
Department of Applied Sciences
Indian Institute of Information Technology Allahabad
Devghat, Jhalwa, Prayagraj, Uttar Pradesh, India

Utkarsh Gaharwar
Tech Invention Lifecare Ltd
Mumbai, Maharashtra, India

Anindita Ghosh
Department of Biological Sciences
Indian Institute of Science Education and Research (IISER)
Bhopal, Madhya Pradesh, India

Aradhna Gupta
Department of Biochemistry
University of Allahabad
Prayagraj, Uttar Pradesh, India

Nikhila Khola
Department of Biochemistry
Central University of Haryana
Mahendergarh, Haryana, India

Anoop Kumar Yadav
Department of Computer Science and Engineering
National Institute of Technology
Patna, Bihar, India

Deepak Kumar
Department of Zoology
University of Allahabad
Prayagraj, Uttar Pradesh, India

Nilesh Kumar
IRCP-Biological Data Sciences
University of Alabama at Birmingham
Birmingham, AL

Pawan Kumar Maurya
Department of Biochemistry
Central University of Haryana
Mahendergarh, Haryana, India

Vijaya Nath Mishra
Neurology Professor & Head
Department of Neurology
Institute of Medical Sciences
Banaras Hindu University
Varanasi, Uttar Pradesh, India

Sonu Pahal
Bioinformatics Program
University of Arkansas for Medical Sciences
University of Arkansas at Little Rock
Little Rock, AR

Vishvender Pahal
All India Institute of Medical Sciences
Raebareli, Uttar Pradesh, India

Anita Pal
Department of Zoology
University of Allahabad
Prayagraj, Uttar Pradesh, India

Sangita Paul
CSIR Institute of Genomics and Integrative Biology
New Delhi, India
and
Academy of Scientific and Innovative Research (AcSIR)
Ghaziabad, Uttar Pradesh, India

Sneha Ranjan
Department of Applied Sciences
Indian Institute of Information Technology Allahabad
Devghat, Jhalwa, Prayagraj, Uttar Pradesh, India

Shivani Rath
Movement Disorder
National Institute of Mental Health and Neurosciences
Bengaluru, Karnataka, India

Ram Sagar
Department of Genetic Medicine
School of Medicine
The Johns Hopkins University
Baltimore, Maryland

Arshdeep Kaur Sethi
Department of Medicine
Sri Guru Ram Das University of Health Science
Sri Amritsar, Punjab, India

Bechan Sharma
Department of Biochemistry
University of Allahabad
Prayagraj, Uttar Pradesh, India

Himanshu Singh
Department of Radiology Memorial Sloan Kettering Cancer Center
New York, NY

Sangeeta Singh
Department of Applied Sciences
Indian Institute of Information Technology Allahabad
Devghat, Jhalwa, Prayagraj, Uttar Pradesh, India

Vineeta Singh
Department of Geriatrics
Institute of Medical Sciences
Banaras Hindu University
Varanasi, Uttar Pradesh, India

Diya Srivastava
Department of Information Technology
Indian Institute of Information Technology Allahabad
Devghat, Jhalwa, Prayagraj, Uttar Pradesh, India

Anushree Tripathi
Department of Computer Science and Engineering
National Institute of Technology
Patna, Bihar, India

Somu Yadav
Department of Biochemistry
Central University of Haryana
Mahendergarh, Haryana, India

Part I

Neurodegenerative Diseases and Synaptic Plasticity

1 Neurodegenerative Diseases (ND)

An Introduction

*Aradhna Gupta and Bechan Sharma**
University of Allahabad
Prayagraj, Uttar Pradesh, India

1.1 INTRODUCTION

Neurodegenerative diseases (NDs) affect 15% of the total world population [1,2]. These diseases arise due to gradually progressive, unrepaired loss of neurons occurring in older ages often leading to death [3]. They are characterized by age-dependent loss of cognitive skills (dementia), motor functions (ataxia), mental impairment, and debilitation. As we age, we are more susceptible to epigenetic changes, like DNA methylations and histone modification, that affect the brain, increasing the risk of ND [4]. The disease/ND progression depends on genetics and environmental factors such as trauma, drugs, and toxins [5]. Neurons primarily derive energy through cellular respiration, a process that converts glucose into adenosine triphosphate (ATP), which is crucial for cell function. This energy sustains the resting membrane potential, facilitates action potentials, and supports various cellular activities vital for communication and information processing in the nervous system.

Neurologists classify the nervous system into two parts: the central nervous system (CNS) and the peripheral nervous system (PNS). The CNS comprises the brain and spinal cord and is responsible for safeguarding, repairing, and maintaining energy balance in neuronal cells. where neurodevelopmental disorders can affect either or both. The PNS includes cranial-spinal nerves and ganglia, which transmit information from the brain to effector organs. Unlike the CNS, which lacks cell repair capabilities, the PNS possesses intrinsic repair and regeneration abilities. NDs can affect either or both of these systems. They are more common in older adults, but they can impact individuals of any age [6].

Organizations such as the World Health Organization (WHO) and the National Health Service in the United Kingdom have stated that "available resources for neurological services are insufficient in most countries of the world compared with global need for neurological care" and that "Neurological services are not sustainable in their current form and redesign is needed" [7].

Neuro-regeneration may be a hope for individuals suffering from these debilitating diseases. The field of neural tissue engineering focuses on generating new neurons, axons, synapses, myelin, and glial cells. Orchestrated and controlled experiments on neural stem cells found in different regions of the brain, like the subependymal zone, gyrus dentatus, and rostral migration pathway, might shed some light on neuroprotection and regeneration [8–12].

Anthrobots are spheroid-shaped multicellular biological robots (biobots) made from adult human lung cells of size 30–500 μm, having cilia and the ability to move with a speed ranging between 5 and 50 μ/s. These anthrobots can fuse to form a large structure called a superbot which has demonstrated the ability to regenerate neurons [13].

DOI: 10.1201/9781003464648-2

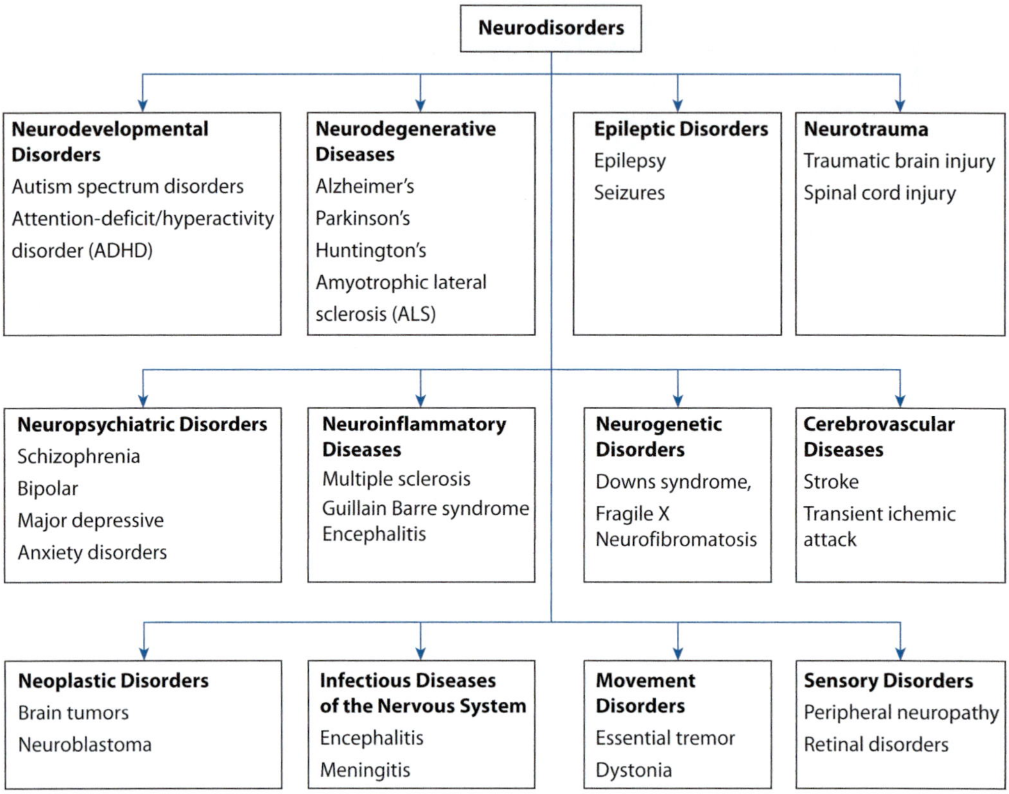

FIGURE 1.1 General classification of neuro diseases based on different types of disorders.

1.2 CLASSIFICATION OF NEURO DISEASES

There are many classifications of NDs. They can be classified on the basis of clinical, anatomical, and protein vulnerability [14–16]. The vulnerable proteins include microtubule associated protein tau (MAPT), Aβ (cleavage of amyloid precursor protein), α-synuclein, prions proteins, transactive response (TAR) DNA binding proteins, and FET proteins. Hereditary disorder associated proteins are ferritin related ND, neuroserpin, and familial cerebral amyloidosis. Cognitive decline and dementia are observed clinically, affecting regions like the neocortical areas, hippocampus, limbic system, and entorhinal cortex. Locomotive disorders encompass hyper- or hypo-kinetic movements and cerebellar dysfunction, involving regions such as subcortical nuclei, diencephalon, cranial nerve nuclei, cerebellar hemispheres, precentral gyrus, and spinal motor neurons. Figure 1.1 illustrates the overall classification of disorders based on these manifestations.

1.3 BRIEF REVIEW OF COMMON NEURO DISEASES

1.3.1 Alzheimer's Disease

Alzheimer's disease (AD) affects the part of the brain that controls emotions, memory, and speech. Cholinergic synapses, which play a crucial role in learning and memory (cognition), are plentiful in the neocortex, limbic system, thalamus, and striatum [17]. There is a loss of cognitive skills and deposition of intracellular neurofibrils tangles (Aβ fibrils) and extracellular p-tau proteins inclusions in the amygdala, cortex, and hippocampus [18,19].

1.3.1.1 Symptoms

Symptoms include trouble in thinking and remembering, even recent events, and brain atrophy. There can be personal changes like depression, mood swings, social withdrawal, anger, distrust, changed sleep patterns, delusions, and wandering.

1.3.1.2 Complications

As this disease affects memory and thought, the patient may not be able to express pain and may suffer from side effects like bowel-related problems leading to uncontrolled bowel movements, constipation, diarrhea, bedsores, and dental problems like tooth decay, plaques, and fractures.

1.3.1.3 Causes

Beta amyloid protein fragments accumulate as plaques along with other cell debris within nerve cells. Simultaneously, tau proteins, responsible for nutrient transport to neurons, undergo shape changes and form neurofibrillary tangles. These processes collectively impair neuronal function and connectivity.

1.3.1.4 Risk Factors

The risk of developing AD rises with advancing age and having a family history. A specific gene variant, apolipoprotein E e4, is predominantly found in patients afflicted with the disease. Downs syndrome XXX (chromosome 21) increases the production of beta amyloid protein whose fragments become plaques that are deposited in the brain, the leading cause of this disease. More females are reported to be suffering from this disease. Mild cognitive impairments slowly may lead to dementia. Alcohol consumption, poor sleep, cardiovascular disease, diabetes, stress, obesity, and less education all have an impact in increasing the chance of this disease.

1.3.1.5 Prevention

Lecture sessions, improved life style, diet, exercise, social activities meditations and regular monitoring of sugar, cholesterol and blood pressure can help in lowering the risk of Alzheimer disease's. Meditation and regular monitoring of sugar, cholesterol and blood pressure can help in lowering the risk of Alzheimer's disease.

1.3.2 Parkinson's Disease

Parkinson's disease (PD) is a persistent degenerative condition that impacts both motor and non-motor neurons. About a 50–60% loss of dopaminergic neurons (nerve cells that produce "dopamine," a chemical messenger resulting in irregular brain activity and movement associated problems like hand tremors/stiffness) in the substantia nigra leads to motor function issues, attributed to the buildup of α-synuclein protein bound to a ubiquitin complex [19,20].

1.3.2.1 Symptoms

Symptoms of PD include rhythmic shaking (tremors) of hands, which may decrease while performing work; slowed movement (bradykinesia); limited range of motion due to painful, rigid movements, accompanied by a stooped body posture and compromised balance; and loss of autonomic movements such as blinking, smiling, and swinging of arms. Speech problems like softness, quickness, or slurring, along with difficulties in writing, are also common.

1.3.2.2 Complications

PD affects cognition, sleep, and the brain parts involved in sensory perception and interoception. In the advanced stage, dementia is seen. There are emotional shifts such as depression, anxiety, fear, and reduced motivation. Additionally, other complications may arise, including issues with swallowing and chewing, including drooling due to difficulty in swallowing; bladder problems like inability to control urine; constipation due to slow digestion; dizziness/light-headedness due to sudden changes in blood pressure; fatigue; pain; difficulty in identifying odors; and decreased sexual desire.

1.3.2.3 Causes

The death of neurons occurs through the aggregation of alpha-synuclein proteins, forming Lewy bodies, which are microscopic markers of PD. Environmental factors like toxins/pesticides exposure, or head injury all contribute in aggravating the disease.

1.3.2.4 Risk Factors

It is commonly observed in individuals around the age of 60. Having a family history of PD raises the likelihood of developing it, with males being more susceptible than females.

1.3.2.5 Prevention

Medications, regular aerobic exercise, and green tea can help prevent PD.

1.3.3 Spinal Cord Injury

The spinal cord is a bundle of nerves within the central canal of bones. It facilitates communication between the brain and the rest of the body, enabling movement and sensation.

1.3.3.1 Symptoms

Signs of spinal cord injury include loss of movement and sensation; impaired bowel or bladder control; breathing difficulties; coughing issues; pain; and paraplegia affecting the trunk, legs, and pelvic organs.

1.3.3.2 Causes

Spinal cord injuries typically result from over bending or bending with a jerk. Causes include sports accidents, falls, violence, infections affecting the spinal cord, arthritis, cancer, inflammation, infections, or disk degeneration in the spine.

1.3.3.3 Complications

Complications of spinal cord injuries include urinary incontinence and uncontrolled bowel movements, loss of sensation below the injured area, changes in blood pressure, muscle tone (flaccidity, spasticity), decreased bone density, inability to cough or breathe, weight loss, muscle thinning, pain, and depression.

1.3.3.4 Risk Factors

Males often suffer from this injury between 16–30 years of age or older than 65, alcoholics, osteoporosis. Osteoporosis or loss of bone mass has been seen in people with spinal cord injuries which may be due to lack of muscle traction, neural factors, elevated levels of adiponectin, leptin. The reasons for males suffering from this diseases are sports, motorcycling and gunshots. (https://spinalcordinjurylawyers.com)

1.3.3.5 Prevention

Don't drink and drive, make sure the water level of a swimming pool is not less than 9 ft deep before diving, prevent falls, take precautions while playing sports.

1.3.4 Brain Tumor

A brain tumor is a growth of cells in the brain or nearby area. It can be benign or malignant. There are many types of brain tumors depending on location.

1.3.4.1 Symptoms

Brain tumor symptoms include headaches (severe and often); feeling of pressure in the head; vomiting; nausea; blurred vision; problems in balance, speech, hearing, and memory; fatigue; hunger; dizziness; and weight gain.

1.3.4.2 Causes and Risk Factors

Risk factors exposure to radiation, causes are cancer that spread from one part of body to other, genetic conditions and mutations in DNA.

1.3.5 Brain Aneurysm

A brain aneurysm, also called a cerebral or intracranial aneurysm, refers to a weakened or thin section of a blood vessel in the brain that expands like a balloon. Aneurysms can be of different sizes and shapes and can rupture, resulting in bleeding into nearby tissues (hemorrhage). This can lead to severe neurological complications or even death.

1.3.5.1 Symptom

A ruptured aneurysm can cause a severe headache if it presses against brain tissues or nerves.

1.3.5.2 Cause

Aneurysms can appear in any part of the brain, but they occur more often near the base of the brain and at branching points of arteries, as these spots are weak. Factors such as genetic predisposition, high blood pressure, smoking, excessive alcohol consumption, and specific medical conditions also play a role.

1.3.5.3 Risk Factors

Aneurysms are more common in adults 30–60 years of age and in women. They are also more common in people who have high blood pressure, alcoholism, inherited connective tissue disorder, aorta contraction, polycystic kidney diseases, and a family history, and in those who use drugs like cocaine.

1.3.5.4 Complications

Complications include changed sodium levels, re-bleeding, excess fluid accumulation, and narrowed blood vessels.

1.3.5.5 Prevention

Once aneurysms develop, they can't be prevented. But to some extent its growth can be controlled by quitting smoking, alcohols, and narcotics. Regular monitoring of blood pressure, and other vital parameters may help in decreasing the risk of disease development.

1.3.6 Epilepsy

Epilepsy is a neurological disorder characterized by recurring seizures due to an imbalance in the release of neurotransmitters, tumors, and stroke and brain damage from illness or injury.

1.3.6.1 Symptoms

Seizures are the hallmark of epilepsy. They can range from momentary lapses of consciousness and staring spells to convulsions and loss of consciousness. Seizures can differ in type and severity from one person to another.

1.3.6.2 Complications and Risk Factors

Seizures can cause related injuries because they can occur at any time and any place and cause cognitive and behavioral issues. They are sometimes associated with other medical conditions, such as migraines, sleep disorders, or developmental disorders. Managing these comorbidities may add complexity to the overall care of individuals with epilepsy.

1.3.6.3 Prevention

Seizures can be prevented with the use of medicines, surgery, therapies that stimulate the brain using a device, a ketogenic diet, enough sleep, and exercise.

1.3.7 Hydrocephalus

Hydrocephalus is defined by the abnormal buildup of cerebrospinal fluid (CSF) in the brain cavities, causing elevated intracranial pressure. It arises from an imbalance between CSF production and absorption, leading to an accumulation that can exert pressure and harm brain tissue. CSF, a clear and colorless fluid, envelops the brain and spinal cord, offering protection and nutrients to these vital structures.

1.3.7.1 Symptoms

Symptoms of hydrocephalus include headaches, nausea, vomiting, cognitive difficulties, vision problems, and in infants, an enlarging head circumference.

1.3.7.2 Cause

Obstruction (blockages that prevent the normal flow of cerebrospinal fluid within the ventricular system) may be due to congenital abnormalities, tumors, infections, or other factors. Impaired CSF absorption can occur in conditions such as arachnoiditis or subarachnoid hemorrhage or overproduction.

1.3.7.3 Risk Factors

Risk factors for hydrocephalus include congenital hydrocephalus, bleeding within the ventricles, brain tumors, meningitis, and other traumatic injury.

1.3.7.4 Complications

Complications of hydrocephalus include learning, developmental or physical disabilities. Serious complications can lead to death.

1.3.7.5 Treatment Options

Treatment options include surgery, medications, and the placement of a shunt, a slender tube that redirects excess cerebrospinal fluid to a different body area for absorption.

1.3.8 Meningitis

Meningitis is the inflammation of membrane (meninges) covering the brain and spinal cord by virus and bacteria. It can also be caused by fungi, parasites, or non-infectious factors like certain medications, autoimmune disorders, or cancers.

1.3.8.1 Symptoms

Symptoms of meningitis include headache, fever, vomiting, nausea, light sensitivity, stiff neck, joint pain, confusion, drowsiness, and seizures.

1.3.8.2 Cause

The cause of meningitis is a bacterial or virus infection in the brain.

1.3.8.3 Risk Factors

Risk factors for meningitis include a weakened immune system and skipping vaccines. Meningitis is most common in children under 5 years old and adults under 20 years old.

1.3.8.4 Complications

Complications of meningitis include hearing loss, problems in walking, memory, learning, brain damage, kidney failure, shock, and seizures.

1.3.8.5 Prevention

Vaccination (for certain types of bacterial meningitis), practicing good hygiene, and avoiding close contact with individuals who have respiratory infections or meningitis can help prevent meningitis.

1.3.9 Multiple Sclerosis

Multiple sclerosis (MS) is a progressive autoimmune neurological disorder that chronically affects the central nervous system (CNS). In MS, the immune system cells target and erode the protective myelin sheath surrounding nerve fibers, disrupting communication between the brain and various body parts. The condition is marked by inflammation and demyelination of oligodendrocytes and axonal neurons in subcortical structures [21,22].

1.3.9.1 Symptoms

Symptoms of multiple sclerosis include difficulty walking, fatigue, numbness or tingling, muscle weakness, coordination and balance problems, vision issues, and bladder or bowel function troubles. These symptoms may fluctuate or progress gradually over time.

1.3.9.2 Cause

The cause of MS is a malfunctioned immune system that damages the insulation of nerve fibers (myelin), which is made of fatty substances, resulting in damaged and exposed nerve fibers that carry messages.

1.3.9.3 Complications

Complications of MS include muscle stiffness/weakness, paralysis, urinary incontinence and uncontrolled bowel movements, cognitive problems, seizures, depression, and anxiety.

1.3.9.4 Risk Factors

Risk factors for MS include having a family history of MS, infections, being female, obesity, deficient vitamin D and sunlight exposure, smoking, and diseases like thyroid (autoimmune hypothyroidism), psoriasis, type 1 diabetes, pernicious anemia, and inflammatory bowel disease. The onset of MS usually occurs around 20–40 years of age.

1.3.9.5 Treatment and Management

Rehabilitation involves physical and occupational therapy, along with lifestyle adjustments like regular exercise, a balanced diet, and stress management. Medications such as disease-modifying therapies (DMTs) are commonly used to lessen the frequency and severity of relapses. Additionally, symptomatic treatments target specific symptoms like fatigue and muscle spasms with appropriate medications. MS is a complex condition, and its course can vary widely among individuals. The management and treatment plan for someone with MS is often tailored to that person's specific symptoms and needs. It's crucial to have regular check-ups with healthcare providers to monitor the disease's progression and make any necessary adjustments to treatment.

1.3.10 Amyotrophic Lateral Sclerosis

Known commonly as Lou Gehrig's disease, amyotrophic lateral sclerosis (ALS) is a progressive neurodegenerative disorder that affects brain cells, resulting in a gradual decline in movement ability. Its amyotrophic and lateral sclerosis due to mutation in superoxide dismutase 1 which involves deposition of misfolded proteins [23]. Misfolded proteins spreading from cell to cell infiltrate healthy tissues and contribute to the progression of the disease [24,25]. Motor neurons in the lower spine control the muscles of the legs and feet, and those of the upper spine control the arms and hands. Brainstem motor neurons control mouth movements during speaking and eating [26].

Duchenne muscular dystrophy is an X-linked genetic disorder, characterized by mutations in the dystrophin gene. Other disease due to gene mutation is Huntington gene which results in brain cell apoptosis in striatum [27] and leucine rich repeat kinase 2 all secrete inflammatory cytokines inducing neuronal and glial senescence [28]. Activation of microglia and astrocytes by these mutant proteins results in increased brain inflammation and neurological symptoms, along with the suppression of normal astrocyte function [29,30]. Microglia cells scavenge damaged neurons and promote healthy brain physiology [31].

1.3.10.1 Symptoms and Complications

In ALS, muscle weakness begins in the hands or feet and gradually affects other parts of the body. Complications include difficulty speaking, breathing and swallowing; muscle cramps and twitching due to the nerve cell damage; and atrophy of muscles leading to paralysis.

1.3.10.2 Cause

In ALS, the damaged motor neurons cannot pass messages; hence, there is no movement. The cause of ALS could be a blend of genetic and environmental factors, or in certain instances, there seems to be a familial or inherited element, while other instances are sporadic with no clear family history.

1.3.10.3 Risk Factors

ALS is most common in those aged 60 to mid-80s. Men and those who smoke are more prone to ALS. In 10% of the cases there may be a relation to genetics.

1.3.10.4 Treatment

At present, there is no cure for ALS, so treatment revolves around symptom management, preserving function, and enhancing quality of life. This often involves a combination of medications, physical therapy, and the use of assistive devices to aid in daily activities.

1.3.11 Stroke

A cerebrovascular accident (CVA), commonly known as a stroke, occurs when there's a sudden disruption in blood flow to the brain or when a blood vessel in the brain ruptures, causing damage to brain tissue. This interruption in blood supply deprives the affected brain area of oxygen and vital nutrients, leading to brain damage or potentially fatal outcomes.

1.3.11.1 Symptoms

The acronym FAST can help identify the major symptoms of a stroke: facial drooping (F), arm weakness (A), and speech difficulties (S). If these symptoms are present, then time (T) is of the essence and one should seek medical attention immediately. Other symptoms of a stroke may involve numbness or weakness on one side of the body, difficulty in speaking or understanding speech, sudden confusion, vision issues, severe headache, nausea, vomiting, and loss of consciousness.

1.3.11.2 Risk Factors

Risk factors for stroke include high blood pressure, smoking, diabetes, elevated cholesterol levels, atrial fibrillation, and a family history of the condition.

1.3.11.3 Prevention

Treatment options such as clot-dissolving medications, thrombectomy, as well as physical, occupational, and speech therapy, can help manage stroke. Implementing lifestyle modifications like adopting a nutritious diet, engaging in consistent physical activity, ceasing smoking, and effectively managing conditions such as high blood pressure and diabetes can substantially lower the risk of experiencing a stroke.

1.3.12 Encephalitis

Encephalitis is inflammation of active brain tissues by infection or an autoimmune response (the body's own immune response attacks the brain).

1.3.12.1 Symptoms

Symptoms of encephalitis include a range of issues such as fevers, seizures, headaches, movement disorders, sensitivity to light and sound, stiffness, and loss of consciousness. Additionally, it can lead to paralysis, impaired speech, double vision, coma, and cognitive impairments.

1.3.12.2 Treatment

Early medication can improve the outcome and reduce the risk of complications.

1.4 REASONS WHY NEURODEGENERATIVE DISEASES OCCUR

There are many reasons for the occurrence of NDS, including accumulation of tau proteins, amyloid β and α-synuclein, oxidative stress, neuroinflammation, and neuroplasticity [32].

As we age, the brain tissues, which consist of postmitotic neurons and oligodendrocytes, become more vulnerable to DNA damage. Unlike proliferating cells, postmitotic cells are particularly susceptible to this damage [33]. Mutations in base excision repair/telomerase inactivation, defects in proteostasis, and DNA damage decrease NAD^+, an essential cofactor in repairs, increasing the vulnerability of ND in ageing [34–36]. Neurons have a very high energy demand and mitochondrial functions decrease with age, so there is a decrease in the energy supply. With age, neurons experience an accumulation of reactive oxygen species (ROS) due to DNA damage, mitochondrial dysfunction, and oxidative stress, which leads to impaired DNA repair mechanisms [37–39]. Since mitochondrial DNA are not protected by histone proteins, they are more prone to oxidative damage [40]. The crosstalk between immune system cells like mast cells, microglia (removes cellular debris), T cells, and oligodendrocytes is vital to protect immune response losses in the brain [41]. As we age, mitochondrial dysfunction leads to an increase in ROS, activating microglia and enhancing NF-κB signaling. This cascade results in the heightened production of pro-inflammatory chemokines and cytokines like IL-1β, IL-6, and tumor necrosis factor, ultimately contributing to neuroinflammation [42,43]. In Alzheimer's and Parkinson's diseases, there is evidence of oxidative damage to proteins, lipids, and nucleic acids, indicating a connection between these conditions and oxidative stress [44].

In Huntington's disease (HD) mitochondrial dysfunction in complex ll and lll and impaired calcium homeostasis has been reported [45]. Neurons are highly sensitive to calcium levels, and any disturbance in calcium homeostasis can result in neuronal damage [46,47]. Elevated calcium concentrations within the endoplasmic reticulum disrupt the delicate balance between the Ca^{2+}-dependent phosphatase calcineurin (CaN) and the calmodulin-dependent protein kinase II (CaMKII), both of which are abundant in synaptic sites. This imbalance is a contributing factor to synaptic loss and the progression of neurodegeneration [48].

The hormone prolactin may regulate/inhibit neurogenesis and promotes mood lowering or depression [49]. It can easily pass blood brain barrier (BBB); its role in CNS and neurodegeneration is vast. Insulin plays a crucial role in learning, memory, and cognition, with abundant insulin receptors identified on dopaminergic neurons within the substantia nigra [50].

Prion protein aggregates are also responsible for the spread of disease from one neuron to another [51]. The neurotropic viruses like zika, rabies, and Japanese encephalitis, all cause

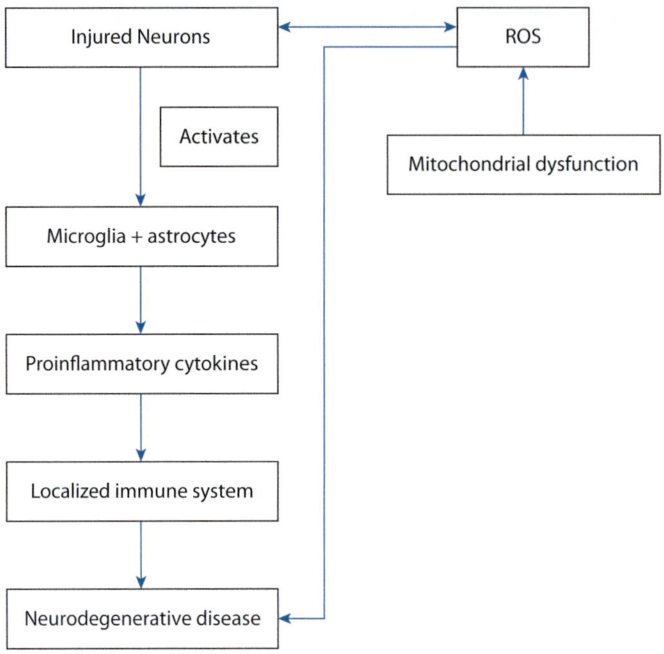

FIGURE 1.2 Neurodegenerative disease by mitochondrial dysfunction.

damage to neurons or virus induced apoptosis [52,53]. Systemic infections to the microbiota gut brain axis contribute to NDs following the release of microbiome microorganisms and their toxins [54]. Figure 1.2 shows a generalized description of how mitochondrial dysfunction can lead to ND.

1.5 COMORBIDITY IN NEURO DISEASES

Patients suffering from these diseases are always found to have other neuropathies and infections. They can occur due to infection in the urinary tract, sepsis, brain injury, or acute muscular dystrophy [55–57]. Chronic peripheral inflammations like diabetes, hypertension, and hyperlipidemia all contribute to ND and ageing [58]. The prevalence of Alzheimer's was diagnosed in those with diabetics and hypertension; and Parkinson's is prevalent in people suffering from inflammatory bowel diseases [59,60]. Social/environmental stress and diagnosis/preclinical course contribute in aggravating the condition of these patients, influencing their health span and quality of life.

1.6 DISEASE SIGNS AND CHIEF COMPLAINTS

While the physiological manifestations of NDs, such as demyelination, dendrite loss, and neuronal death [61,62], may differ, they all stem from a common etiology at the cellular and molecular levels [63,64]. The two most common old age-related NDs are Alzheimer's and Parkinson's, and it is expected that by 2040 the number of Parkinson's patients will be doubled after the age of 65 [65,66]. MRI studies have shown that phenotypic cognitive impairments take place at the very later stage, while the structural changes appear at the very early stage [67].

Chief complaints of patients suffering from ND are disrupted circadian rhythms, motor dysfunctions, cognitive impairment insomnia/hypersomnia, and sleep apnea [68].

1.7 WASTE CLEARANCE VIA THE GLYMPHATIC SYSTEM IN ND

The lymphatic system is crucial for eliminating waste and cellular debris, but brain cells lack this system [69]. Instead, the brain relies on a specialized pathway for clearance, which is dependent on astrocyte-specific water channels called AQP4. As individuals age, the efficiency of this channel tends to decrease, resulting in an increase in AQP4 depolarization. This channel is crucial for transporting cerebrospinal fluid into the brain and facilitating the movement of interstitial fluid back into perivenous spaces. Its deletion or impairment can disrupt the normal flow of fluids. Referred to as the glia-dependent lymphatic system or glymphatic system, it operates primarily during sleep. As a result, disrupted sleep can cause waste to accumulate in the brain [70–72].

1.8 MEDICATIONS AND SELF-CARE

Since these NDs are irreversible, understanding their mechanism of progression will help us in designing more accurate and precise therapeutic treatment approaches. Palliative medicines offer only symptomatic relief [73]. Medications for Parkinson's are based on dopamine neurochemistry; for Alzheimer's, acetylcholinesterase inhibitors are used. Limited calorie intake and increased physical activity might protect us from NDs [74]. Intermittent fasting can improve the brain functions and decrease NDs associated with inflammatory responses independent of age [75,76]. Ketogenic diets can ameliorate the energy crisis characterized by a deteriorated brain's glucose metabolism [77].

Phytochemicals obtained from different plants like ashwagandha, brahmi, Indian pennywort, velvet bean, ginseng, Korean ginseng, henbane, black henbane, or stinking nightshade, amla, garden heliotrope, stinking gum, dhatura, oat, sugar apple, sway, muskrat root, dwarf morning glory, slender dwarf morning glory, and rosemary, are also known to improve or delay the course of disease [78–111]. Table 1.1 shows the effects of different phytochemicals obtained from plants.

TABLE 1.1
Effects of Phytochemicals Obtained from Plants

Plant Name	Family	Part Used	Phytochemical	Neuro Effect
Withania somnifera (ashwagandha)	Solanaceae	Root	Withanolide A, withanoside IV, VI	Reparation of neuronal axons and dendrites, along with the rebuilding of pre and post-neuronal connections, and the mitigation of stress
Bacopa monnieri (brahmi)	Plantaginaceae	Leaf	Brahmine, herpestine, saponins (bacosides A, A3, and B and bacopasaponin A to F), D-mannitol, betulinic acid, β-sitosterol, and stigmasterols	Possessing neuroprotective qualities, it reinstates the functioning of cholinergic enzymes, leading to enhanced cognitive functions.
Centella asiatica (Indian pennywort)	Umbelliferae (Apiceae)	Whole part	Triterpenoids, volatile fatty acids, glycosides, flavonoids, alkaloids, tannins, and asiaticoside	Acting as an anti-inflammatory agent, it inhibits the excessive activation of the p38 MAPK pathway
Mucuna pruriens (velvet bean)	Fabaceae	Whole plant	L-dopa and hallucinogenic tryptamines, and anti-nutritional factors (phenols and tannins)	L-dopa serves as a precursor to dopamine, capable of crossing the blood-brain barrier and restoring neurotransmission.
Papaver somniferum (opium poppy)	Papaveraceae	Latex capsule seeds	Phenanthrenes (thebaine morphine, codeine, and sanguinarine)	Diminish brain activity (painkillers) and elevate mood by acting on specific neurons

(Continued)

TABLE 1.1 ***(Continued)***

Effects of Phytochemicals Obtained from Plants

Plant Name	Family	Part Used	Phytochemical	Neuro Effect
Loranthus longifolia	Loranthaceae	Leafs	Alkaloids, flavonoids, tannins, terpenoids, reducing sugars, carbohydrates and cardiac glycosides in aqueous extract	Protection for the nervous system against anxiety and depression.
Aconitum (aconite, monkshood)	Ranunculaceae	Tubers	Aconitine, mesaconitine	Disorders affecting neurons, causing pain and inflammation.
Panax ginseng (Asian ginseng)	Araliliaceae	Roots and rhizomes	Ginsenosides (Rg1), ginseng (polysaccharides, polypeptides, panaxosides)	Acting as an antidepressant through the enhancement of the BDNF-TrkB signaling pathway, it raises acetylcholine levels, providing anti-depressant, anti-Alzheimer's, and anti-Parkinson's effects while offering neuronal protection.
Hyoscyamus niger (henbane)	Solanaceae	Leaves	Hyoscyamine, scopolamine, and tropane alkaloids	Inhibits acetylcholine function in the brain and acts as an antagonist to muscarinic receptors.
Embilica officinalis (amla)	Euphorbiaceae	Root	Gallic acid, ascorbic acid, ellagic acid, rutin, quercetin, and catechol, tannins, epigallocatechin-3-gallat and polyphenols	Protection for the nervous system, sedation for sleep disorders, and relief for anxiety and epilepsy.
Valeriana officinalis (garden heliotrope)	Caprifoliaceae	Fruit	Actinidine, chatinine, shyanthine, valerianine, and valerine	Augmented GABA receptor activity prevents excitotoxicity and neuronal demise.
Ferula asafoetida (stinking gum)	Apiaceae	Rhizome, tap root	Ferulic acid, umbelliferone, asaresinotannols, farnesiferols A, B, and C, glucose, galactose, l-arabinose, rhamnose, and glucuronic acid and volatile oil (3-17%) consisting of disulfides (2-butyl propenyl disulfide) with monoterpenes (α- and β-pinene, etc.), free ferulic acid, valeric acid, and traces of vanillin	Sedative agent with an antioxidative property.
Datura metel (datura)	Solanaceae	Seeds, leafs	Tropane alkaloids, tannins, flavonoids, saponins and withanolides	Safeguarding the nervous system against seizures.
Avena sativa (oat)	Poaceae	Seed	*β*-glucans, avenanthramides, tocols, sterols, and avenacosides	Elevated production of brain growth factors (neurotrophins) and the vasodilatory molecule nitric oxide, crucial for regulating cerebral blood flow, along with the modulation of neurotransmission and the inhibition of enzymes like MAO-B and AChE, responsible for oxidizing or hydrolyzing various neurotransmitters

(Continued)

TABLE 1.1 ***(Continued)***
Effects of Phytochemicals Obtained from Plants

Plant Name	Family	Part Used	Phytochemical	Neuro Effect
Annona squamosa (sugar apple)	Annonaceae	Seeds, bark, leaves	Diterpenoid alkaloid atisine, oxophoebine, reticuline, isocorydine, methylcorydaldine and flavonoid quercetin-3-O-glucoside	Exhibits anti-inflammatory impact by decreasing NF-κβ and mitigates apoptosis in neural cells by reducing caspase 3.
Acoros calamus (sweet flag)	Acoraceae	Leaves, rhizome	Phenylpropanoids (asarone and eugenol), sterols, triterpene glycosides, triterpenoid saponins, sesquiterpenoids, monoterpenes, and alkaloids	Protection for the nervous system and improved memory capabilities.
Evolvulus alsinoides (dwarf morning-glory)	Convolvulaceae	Whole plant	Scopoletin, umbelliferone, scopolin and 2-methyl-1,2,3,4-butanetetrol	Enhances the formation of spatial memory and inhibits acetylcholinesterase (AChE).
Rosmarinus officinalis (rosemary)	Lamiaceae	leaves	Rosmarinic acid, camphor, caffeic acid, ursolic acid, betulinic acid, carnosic acid, and carnosol	Reduced lipid peroxidation observed in the cerebral tissues of patients experiencing ischemia.ba
Aegle marmelos (bael fruit)	Rutaceae	Leaves, bark, roots, fruit, seeds	Marmenol, marmin, marmelosin, marmelide, psoralen, alloimperatorin, rutaretin, scopoletin, aegelin, marmelin, fagarine, anhydromarmelin, limonene, â-phellandrene, betulinic acid, marmesin, imperatorin, marmelosin, luvangentin, and auroptene	Reduced oxidative stress and safeguarding neural health.
Rauwolfia serpentine (Indian snakeroot)	Apocynaceae		Alkaloids of the indole alkaloid (ajmaline, ajmalicine, reserpine, serpentine), and others	Psychiatric disorders such as schizophrenia and bipolar disorder, neurological conditions like epilepsy and seizures, as well as difficulties with sleep such as insomnia and other sleep-related problems.

1.9 CONCLUSION

Since NDs are irreversible, understanding their mechanism of progression will help us in designing better accurate precise therapeutic approach treatments. The pharmaceutical companies should focus on drugs that stop the disease progression. Though animal models gave us a sound understanding of these diseases, they are very limited while humans exhibit heterogeneity in etiology and pathophysiology. Studying pluripotent stem cells will provide valuable insights into the biochemical and molecular underpinnings of these diseases. This research can lead to a deeper understanding of their mechanisms and potential therapeutic interventions. Life style changes can delay the progression of NDs and may protect neuroglial cells from degeneration, extending cognitive longevity. The palliative treatments are best for these disorders, healthcare providers should anticipate the next situation and act accordingly. Anthrobots may be a new ray of hope for people suffering from these neurodegenerative diseases.

ACKNOWLEDGMENT

The authors express gratitude to the University of Allahabad for offering facilities to conduct the current study.

CONFLICTS OF INTEREST

There is no conflict of interest to be disclosed.

REFERENCES

1. Van Schependom, J., and D'haeseleer, M. Advances in Neurodegenerative Diseases. J. Clin. Med., 2023; 12(5): 1709. doi:10.3390/jcm12051709.
2. Feigin, V. L., Vos, T., Nichols, E., O Owolabi, M., Carroll, W. M., Dichgans, M., Deuschl, G., Parmar, P., Brainin, M., and Murray, C. The Global Burden of Neurological Disorders: Translating Evidence into Policy. Lancet Neurol., 2019; 19: 255–265.
3. Katsnelson, A., De Strooper, B., and Zoghbi, H. Y. Neurodegeneration: From Cellular Concepts to Clinical Applications. Sci. Transl. Med., 2016; 8(364): 364ps18.
4. Cacabelos, R., Cacabelos, P., and Carril, J. C. (2019). Epigenetics and Pharmacoepigenetics of Age-Related Neurodegenerative Disorders. Pharmacoepigenetics. Academic Press. 903–950. doi:10.1016/b978-0-12-813939-4.00038-3.
5. Dean, D. C., 3rd et al. Brain Differences in Infants at Differential Genetic Risk for Late-Onset Alzheimer Disease: A Cross- Sectional Imaging Study. JAMA Neurol., 2014; 71: 11–22.
6. Cravello, L., et al. Chronic Pain in the Elderly With Cognitive Decline: A Narrative Review. Pain Ther., 2019; 8: 53–65. doi:10.1007/s40122-019-0111-7.
7. Dorsey, E. R., Glidden, A. M., Holloway, M. R., Birbeck, G. L., and Schwamm, L. H. Teleneurology and mobile Technologies: The Future of Neurological Care. Nat. Rev. Neurol., 2018; 14: 285–297.
8. Gage, F. H. Neurogenesis in the Adult Brain. J. Neurosci., 2002; 22: 612–613.
9. Reynolds, B. A., and Weiss, S. Generation of Neurons and Astrocytes from Isolated Cells of the Adult Mammalian Central Nervous System. Science, 1992; 255: 1707–1710.
10. Shihabuddin, L. S., Horner, P. J., Ray, J., and Gage, F. H. Adult Spinal Cord Stem Cells Generate Neurons after Transplantation in the Adult Dentate Gyrus. J. Neurosci. Off. J. Soc. Neurosci., 2000; 20: 8727–8735.
11. Pencea, V., Bingaman, K. D., Freedman, L. J., and Luskin, M. B. Neurogenesis in the Subventricular Zone and Rostral Migratory Stream of the Neonatal and Adult Primate Forebrain. Exp. Neurol., 2001; 172: 1–16. doi:10.1006/exnr.2001.7768.
12. Mueller, B. K., Mueller, R., and Schoemaker, H. Stimulating Neuroregeneration as a Therapeutic Drug Approach for Traumatic Brain Injury. Br. J. Pharmacol., 2009; 157(5): 675–85. doi:10.1111/j.1476-5381.2009.00220.x.
13. Gumuskaya, G., Srivastava, P., Cooper, B. G., Lesser, H., Semegran, B., Garnier, S., and Levin, M., Motile Living Biobots Self-Construct from Adult Human Somatic Progenitor Seed Cells. Adv. Sci. 2023; 11: 2303575. doi:10.1002/advs.202303575.
14. Kovacs, G. G. (2015). Introduction: classification of neurodegenerative diseases. In: GG Kovacs (Ed.), Neuropathology of Neurodegenerative Diseases: A Practical Guide. Cambridge University Press.
15. Kovacs, G. G. Current Concepts of Neurodegenerative Diseases. EMJ Neurol., 2014; 1: 78–86.
16. Kovacs, G. G. (2018). Concepts and Classification of Neurodegenerative Diseases. Neuropathology, 301–307. doi:10.1016/b978-0-12-802395-2.00021-3.
17. Hampel, H., Mesulam, M.-M., Cuello, A. C., Farlow, M. R., Giacobini, E., Grossberg, G. T., Khachaturian, A. S., Vergallo, A., Cavedo, E., and Snyder, P. J. The Cholinergic System in the Pathophysiology and Treatment of Alzheimer's Disease. Brain J. Neurol., 2018; 141: 1917–1933.
18. Querfurth, H. W., and LaFerla, F. M. Alzheimer's Disease. N. Engl. J. Med., 2010; 362: 329–344. doi:10.1056/nejmra0909142.
19. Hyman, B. T., and Van Hoesen, G. W. Neuron Numbers in Alzheimer's Disease: Cell-Specific Pathology. Neurobiol. Aging, 1987; 8: 555–556.
20. Ottolini, D., Cali, T., Szabo, I., and Brini, M. Alpha-Synuclein at the Intracellular and the Extracellular Side: Functional and Dysfunctional Implications. Boil. Chem., 2017; 398: 77–100.

21. Haider, L., et al. Oxidative Damage in Multiple Sclerosis Lesions. Brain. 2011; 134: 1914–1924. doi:10.1093/brain/awr128.
22. Moccia, M., et al. Grey: White Matter Ratio at Diagnosis and the Risk of 10-Year Multiple Sclerosis Progression. Eur. J. Neurol., 2017, 24, 195–204.
23. Malaspina, A., Puentes, F., and Amor, S. Disease Origin and Progression in Amyotrophic Lateral Sclerosis: An Immunology Perspective. Int. Immunol., 2014;27:117–129. doi:10.1093/intimm/dxu099.
24. Hatters, D. M. Protein Misfolding Inside Cells: The Case of Huntingtin and Huntington's Disease. IUBMB Life, 2008; 60: 724–728.
25. Hussain, R., Zubair, H., Pursell, S., and Shahab, M. Neurodegenerative Diseases: Regenerative Mechanisms and Novel Therapeutic Approaches. Brain Sci., 2018; 8(9): 177. doi:10.3390/brainsci8090177.
26. Jones, C. M., and Coleman, S. (2007). Neurodegenerative Diseases. Palliative Care. Elsevier. 382–395. doi:10.1016/b978-141602597-9.10026-2.
27. Sciacca, G., and Cicchetti, F. Mutant Huntingtin Protein Expression and Blood-Spinal Cord Barrier Dysfunction in Huntington Disease. Ann. Neurol., 2017; 82: 981–994.
28. Musi, N., et al. Tau Protein Aggregation is Associated with Cellular Senescence in the Brain. Aging Cell. 2018; 17: e12840.
29. Crotti, A., et al. Mutant Huntingtin Promotes Autonomous Microglia Activation via Myeloid Lineage-Determining Factors. Nat. Neurosci., 2014; 17(4): 513–521. doi:10.1038/nn.3668.
30. Hsiao, H. Y., Chen, Y. C., Chen, H. M., Tu, P. H., and Chern, Y. A Critical Role of Astrocyte-Mediated Nuclear Factor-kappaB-Dependent Inflammation in Huntington's Disease. Hum. Mol. Genet. 2013; 22(9): 1826–1842. doi:10.1093/hmg/ddt036.
31. Tremblay, M. E., Stevens, B., Sierra, A., Wake, H., Bessis, A., and Nimmerjahn, A. The Role of Microglia in the Healthy Brain. J. Neurosci., 2011; 31(45): 16064–16069. doi:10.1523/jneurosci.4158-11.2011.
32. Vaquer-Alicea, J., and Diamond, M. I. Propagation of Protein Aggregation in Neurodegenerative Diseases. Ann. Rev. Biochem., 2016; 88(1). doi:10.1146/annurev-biochem-061516-045049.
33. Madabhushi, R., Pan, L., and Tsai, L. H. DNA Damage and its Links to Neurodegeneration. Neuron, 2014; 83: 266–282.
34. Leandro, G. S., Sykora, P., and Bohr, V. A. The Impact of Base Excision DNA Repair in Age-Related Neurodegenerative Diseases. Mutat. Res., 2015; 776: 31–39.
35. Fang, E. F., et al. NAD(+) in Aging: Molecular Mechanisms and Translational Implications. Trends Mol. Med., 2017; 23: 899–916.
36. Eitan, E., Hutchison, E. R., and Mattson, M. P. Telomere Shortening in Neurological Disorders: An Abundance of Unanswered Questions. Trends Neurosci., 2014; 37: 256–263.
37. Maynard, S., Fang, E. F., Scheibye-Knudsen, M., Croteau, D. L., and Bohr, V. A. DNA Damage, DNA Repair, Aging, and Neurodegeneration. Cold Spring Harb. Perspect. Med., 2015; 5: a025130. doi:10.1101/cshperspect.
38. Singh, A., Kukreti, R., Saso, L., and Kukreti, S. Oxidative Stress: A Key Modulator in Neurodegenerative Diseases. Molecules, 2019; 24(8): 1583. doi:10.3390/molecules24081583.
39. Collin, F. Chemical Basis of Reactive Oxygen Species Reactivity and Involvement in Neurodegenerative Diseases. Int. J. Mol. Sci., 2019; 20(10): 2407. doi:10.3390/ijms20102407.
40. Richter, C., Park, J. W., and Ames, B. N. Normal Oxidative Damage to Mitochondrial and Nuclear DNA is Extensive. Proc. Natl. Acad. Sci., 1988; 85: 6465–6467. doi:10.1073/pnas.85.17.6465.
41. Amor, S., and Woodroofe, M. N. Innate and Adaptive Immune Responses in Neurodegeneration and Repair. Immunology, 2014; 141: 287–291.
42. Cunningham, C. Microglia and Neurodegeneration: The Role of Systemic Inflammation. Glia, 2013; 61: 71–90.
43. Voet, S., Srinivasan, S., Lamkanfi, M., and van Loo, G. Inflammasomes in Neuroinflammatory and Neurodegenerative Diseases. EMBO Mol. Med., 2019; e10248. doi:10.15252/emmm.201810248.
44. Wang, X., Wang, W., Li, L., Perry, G., Lee, H. G., and Zhu, X. Oxidative Stress and Mitochondrial Dysfunction in Alzheimer's Disease. Biochim. Biophys. Acta., 2014; 1842(8): 1240–1247. doi:10.1016/j.bbadis.2013.10.015.
45. Schapira, A. H., Olanow, C. W., Greenamyre, J. T., and Bezard, E., 2014. Slowing of Neurodegeneration in Parkinson's Disease and Huntington's Disease: Future Therapeutic Perspectives. Lancet, 384(9942): 545–555. doi:10.1016/S0140-6736(14)61010-2.
46. Bezprozvanny, I., and Mattson, M. P. Neuronal Calcium Mishandling and the Pathogenesis of Alzheimer's Disease. Trends Neurosci., 2008; 31: 454–463.
47. Pchitskaya, E., Popugaeva, E., and Bezprozvanny, I. Calcium Signaling and Molecular Mechanisms Underlying Neurodegenerative Diseases. Cell Calcium, 2018; 70: 87–94. doi:10.1016/j.ceca.2017.06.008.

48. Popugaeva, E., Pchitskaya, E., and Bezprozvanny, I. Dysregulation of Neuronal Calcium Homeostasis in Alzheimer's Disease - A Therapeutic Opportunity? Biochem. Biophys. Res. Commun., 2017; 483: 998–1004.
49. Torner, L. Actions of Prolactin in the Brain: From Physiological Adaptations to Stress and Neurogenesis to Psychopathology. Front. Endocrinol., 2016; 7: 25.
50. de la Monte, S. M., Tong, M., Bowling, N., and Moskal, P. Si-RNA Inhibition of Brain Insulin or Insulin-Like Growth Factor Receptors Causes Developmental Cerebellar Abnormalities: Relevance to Fetal Alcohol Spectrum Disorder. Mol. Brain, 2011; 4: 13.
51. Lim, J., and Yue, Z. Neuronal Aggregates: Formation, Clearance, and Spreading. Dev. Cell, 2015; 32(4): 491–501. doi:10.1016/j.devcel.2015.02.002.
52. Hotta, H. [Neurotropic Viruses–Classification, Structure and Characteristics]. Nihon Rinsho, 1997; 55(4): 777–782. [Japanese].
53. Palazzo, E., et al. Role of Nacylethanolamines in the Neuroinflammation: Ultramicronized Palmitoylethanolamide in the Relief of Chronic Pain and Neurodegenerative Diseases. Neuropsychiatry, 2019;9:2035–2046.
54. Stephenson, J., Nutma, E., van der Valk, P., and Amor, S. Inflammation in CNS Neurodegenerative Diseases. Immunology, 2018; 154: 204–219. doi:10.1111/imm.12922
55. Martin, L. J. (2002). Neurodegenerative Disorders. Encyclopedia of the Human Brain. Academic Press. 441–463. doi:10.1016/b0-12-227210-2/00240-5
56. Iwashyna, T. J., Ely, E. W., Smith, D. M., and Langa, K. M. Long-Term Cognitive Impairment and Functional Disability Among Survivors of Severe Sepsis. JAMA, 2010; 304: 1787–1794.
57. Zheng, K. S., Dorfman, B. J., Christos, P. J., Khadem, N. R., Henchcliffe, C., Piboolnurak, P., and Nirenberg, M. J. Clinical Characteristics of Exacerbations in Parkinson Disease. Neurologist, 2012; 18: 120–124.
58. Süβ, P., Lana, A. J., and Schlachetzki, J. C. M. Chronic Peripheral Inflammation: A Possible Contributor to Neurodegenerative Diseases. Neural Regen. Res., 2021;16(9): 1711–1714. doi:10.4103/1673-5374.306060.
59. Barnes, D. E., and Yaffe, K. The Projected Effect of Risk Factor Reduction on Alzheimer's Disease Prevalence. Lancet Neurol., 2011; 10: 819–828.
60. Peter, I., Dubinsky, M., Bressman, S., Park, A., Lu, C., Chen, N., and Wang, A. Antitumor Necrosis Factor Therapy and Incidence of Parkinson Disease Among Patients with Inflammatory Bowel Disease. JAMA Neurol., 2018;75: 939–946.
61. Zis, P., McHugh, P. C., Manca, M., Sarrigiannis, P. G., Rao, D. G., and Hadjivassiliou, M. Increased Oxidative Stress as a Risk Factor in Chronic Idiopathic Axonal Polyneuropathy. J. Mol. Neurosci., 2018; 66(4): 547–551. doi:10.1007/s12031-018-1200-5.
62. Yadav, S., Mansoori, A. N., and Sisodiya, J. S. Current Pathologic Determinants of Complex Neurodegenerative Diseases: A Review. Int. J. Pharm. Technol., 2013; 5(2):2607–2621.
63. Vadakkan, K. I. Neurodegenerative Disorders Share Common Features of "Loss of Function" States of a Proposed Mechanism of Nervous System Functions. Biomed. Pharmacother. Biomed. Pharmacother., 2016; 83: 412–430.
64. Gitler, A. D., Dhillon, P., and Shorter, J. Neurodegenerative Disease: Models, Mechanisms, and a New Hope. Dis. Model Mech., 2017;10: 499–502.
65. Kowal, S. L., Dall, T. M., Chakrabarti, R., Storm, M. V., and Jain, A. The Current and Projected Economic Burden of Parkinson's Disease in the United States. Mov. Disord., 2013; 28(3): 311–318. doi:10.1002/mds.25292.
66. Yacoubian, T. A. (2017). Neurodegenerative Disorders. Drug Discovery Approaches for the Treatment of Neurodegenerative Disorders. Academic Press. 1–16. doi:10.1016/b978-0-12-802810-0.00001-5
67. Hou, Y., Dan, X., Babbar, M., Wei, Y., Hasselbalch, S. G., Croteau, D. L., and Bohr, V. A. Ageing as a Risk Factor for Neurodegenerative Disease. Nat. Rev. Neurol., 2019; 15(10): 565–581. doi:10.1038/s41582-019-0244-7
68. Hölscher, C. Brain Insulin Resistance: Role in Neurodegenerative Disease and Potential for Targeting. Expert Opin. Investig. Drugs, 2020; 29: 333–348. doi:10.1080/13543784.2020.1738383
69. Iliff, J. J., and Nedergaard, M. Is There a Cerebral Lymphatic System? Stroke, 2013; 44: S93–S95.
70. Jessen, N. A., Munk, A. S., Lundgaard, I., and Nedergaard, M. The Glymphatic System: A beginner's Guide. Neurochem. Res., 2015; 40: 2583–2599.
71. Kress, B. T., et al. Impairment of Paravascular Clearance Pathways in the Aging Brain. Ann. Neurol., 2014; 76, 845–861.
72. Iliff, J. J., et al. A Paravascular Pathway Facilitates CSF Flow through the Brain Parenchyma and the Clearance of Interstitial Solutes, Including Amyloid Beta. Sci. Transl. Med., 2012; 4: 147ra111.

73. Cacabelos, R. Parkinson's Disease: From Pathogenesis to Pharmacogenomics. Int. J. Mol. Sci., 2017; 18: 551.
74. Babbar, M., and Sheikh, M. S. Metabolic Stress and Disorders Related to Alterations in Mitochondrial Fission or Fusion. Mol. Cell. Pharmacol., 2013; 5: 109–133.
75. Vasconcelos, A. R., et al. Effects of Intermittent Fasting on Age-Related Changes on Na,K-ATPase Activity and Oxidative Status Induced by Lipopolysaccharide in Rat Hippocampus. Neurobiol. Aging, 2015;36(5):1914–1923. doi:10.1016/j.neurobiolaging.2015.02.020.
76. Rekatsina, M., et al. Pathophysiology and Therapeutic Perspectives of Oxidative Stress and Neurodegenerative Diseases: A Narrative Review. Adv. Ther. 37: 113–139 (2020). doi:10.1007/s12325-019-01148-5.
77. Jensen, N. J., Wodschow, H. Z., Nilsson, M., and Rungby, J. Effects of Ketone Bodies on Brain Metabolism and Function in Neurodegenerative Diseases. Int. J. Mol. Sci., 2020; 21(22): 8767. doi: 10.3390/ijms21228767.
78. Cardinali, D. P. Melatonin: Clinical Perspectives in Neurodegeneration. Front. Endocrinol., 2019; 10: 480. doi:10.3389/fendo.2019.00480.
79. Mannangatti, P., and Naidu, K. N. (2016). Indian Herbs for the Treatment of Neurodegenerative Disease. The Benefits of Natural Products for Neurodegenerative Diseases. Springer. 323–336. doi: 10.1007/978-3-319-28383-8_17.
80. Nabi, M., and Tabassum, N. Role of Environmental Toxicants on Neurodegenerative Disorders. Front. Toxicol., 2022; 4: 837579. doi:10.3389/ftox.2022.837579.
81. Alasfar, R. H., and Isaifan, R. J. Aluminum Environmental Pollution: The Silent Killer. Environ. Sci. Pollut. Res. Int., 2021; 28(33): 44587–44597. doi:10.1007/s11356-021-14700-0.
82. Klotz, K., Weistenhöfer, W., Neff, F., Hartwig, A., van Thriel, C., and Drexler, H. The Health Effects of Aluminum Exposure. Dtsch. Arzteblatt Int., 2017; 114(39): 653–659. doi:10.3238/arztebl.2017.0653.
83. Verstraeten, S. V., Aimo, L., and Oteiza, P. I. Aluminium and Lead: Molecular Mechanisms of Brain Toxicity. Arch. Toxicol., 2008; 82(11): 789–802. doi:10.1007/s00204-008-0345-3.
84. Chung, J. Y., Yu, S. D., and Hong, Y. S. Environmental Source of Arsenic Exposure. J. Prev. Med. Public Health, 2014; 47(5): 253–257. doi:10.3961/jpmph.14.036.
85. Finnegan, P. M., and Chen, W. Arsenic Toxicity: The Effects on Plant Metabolism. Front. Physio., 2012; 3: 182. doi:10.3389/fphys.2012.00182.
86. Beard, E., Lengacher, S., Dias, S., Magistretti, P. J., and Finsterwald, C. Astrocytes as Key Regulators of Brain Energy Metabolism: New Therapeutic Perspectives. Front. Physiol., 2022; 12: 825816. doi:10.3389/fphys.2021.825816. Erratum in: Front Physiol. 2022 Feb 25;13:867827.
87. De Keyser, J., Mostert, J. P., and Koch, M. W. Dysfunctional Astrocytes as Key Players in the Pathogenesis of Central Nervous System Disorders. J. Neurol. Sci., 2008; 267(1–2): 3–16. doi:10.1016/j.jns.2007.08.044.
88. Bhattacharya, S. K., and Muruganandam, A. V. Adaptogenic Activity of *Withania somnifera*: An Experimental Study Using a Rat Model of Chronic Stress. Pharmacol. Biochem. Behav., 2003; 75(3): 547–555.
89. Nathan, P. J., et al. The Acute Effects of an Extract of Bacopa Monniera (Brahmi) on Cognitive Function in Healthy Normal Subjects. Hum. Psychopharmacol., 2001; 16(4): 345–351.
90. Aguiar, S., and Borowski, T. Neuropharmacological Review of the Nootropic Herb Bacopa monnieri. Rejuvenation Res., 2013; 16(4): 313–326. doi:10.1089/rej.2013.1431.
91. Das, A. J. Review on Nutritional, Medicinal and Pharmacological Properties of Centella asiatica (Indian Pennywort). J. Biol. Active Prod. Nat., 2011; 1(4): 216–228.
92. Chen, S., et al. Asiaticoside Attenuates Memory Impairment Induced by Transient Cerebral Ischemia-Reperfusion in Mice through Anti-Infl Ammatory Mechanism. Pharmacol. Biochem. Behav., 2014; 122: 7–15.
93. Lampariello, L. R., Cortelazzo, A., Guerranti, R., Sticozzi, C., and Valacchi, G. The Magic Velvet Bean of Mucuna pruriens. J. Tradit. Complement Med., 2012; 2(4): 331–339. doi:10.1016/s2225-4110(16)30119-5.
94. Ballantyne, J. C., and Mao, J. Opioid Therapy for Chronic Pain. N. Engl. J. Med., 2012; 349(20): 1943–1953.
95. Ameri, A. The Effects of Aconitum Alkaloids on the Central Nervous System. Prog. Neuro., 1998; 56(2): 211–235. doi:10.1016/s0301-0082(98)00037-9.
96. Hou, W., Wang, Y., Zheng, P., and Cui, R. Effects of Ginseng on Neurological Disorders. Front. Cell. Neurosci., 2020; 14: 55. doi:10.3389/fncel.2020.00055.
97. Volgin, A. D., Yakovlev, O. A., Demin, K. A., Alekseeva, P. A., Kyzar, E. J., Collins, C., Nichols, D. E., and Kalueff, A. V. Understanding Central Nervous System Effects of Deliriant Hallucinogenic Drugs through Experimental Animal Models. ACS Chem. Neurosci., 2018;10:143–154.

98. Malva, J. O., Santos, S., and Macedo, T. Neuroprotective Properties of *Valeriana officinalis* Extracts. Neuro. Res., 2004; 6(2): 131–140. doi:10.1007/bf03033215.
99. Chahal, A. K., Chandan, G., Kumar, R., Chhillar, A. K., Saini, A. K., and Saini, R. V. Bioactive Constituents of Emblica officinalis Overcome Oxidative Stress in Mammalian Cells by Inhibiting Hyperoxidation of Peroxiredoxins. J. Food Biochem., 2020; 44(2): e13115. doi:10.1111/jfbc.13115.
100. Husain, I., et al. Exploring the Multifaceted Neuroprotective Actions of *Emblica officinalis* (Amla): A Review. Metab. Brain Dis., 2019; 34(4): 957–965. doi:10.1007/s11011-019-00400-9.
101. Mahendra, P., and Bisht, S. Ferula asafoetida: Traditional Uses and Pharmacological Activity. Pharmacogn. Rev., 2012; 6(12): 141–146. doi:10.4103/0973-7847.99948.
102. Alam, W., Khan, H., Khan, S. A., Nazir, S., and Akkol, E. K. Datura Metel: A Review on Chemical Constituents, Traditional Uses and Pharmacological Activities. Curr. Pharm. Des., 2021; 27(22): 2545–2557. doi:10.2174/1381612826666200519113752.
103. Kennedy, D. O., Bonnländer, B., Lang, S. C., Pischel, I., Forster, J., Khan, J., Jackson, P. A., and Wightman, E. L. Acute and Chronic Effects of Green Oat (*Avena sativa*) Extract on Cognitive Function and Mood during a Laboratory Stressor in Healthy Adults: A Randomised, Double-Blind, Placebo-Controlled Study in Healthy Humans. Nutrients, 2020; 12(6): 1598. doi:10.3390/nu12061598.
104. Sharma, V., Sharma, R., Gautam, D. S., Kuca, K., Nepovimova, E., and Martins, N. Role of Vacha (*Acorus calamus* Linn.) in Neurological and Metabolic Disorders: Evidence from Ethnopharmacology, Phytochemistry, Pharmacology and Clinical Study. J. Clin. Med., 2020; 9(4): 1176. doi:10.3390/jcm9041176.
105. Yadav, M. K., Singh, S. K., Singh, M., Mishra, S. S., Singh, A. K., Tripathi, J. S., and Tripathi, Y. B. Neuroprotective Activity of *Evolvulus alsinoides* & *Centella asiatica* Ethanolic Extracts in Scopolamine-Induced Amnesia in Swiss Albino Mice. Open Access Maced. J. Med. Sci., 2019; 7(7): 1059–1066. doi:10.3889/oamjms.2019.247.
106. Vallverdú-Queralt, A., Regueiro, J., Martínez-Huélamo, M., Rinaldi Alvarenga, J. F., Leal, L. N., and Lamuela-Raventos, R. M. A Comprehensive Study on the Phenolic Profile of Widely Used Culinary Herbs and Spices: Rosemary, Thyme, Oregano, Cinnamon, Cumin and Bay. Food Chem., 2014; 154: 299–307.
107. Di Cesare Mannelli, L., et al. Anti-Neuropathic Effects of *Rosmarinus officinalis* L. Terpenoid Fraction: Relevance of Nicotinic Receptors. Sci. Rep., 2016; 6: 34832. doi:10.1038/srep34832.
108. de Oliveira, J. R., Camargo, S. E. A., and de Oliveira, L. D. *Rosmarinus officinalis* L. (Rosemary) as Therapeutic and Prophylactic Agent. J. Biomed. Sci., 2019; 26: 5. doi:10.1186/s12929-019-0499-8.
109. Bansal, Y., and Bansal, G. Analytical Methods for Standardization of *Aegle Marmelos*: A Review. J. Pharm. Educ. Res., 2011; 2(2): 37–44.
110. Tandon, P. N. Indian *Rauwolfia* Research Led to the Evolution of Neuropsychopharmacology & the 2000 Nobel Prize (Part II). Indian J. Med. Res., 2021; 154(1): 169–174. doi:10.4103/ijmr.ijmr_1674_21.
111. Gupta, A., Pandey, R., Singh, S., and Sharma, B. A Review on Toxins Mediated Neuro Diseases and Ameliorative Role of Certain Phytochemicals. J. Pharmacol. Clin. Toxicol., 2023; 11(3): 1179002E.

2 Synaptic Plasticity and Its Uses

*Sneha Ranjan, Shrey Dwivedi, and Sangeeta Singh**
Indian Institute of Information Technology Allahabad
Prayagraj, Uttar Pradesh, India

2.1 INTRODUCTION

The basic goal of neuroscience is to understand how the brain stores memory and develops learning ability in humans to survive in challenging environments. Neurons execute the uptake, processing, and storage of information sent by their neighbouring neurons. This information is then integrated into a signal that is fed into the neural network. Reception and transmission of this information involves 10,000 to 100,000 excitatory neurons. But the above attributes of neurons are imparted to them by their synaptic connection arrays, which essentially establish function-specific dedicated neural circuits. Synapses arbitrate the point-to-point communication and build circuits between neurons via asymmetric specialized intercellular junctions. These cellular junctions are flanked by the pre-synaptic neuron—axonal end of the information transmitting neuron and the dendritic (also called spines) segment of a post-synaptic neuron (the information receiving neuron). It is difficult to anticipate the computational nature of synapses on the basis of their site of formation or ultrastructure given their variations [1,2]. Furthermore, neuronal input and output synapses also vary, which in turn is determined by trans-synaptic interactions among neurons in synapse with each other. Synaptic diversity includes variation in neuro-modulatory receptors, the type of neurotransmitter, its chances for release, and configuration of the post-synaptic receptor. This translates to diversity in composition of proteins found in the synapses that make up the "synaptome" of neural circuits [3,4].

Around 70 years ago, Donald Hebb postulated a theory for understanding the synaptic connections between neurons. He postulated that if cell A takes part most frequently or persistently in the activity or firing of cell B, the bond between cell A and cell B would naturally grow and become stronger [5]. His theory has been presented more concisely by Siegrid Lowel: "Cells that fire together, wire together." The idea behind Hebb's postulate is that he observed learning-related changes based on the strength of the synaptic connection between neurons and an assembly of the neuronal family. Hebb's postulate revolves around the idea of efficient credit assignment, a crucial procedure for assigning synapses that are accountable for specific responses, be they related to escape, reward, or other adaptive behaviors [6]. These learning-related changes and the adaptivity of new information are what describe the term "plasticity" whereas "synaptic plasticity" refers to the transmission of information between two neurons to strengthen or weaken the synaptic connections. This strengthening and weakening of synaptic connections are driven by two factors: long-term potentiation (LTP) and long-term depression (LTD) [7], discussed later in this chapter. Synaptic plasticity can be either short-term or lifelong depending on the occurrence of memories that are stored in brain circuits from milliseconds to minutes or throughout life [8]. A special class of phosphoprotein family, "synapsins," located in the pre-synaptic terminal releases a neurotransmitter such as acetylcholine (Ach), which is further taken up by the post-synaptic receptor into the synapse. The synapsins are phosphorylated by the activation of protein kinase to release neurotransmitters (Ach) in the post-synaptic terminal, the only mechanism that permits synaptic plasticity [9]. Ach is diffused across the synapses by the synaptic vessels when an action potential is generated causing a calcium (Ca^{2+}) ion-gated channel to activate as shown in Figure 2.1. Stimulation of NMDA (n-methyl-d-aspartate) receptors and protein-kinase C (PkC) are two examples of the signaling cascades that are triggered at the post-synaptic terminal through the association of Ach

DOI: 10.1201/9781003464648-3

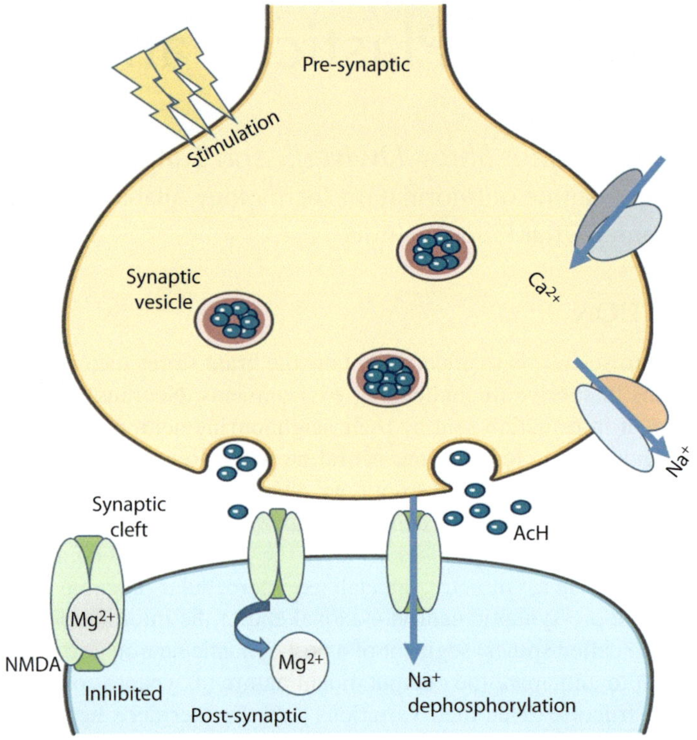

FIGURE 2.1 The structure of synaptic plasticity.

and the receptor [10]. Glutamate receptors of the NMDA type are initially at rest because the magnesium (Mg^{2+}) ion is blocking them. The Mg^{2+} block is lifted as soon as the membrane depolarizes as a result of an action potential, allowing Ca^{2+} ions to enter cells via the NMDA receptor [11]. The influx of Ca^{2+} ions into the cells causes the NMDA receptor to become activated, which initiates lifetime potentiation as shown in Figure 2.1. A more thorough explanation of this process will be covered later in this chapter.

2.2 SYNAPTIC PLASTICITY—CLASSIFICATION AND ATTRIBUTES UNDER HOMEOSTATIC CONDITIONS

2.2.1 Classification of Synaptic Plasticity

The complex interactions between Ca^{2+} signaling, downstream biochemical processes, and NMDA receptors highlight their underlying influence on the flexibility and strength of synaptic connections [12]. These interactions are essential for the brain to utilize its learning and memory abilities properly. Synaptic plasticity is classified into two categories, i.e., LTP and LTD. LTP increases the strength of neuronal connection by maintaining a connection between two neurons and forms long-lasting memory whereas for LTD it is vice versa [13]. These two mechanisms are activated by NMDA receptor through Ca^{2+} ion influx. NMDA is a type of glutamate receptor that acts as an ion channel as it allows the passage of various ions such as Ca^{2+}, sodium (Na^{+}), and potassium ions whereas it also acts as a receptor [14]. Depolarization of neurons is necessary to eliminate the Mg^{2+} ion that blocks NMDA at resting membrane potential. When the electric current across the membrane becomes negative because the cell is receiving more Na^{+} ions, the membrane is depolarized removing the Mg^{2+} ion inhibition. As soon as this inhibition is removed, the NMDA receptor allows an influx of Ca^{2+} ions into the cells. This inflow triggers intracellular signaling cascades in which Ca^{2+}/calmodulin protein kinase II (CaMKII) is activated. CaMKII is a crucial enzyme localized in the post-synaptic density and has a unique feature to undergo autophosphorylation even when the Ca^{2+} level declines [15–17]. Activation of CaMKII further leads to phosphorylation of the α-amino-3-hydroxyl-5-methyl-4-isoxazolaproprionic acid (AMPA) receptor increasing its sensitivity towards the glutamate (neurotransmitter) receptor, an alteration that enhances the number of AMPA receptors to enter into the post-synaptic membrane. The insertion of the AMPA receptor strengthens the synaptic connection between two neurons leading to LTP. These alterations show a biological process that drives fresh memory formation and stores memory throughout life. On the other hand, in LTD low levels of Ca^{2+} ions or other signaling cascades activate phosphatases, an enzyme that removes the phosphate group from proteins and counteracts the activity of kinases. Activation of phosphatases dephosphorylates the AMPA receptor, and also its sensitivity towards glutamate receptor is reduced, which then weakens the synaptic connection as shown in Figure 2.2 [18–20]. The trigger for intracellular signaling cascades in both scenarios is the Ca^{2+} influx via NMDA receptors. In turn, these cascades alter the post-synaptic environment, resulting in long-lasting modifications to synaptic strength and supporting the complex mechanisms involved in memory and learning.

Neurotrophins determine the strength and adaptability of neural connections through their complex interactions with receptors and the complex signaling cascades. These dynamic processes that neurotrophins orchestrate play a major role in the intricate mechanisms that underlie learning and memory in the brain. Certain neurotrophins such as the brain-derived neurotrophic factor (BDNF) also participate in the induction of LTP and LTD. Neurotrophins are released and discharged in a

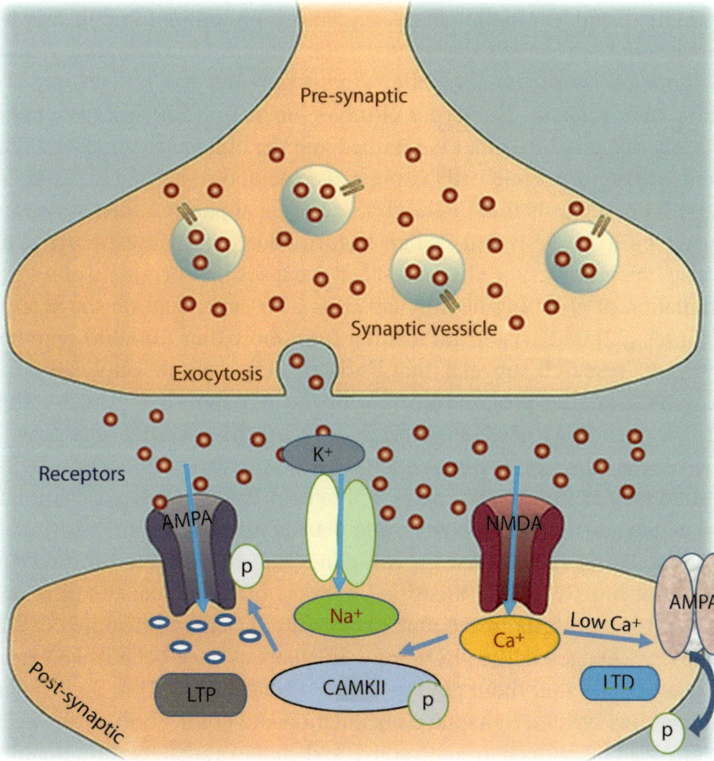

FIGURE 2.2 The mechanism of LTP and LTD is shown. Because of the dephosphorylation of the AMPA receptor, low Ca^{2+} ion inflow causes LTD, but its phosphorylation causes LTP.

manner dependent on activity and bind to their receptors; for example, BDNF binds to the TrKB receptor on the post-synaptic membrane and causes phosphoinositide-3 kinase or mitogen-activated protein kinase (MAPK) pathways to initiate intracellular signaling cascades.This interaction phosphorylates the AMPA receptor and enhances its insertions into the post-synaptic membrane as well as its sensitivity towards glutamate receptors, leading to LTP by strengthening the synaptic connection between two neurons [21,22]. In contrast, LTD has different signaling pathways to follow. Neurotrophins such as P75NTR are activated, and in response to this, phospholipase C is also activated triggering an intracellular signaling cascade. Furthermore, this enzyme activity induces the production of diacylglycerol (DAG) and inositol triphosphate (IP3). While DAG activates protein kinase C, IP3 induces the release of Ca^{2+} ions, which together are involved in the internalization of the AMPA receptor into the post-synaptic membrane, and reduce AMPAsensitivity towards glutamate receptors weakening the synapse [23,24].

2.2.2 Changes and Processes during Development

Cells embark on synapse assembly with the onset of embryogenesis in humans, and this carries over to the postnatal cortex spanning many years. After this massive period of synaptogenesis, there is a decline in the number of synapses over the first two decades of life, which finally levels out by the end of the third decade [25]. In the cortex's shrinking phase, an incredible number of developmental synapses are eliminated, accounting for thousands of synapses each second, or more than 40% of all synapses [26]. The synaptic connections of most neurons in the brain, along with their corresponding axonal and dendritic components, are not stable. The half-life of the hippocampal region's synaptic connections might be shorter than that of a delayed, lifetime potentiation [27]. Microglia play a significant role in the elimination of synapses, which is initiated by the interaction between neuronal CD47 and microglial signal-regulatory protein-α [28,29]. It is well-exemplified that a tug-of-war occurs between synapses during the process of developing synapse elimination. The traditional homogeneity of the cerebellar circuit provides the best example of the eradication of the climbing fiber synapse. Numerous climbing fibers initially innervate cerebellar Purkinje cells; as a result, only one climbing fiber is retained and the others are removed. The mechanism is activity-dependent wherein Purkinje cells express the metabotropic-glutamate receptor (mGluR1) and P-KCγ. [30,31]. This suggests that mGluR1 stimulation at parallel-fiber synapses causes P-KCγ and phospholipase Cb4 to be activated, which in turn causes a backward signal to be generated, which in turn causes supernumerary climbing-fiber synapses that originate at a Purkinje cell to be eliminated. Elimination of climbing-fiber synapses is also dependent on the activity of numerous parallel fiber synapses [32] that arise from feeding from more than 100,000 granule cells for every Purkinje cell. Previous research suggests that Purkinje cells may be implicated in the retrograde signaling pathway by secreting progranulin [33] or the semaphorins Sema3A and Sema7A [34]. However, it is still unclear how these intermediaries fit in with mGluR1 signaling. The process by which embryonic neurons develop involves axons that finally form connections with their synaptic destinations by following conventional routes. Canonical chemotropic navigational cues, such as secreted proteins called netrins, semaphorins, and plasma membrane proteins that manipulate neuronal growth by remodeling of neuronal growth cones, are responsible for directing axon growth [35,36]. This family of hints was first identified because of its critical function in the creation of new neurons, but it also manifests in the mature nervous system, with some of its members being amplified at synapses. Changes in the physiology of synapses have been linked to semaphorin signaling pathways, suggesting that these pathways may influence the form and transmission of synapses in the adult nervous system. Synaptic consolidation and eradication during activity-dependent plasticity are determined by both chemo-attractive and chemo-repellent directional cues, according to reorganization of the dendritic cytoskeleton during memory formation [37–39]. Growth cone collapse during development and the adult brain are the two parallel manifestations of semaphoring IIIA (Sema3A), a repulsive directional indication [40]. With the exception of Sema3s, very few semaphorins are membrane-bound in general. They are secreted and attached to a transmembrane complex of holoreceptors, namely neuropilin-1 (Nrpl) and Plexin A1-4, which regulates the formation of the dendritic spine and neuronal morphology [41]. Sema3A causes dendritic spine collapse in the developing nervous system by activating collapsin response mediator protein 2 (CRMP2), which is mediated by Nrpl/Plexin A1-4 [42]. CRMP2 signaling has been linked to synaptic function, Alzheimer's disease (AD) and other neurodegenerative diseases [43]. Microtubule breakdown, cytoskeletal remodeling, and dendritic spine collapse are coordinated by CRMP2 phosphorylation, and these events may subsequently promote the reinforcement of auxiliary existing synapses [44]. Astrocytes can modify synaptic plasticity in both the short- and long-term through a variety of mechanisms, including glio-transmission, structural plasticity, spine removal, and neurotransmitter and ion uptake. Microglia may increase synapse turnover and homeostasis by continuously examining synapses with dynamic processes. Oligodendrocytes facilitate activity-dependent learning by introducing new myelination or modifying existing myelin characteristics, which ultimately results in modified circuit activity. Neurons are not the only cells that can communicate with one another; these cells can also link with one another. Gap junctions, for instance, provide astrocyte-to-astrocyte communication, which helps maintain synchronization across the circuit. Research conducted *in vitro* on human neocortical neurons has demonstrated that high-frequency stimulation elicited potentiation in synaptic caliber, while low-frequency stimulation de-potentiated it [45]. This is reciprocal; that is, a stimulus can cause one type of plasticity to react negatively to another. For instance, high-frequency stimulation can stimulate synapses that follow LTD, and vice versa. The crucial events for LTP and LTD, respectively, are the insertion and removal of the AMPA receptor [46]. By analyzing extracellular field potentials and intracellular excitatory post-synaptic potentials, these conclusions were reached. The increase and decrease in electrophysiological reactions in the cases of LTP and LTD, respectively, provide more clarification [47]. Reports on *in vitro* synaptic plasticity induction models align with findings from *in vivo* investigations.

2.3 SYNAPTIC PLASTICITY—ATTRIBUTES UNDER PATHOLOGICAL CONDITIONS

2.3.1 Neurodegeneration and Effect of Major NDs on Plasticity of Synapses

2.3.1.1 Alzheimer's Disease

Alzheimer's disease is a type of neurodegenerative disease that causes memory loss and cognitive dysfunction. According to the Alzheimer's Association in 2022, around 6.5 million people were affected by this disease, and it is believed it will cross 13.8 million after 2060 as it is increasing every 65 seconds [48]. Amyloid-beta (Aβ) and tau protein buildup forming fibrils and oligomers is a hallmark of AD development [49]. When tau protein is over-phosphorylated, intractable twisted fibers known as neurofibrillary tangles (NFTs) are created inside brain cells. Aβ is created through the progressive cleavage of amyloid precursor protein (APP) by the enzyme ϒ-secretase. [50]. Numerous neurological conditions, including synaptic dysfunction, neuronal migration, and improper neural connection, are brought on by the activation of these two proteins. Synaptic plasticity is disrupted by the accumulation of Aβ in brain cells and causes AD. Production of Aβ triggers a series of events that disrupt Ca^{2+} homeostasis. Events included are (i) increased release of Ca^{2+} ion from endoplasmic reticulum through interaction with inositol, 1 4 5, triphosphate, and Ryanodine receptors [51], (ii) irregular flow of Ca^{2+} ion due to Aβ-derived diffusible ligands (ADDLs) by influencing NMDA receptor, as tabulated in Table 2.1 (iii) Aβ induces the formation of pores in the plasma membrane for permeability of Ca^{2+} ion and exocytosis of AMPA and NMDA receptors, and (iv) high expression of Aβ results in the dephosphorylation of calcineurin and phosphorylation of CaMKII [52,53]. These events highlight the bidirectional relationship, implying that NMDAR activity is influenced by Aβ accumulation as well as Aβ production, which collectively contribute to the reduction of LTP and the subsequent increase in LTD, causes the weakening of synaptic

TABLE 2.1
Receptors Involved in Different Neurodegenerative Diseases

Type of NDs	Receptor/Gene Involved	Effect on Synaptic Plasticity	References
AD	NMDA, AMPA, SORTA, PSD95	Induce LTP, in synaptic plasticity Localized in post-synaptic membrane, regulate both LTP and LTD Reduces Aβ production, results in LTP Regulate synaptic transmission and plasticity	[91–93]
TBI	NMDA, AMPA, PSD95, mGLURs	Excessive influx of Na^+ ion causes post-traumatic neural death Increased synaptic Glu causes post-trauma overload of calcium Activation of post-synaptic signaling pathway causes post-traumatic cellular injury Act on pre-synaptic membrane, enhances synaptic strength	[94–97]
PD	dopamine D1/5 VPS35, LRRK2, SH3GL2, PRKN, SYNJ1, DNAJC6	Drop in LTP Defective endocytosis of synaptic vesicles	[56,66]
HD	NMDARs BDNF, Glu, Ach	Poor induction and sustenance of LTP, low PTP in STP	[78,79]
MS	NMDARs, elevated IL-1β	Abnormal LTP (early phase), inhibition of LTD and upregulated LTP (late phase)	[85,86]

connection by impairing neurons over time. This complex interaction highlights the intricacy of the molecular and cellular changes that take place in AD, which eventually lead to synaptic dysfunction and cognitive decline.

2.3.1.2 Traumatic Brain Injury

Traumatic brain injury (TBI) is brought on by outside forces that range from a mild headache to severe damage to brain cells and tissues also known as intracranial damage. Synaptic plasticity is a crucial and multifaceted process in the pathology of TBI. In the cerebrospinal fluid (CSF) of TBI patients, extracellular glutamate levels appear to rise after the trauma and continue abnormally for as long as one week. As described in the pathology of AD, TBI also triggers a cascade of events. Particularly, the excitation of glutamate levels activates two types of inotropic receptors, which are AMPA and NMDA receptors crucial for synaptic plasticity and its transmission as tabulated in Table 2.1. AMPA and NMDA receptors are primarily involved in the influx of Na^+ and Ca^{2+} ions into the post-synaptic neuron. An influx of Na^+ ions by AMPA leads to depolarization of post-synaptic neurons. When the glutamate level is at its peak, the four subunits of the AMPA receptor glutamate ionotropic receptor (GLuR1-4) undergo alteration that leads to aggravation of Ca^{2+} overload and prolongs glutamate toxicity. Therefore, the receptor's Ca^{2+} permeability increases when intracellular Ca^{2+} is released due to the expression of AMPA receptors lacking (GLuR2) subunit. In contrast, the excitotoxicity of TBI is exacerbated due to GLuR2-deficient AMPA receptors. When the NMDA receptor binds to post-synaptic density protein (PSD95) —a protein that is crucial for the regulation of synapses—a complex forms at the synapses and further interacts with PKC, which phosphorylates the GLuR2 subunit of the AMPA receptor. Traumatic brain injury-induced excitotoxicity highlights the complex chemical processes behind synaptic dysfunction and subsequent neuronal death. The complex relationship amongst changed AMPA receptor subtypes, glutamate dysregulation, and the subsequent effects on intracellular Ca^{2+} levels contributes to the pathophysiology of excitotoxic neuronal death following TBI. Gaining an understanding of these mechanisms is essential to creating focused treatment plans that lessen the negative consequences of glutamate dysregulation and shield TBI patients' neurons from damage [54].

2.3.1.3 Parkinson's Disease

In Parkinson's disease (PD), dopaminergic (DA) neurons in the mid-brain's extrapyramidal zone weaken with widespread accretion of Lewy bodies or alpha-synuclein proteins. DA injury is the prime sign of PD, but the subsequent neurodegeneration is also accompanied by motor and non-motor symptoms [55]. An association between hippocampal neuroplasticity and DA transmission has been postulated by reports in case of non-motor symptoms, but the role of systems that do not involve DA cannot be ruled out. Cognitive decline in PD is correlated with alterations in synaptic mechanisms. Neurotoxic models of PD (6-hydroxydopamine (6-OHDA)-lesioned rats and 1-methyl-4-phenyl-1,2,3,6-tetrahydropyridine (MPTP)-lesioned mice) have been reported to exhibit reduced LTP [56–58]. Lower levels of LTP were also reported in the hippocampal cornu ammonis 1 (CA1) region of alpha-synuclein transgenic mice [56]. After alpha-synuclein preformed fibrils (PFF) treatment, a fall in hippocampal neurons' excitatory post-synaptic current along with compromised excitation was also reported [59]. DA neurons control dopamine D1/5 receptors [60] and upregulated synthesis of synaptic proteins [61] for the upkeep of hippocampal LTP (Table 2.1). Both of these factors partake in functional preservation of the hippocampus by linking hippocampal synaptic plasticity and DA input. Clinical [62] and preclinical stages [56,63] of PD have shown improvements in cognitive functions via L-DOPA (levodopa) and 1-3, 4-dihydroxyphenylalanine-mediated hippocampal LTP restoration. Besides DA signaling, synaptic plasticity is adversely affected by neurotransmitters like serotonin [64] and noradrenaline (NA) [65]. Genetic studies on PD have revealed genes responsible for interfering with endocytosis of synaptic vesicles to be a causal factor for the disease. These include vacuolar protein sorting ortholog 35 (VPS35), leucine-rich repeat kinase 2 (LRRK2), endophilin A1 (SH3GL2), parkin (PRKN), synaptojanin 1 (SYNJ1), and auxilin (DNAJC6) as tabulated in Table 2.1 [66].

2.3.1.4 Huntington's Disease

Huntington's disease (HD) is an autosomal dominant neurodegenerative condition that displays impaired motor and cognitive functions [67]. Behavioral alterations in HD such as dystonia, chorea, and emotional decline are also displayed [57]. HD-related depressive-like behavior has been reported to be caused by hippocampal serotonergic disturbance [68], while cognitive degeneration is tied to aberrant extracellular glutamate (Glu) interactions [69]. Deficits in lifelong and short-term functional plasticity vary in mouse and human models of HD [70,71]. Researches on lifelong plasticity dysfunction hint at poor induction and sustenance of LTP, while post-tetanic potentiation (PTP) is reduced in short-term plasticity [72]. In their hippocampi, proline rich tyrosine kinase 2 (Pyk2) deficient, Q175FDN knock-in, and Hu97/18-TG mice were demonstrated to exhibit LTP interference [72–74]. Experimental data suggested that onset of symptoms in R6/1-TG mice, R6/2-TG, and Pyk2-deficient was accompanied by NMDARs-mediated LTP at their hippocampal CA1 synapse (Table 2.1) [73,75,76]. These conditions lead to learning and memory-related deficiencies [47,77]. Hippocampal dysfunction is amplified by lifelong impairment caused by defects in multiple signaling pathways [72]. Neurotrophic factors like brain-derived neurotrophic factor (BDNF) [78], and neurotransmitters like Glu, DA and acetylcholine (Ach) [79] have also been implicated in HD-related synaptic plasticity as tabulated in Table 2.1.

2.3.1.5 Multiple Sclerosis

Multiple sclerosis (MS) is characterized by an autoimmune outbreak on axonal myelin sheaths, which triggers inflammation and degeneration of neurons at the site(s) of damage. Oligodendrocytes (OL) are subsequently lost due to the inflammatory micro-environment, which culminates in central nervous system (CNS) neurons' demyelination [80] and cognitive impairment [81]. Disturbances in synaptic plasticity have been reported in MS patients implying that cognitive aberrations may be a consequence of synaptic imbalance [82]. MS patients' hippocampi exhibited decreased synapse numbers and an adverse effect on entities that participate in synaptic plasticity—integrity, learning, and memory; Glu homeostasis; as well as axonal transport [83]. In animal models for MS like experimental autoimmune encephalomyelitis (EAE) mice, early as well as late phases of the condition were reported to have shown defects in cognitive performance as well as in hippocampal LTP [84]. This is due to elevated levels of interleukin-1β (IL-1β) and abnormal association of NMDARs as tabulated in Table 2.1 [85]. On the other hand, late phase entails LTD inhibition with heightened LTP in the CA1 area of the hippocampus [86]. Pathophysiological changes in synaptic plasticity and LTP flux were further found to be deeply impacted by cytokines like IL-1β [85], invasive T cells [87], and activated microglia [85,87]. LTP stimulation and sustenance along with synaptic activity are determined by ionotropic receptors like NMDA [88] and AMPAR [89], which are expressed under IL-1β-mediated regulation. During the remitting phase of EAE any changes—normal or pathological—in synaptic plasticity of the hippocampus is accompanied by high IL-1β and nicotinamide adenine dinucleotide phosphate oxidase with constant activation of glial cells [90]. It is difficult to determine when the LTP disruption triggers, as some studies suggest early-phase initiation while many others hint at late phase. The prevailing postulate considers that late phase aggravates gliosis, loss of axons, and disintegration of tissue in contrast with the milder early phase of MS advancement [84].

2.4 IMPORTANCE AND CLINICAL RELEVANCE FOR NEURODEGENERATIVE THERAPY

There are various challenges to performing clinical trials of therapeutic processes for NDs due to subjects having impaired cognition and physical dysfunction and the progressive nature of NDs. Therapeutic approaches to prevent, slow the progression of, and cure NDs include physical, chemical, pharmacological, and lifestyle interference. We discuss some of these approaches here.

As discussed earlier, AD is a loss of memory function and cognitive dysfunction having no cure in the medicinal field, but there are still some approaches that can slow down or reduce the symptoms of AD such as immunotherapy, NMDAR modification, and gene therapy Trials for Aβ and tau immunotherapy have been found to reduce the production of these proteins in patients' brains, and Si-RNA gene therapy can silence the APP gene, which reduces the formation of Aβ as tabulated in Table 2.2. Both of these therapies lower the risk of AD. On the other hand, NMDAR modification increases the downstream effect of D-serine, a main agonist of NMDAR.

Therapeutic approaches for TBI are exosome therapy, cell-based therapy, and neurotherapy. Exosome therapy follows a cell-free approach in which a cell-free exosome is derived from mesenchymal stem cells (MSCs), whereas cell-based therapy involves transplantation of bone marrow mesenchymal stem cells (BM-MSCs) and reduces inflammation by causing neuroprotection. Deep brain stimulation (DBS) provides electrical stimulation at constantly high frequencies to the basal ganglia, and this therapy affects the physiological and pathological circuits of neurons [98]. It can mimic the brain lesions' effects at specific sites [sub-thalamic nucleus (STN) and the globus pallidus pars interna (GPi)] and inhibit the activity of neurons. This has been postulated to work by elevating inhibitory GABAergic signals to the nucleus, deactivation of abnormal beta oscillations, obfuscation of coded information, and downregulation of the synthesis and release of neurotransmitters and hormones. These mechanisms can engage the cementing of neural plasticity in a variety of forms preceded by neural circuits' activation or inhibition [99]. The simultaneous recording and storage of many targets and their local field potentials using internal pulse generators (IPGs) facilitates clinicians to better illuminate the DBS-induced symptomatic ablation in PD.

In PD, neuro-inflammation has been targeted through the phosphodiesterase 10A (PDE10A) enzyme, which brings about the hydrolysis of cyclic AMP (cAMP) and cyclic GMP (cGMP) in spiny neurons. PDE10A is associated with synaptic plasticity, ion conductance, and other brain functions and DA signaling regulation [100]. In both early L-DOPA-treated PD patients and *de novo* conditions,

TABLE 2.2
Different NDs and Their Therapeutic Approaches

Target NDs	Therapeutic Approaches	Target	Outcome(s)	References
Alzheimer's disease	Aβ-immunotherapy	Aβ-oligomers	An anti-Aβ vaccine is developed to reduce Aβ burden	[106]
	NMDAR modification	D-amino acid oxidase (DAAO)	Increases the downstream effect of D-serine (main co-agonist of NMDAR)	[107]
	Tau-immunotherapy	Microtubule-binding region of tau	Decreases tau aggregation	[106]
	Si-RNA gene therapy	APP gene	Silencing APP gene reduces the formation of Aβ	[108]
Traumatic brain injury	Exosome therapy	Cell-free exosomes derived from MSCs	Increase angiogenesis and neurogenesis and decrease inflammation in the brain	[109]
	Cell-based therapy	BM-MSCs	Transplantation of BM-MSCs reduces brain damage and inflammation	[110]
	Transcranial brain stimulation (neurotherapy)	Brain	Provide neuroprotection and prevent neuroinflammation	[111]
Parkinson's disease	Deep brain stimulation	Sub-thalamic nucleus (STN) and the globus pallidus pars interna (GPi)	Neural plasticity establishment Symptomatic ablation	[98,99]
	Papaverine-mediated neuro-inflammation reversal	DA activity, PDE10A enzyme inhibition	Drop in both inflammatory gene expression and microglial activation	[100–102,112]
Huntington's disease	Synaptic modulation	p38-alpha enzyme inhibition	Synaptic defect rescue	[103,104]
	GMP-grade hESC-NSC transplantation	Mice hippocampi	Reversal of synaptic alterations Motor deficit recovery Late stage cognitive decline improvement	[105]

striatal dopamine transporter (DAT) deficiency was reported to be triggered by lack of PDE10A. This may establish a possible correlation between DA activity and PDE10A [101]. Papaverine, a PDE10A inhibitor, has been reportedly tied to upregulation in expression of protein kinase A (PKA) signaling components, including BDNF, glial cell line-derived neurotrophic factor (GDNF), and B cell lymphoma 2. Papaverine's inhibitory effect on inflammatory gene expression and microglial activation further validates its suitability to mitigate PD-related neuroinflammation [102].

Synaptic modulation has been targeted for HD via neflamapimod, which is a small molecule reported to have brain penetrability with p38-alpha enzyme inhibition ability. The latter controls inflammation, and on being upregulated to exceedingly high levels can debilitate nerve cell communication [103,104]. Neflamapimod can rescue synaptic defects in mouse models, and its therapeutic application can also be extended to NDs like HD [103,104]. The clinical trial NCT03980938 has been studying the putative rescue of functional abnormality in the hippocampus of HD patients in the early stages. Many stem cell therapies have been tried to target HD. For instance, transplanting human embryonic stem cell-derived neural stem cells (hESC-NSCs) of goods and manufacturing practice grade into Q140 HD knock-in and R6/2 mice is one such treatment [105]. R6/2 animals were shown to reverse synaptic changes, recover from motor deficits, and integrate stem cells into neural circuits, whereas Q140 knock-in mice showed improvement in late stages of cognitive decline. The transplants reduced the aggregation of mutant huntingtin (mHTT) and increased the expression of BDNF [105].

2.4.1 Modulation of Synaptic Plasticity—Optogenetics and Chemogenetics

The mechanisms by which modifications in brain activity lead to learning, memory, and neurological diseases have already been covered in this chapter. Optogenetics and chemogenetics are two techniques through which researchers investigate this modulation. In optogenetics, opsins present in photoreceptor cells of the retina are activated by light at particular wavelengths. Researchers used different engineered methods such as site-directed mutagenesis, chimeric opsins, directed evolution, etc., to develop several families of opsins like channel rhodopsins (ChRs) and halorhodopsins (HRs) as shown in Figure 2.3. ChRs are light-sensitive proteins that are activated in response to specific wavelengths that permit the depolarization of neurons by Na^+ ion influx. In contrast, HRs lead to hyperpolarization of neurons by Cl^- ion influx. These two proteins expressed themselves at the site of the targeted neuronal membrane through the use of viral vectors like adeno-associated virus (AAV) or lentivirus. With extraordinary temporal and spatial precision, researchers can precisely activate or inhibit neurons by applying light at certain wavelengths to the target area. Due to its ability to mimic natural firing patterns and manipulate different neuronal populations within a circuit, this bidirectional control over neuronal activity is very helpful for researching synaptic plasticity [113].

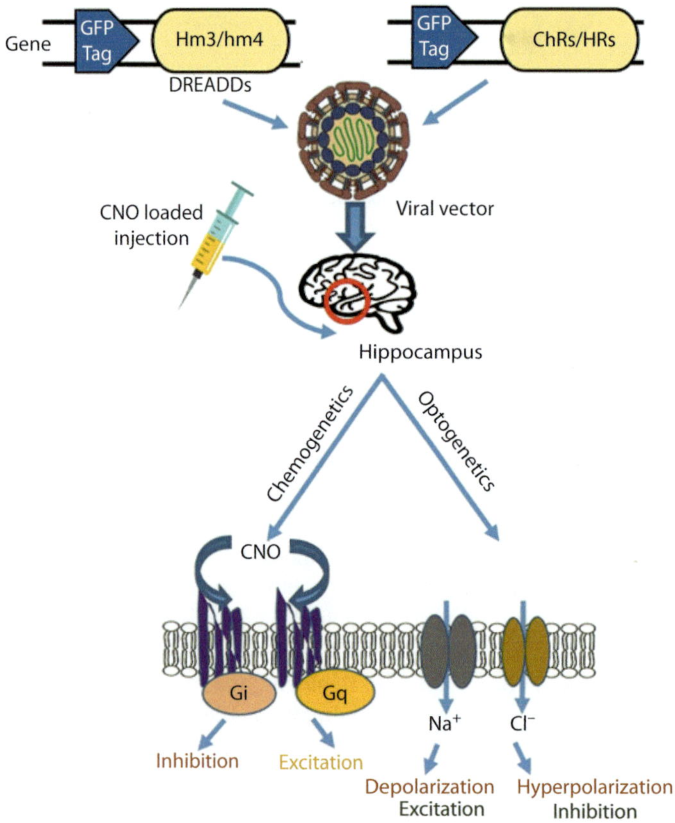

FIGURE 2.3 Chemogenetics and optogenetics' mechanisms are shown. In optogenetics, inhibition results from hyperpolarization and excitement from depolarization in response to axon potential, whereas in chemogenetics, inhibition and excitation are caused by the expression of CNO ligands on the DREADDs receptors.

On the other hand, in chemogenetics, designer receptors exclusively activated by designer drugs (DREADDs) are targeted in the same manner as optogenetics at the targeted neuronal membrane as shown in Figure 2.3. DREADDs are a synthetically designed class of G-protein coupled receptors. Their activity can be precisely altered by the researchers to respond selectively to artificial ligands such as clozapine-N-oxide (CNO). Unlike optogenetics, this uses light's fleeting effects to establish temporal control; chemogenetics uses the drug's pharmacokinetics to produce longer-lasting effects. Chemogenetics provides spatial control as well, enabling bidirectional modulation of particular neuronal populations. By enabling researchers to analyze the intricate workings of neural circuits, examine the molecular processes underlying learning and memory, and look into the functions of particular neuronal populations in a range of physiological and pathological situations, these techniques have greatly advanced our understanding of synaptic plasticity [114].

2.4.2 Imaging Techniques

Imaging techniques are essential for the research of synaptic plasticity because of their ability to view and analyze dynamic changes in the structures and functions of neurons. Two of the techniques are discussed here in this chapter

2.4.2.1 Two-photon Microscopy

The cutting-edge imaging method known as two-photon microscopy is especially useful for examining synaptic plasticity in living tissues because of its remarkable deep-tissue penetration and spatial resolution. In contrast to conventional fluorescence microscopy, two-photon microscopy achieves imaging depth and less damage by simultaneously absorbing two lower-energy photons that stimulate fluorophores. Dendritic spine and synaptic activity dynamic changes are frequently visualized *in vivo* using this approach. Researchers can witness changes in synapse structure and function in real time during learning, memory formation, or other behavioral activities with two-photon microscopy, which can take high-resolution images in living animals. The spatiotemporal dynamics of synaptic plasticity within intact brain circuits have been understood in large part by this imaging technology [115].

2.4.2.2 Ca^{2+} Imaging

Ca^{2+} imaging is a potent method for tracking variations in intracellular Ca^{2+} levels to gain insights into the activity of neurons and synapses during plasticity. Ca^{2+} concentration variations cause fluorescent Ca^{2+} indicators, such as GCaMP, fused with Ca^{2+}-calmodulin, M-13 (a sequence of light chain kinase from myosin), and green fluorescent protein (GFP) to fluctuate in fluorescence intensity. These indicators are chemically or genetically loaded into cells. Because Ca^{2+} is essential for both intracellular signaling and synaptic transmission, this approach is especially useful for researching synaptic plasticity. Ca^{2+} imaging enables researchers to see these changes at the level of individual synapses. As neurons become more active, there is an influx of Ca^{2+} ions into the cell. The spatiotemporal patterns of Ca^{2+} fluctuations can be captured by researchers using sophisticated microscopes and dyes that are sensitive to Ca^{2+}, providing insight into the location and timing of synaptic plasticity. Our comprehension of the fundamental mechanisms regulating synaptic plasticity in both health and disease is aided by this real-time, non-invasive method, which also gives vital information about the activity of individual synapses [116].

2.5 FUTURE PROSPECTS

Future developments in synaptic plasticity research could lead to revolutionary understandings of the intricate processes of the brain and provide opportunities for more precise treatments, cutting-edge technology, and a better comprehension of cognitive functions. Treatment plans for ailments like Parkinson's disease, Alzheimer's disease, and other neuropsychiatric illnesses may be completely changed using precision medicine techniques based on a person's synaptic profile. By customizing therapies to each patient's distinct synapse properties, treatment outcomes might be maximized and adverse effects could be reduced. Still another fascinating area of research is the development of optogenetics and chemogenetics. These methods, which provide accurate regulation of synaptic plasticity and neuronal activity, are still developing. The ability to analyze brain circuits in unprecedented depth may be made possible by future improvements in their accuracy, adaptability, and spatiotemporal resolution. The significance of particular synapses in intricate actions and cognitive processes may become clearer as a result. Gaining such knowledge could be very beneficial for improving brain-machine interface technology and developing tailored neuromodulation medicines. Advances in neuro-prosthetics and improved brain-machine interfaces for people with neurological impairments may result from the merging of synaptic plasticity research and neuro-engineering.

The construction of artificial synapses and bioengineered materials is being investigated by researchers, who draw inspiration from the principles of natural synaptic action. For neural tissue engineering, regenerative medicine, and the creation of novel therapeutic approaches, these projects have great potential. It may be possible to create artificial brain circuits and devices that can duplicate or improve on natural synaptic processes as a result of the confluence of synthetic biology and synaptic plasticity research. The exploration of large-scale synaptic alterations under various experimental settings could be made more efficient and thorough with the use of high-throughput imaging and automated analysis methods. By providing crucial insights into the dynamic nature of synapses under many physiological and pathological situations, these imaging technologies are essential for understanding the spatiotemporal patterns of synaptic plasticity in live tissues.

A comprehensive understanding of the molecular mechanisms controlling synaptic function will also be provided by the integration of multi-omics data, which includes transcriptomics, proteomics, and genomes, with synaptic plasticity research. Through the use of multi-dimensional data, systems biology techniques have the potential to reveal complex regulatory networks that facilitate synaptic plasticity. It will be possible to better understand how molecular processes lead to synaptic adaptations by using this integrative method, which will be crucial in removing the complexity from synaptic signaling pathways.

Finally, there is great promise for the future of synaptic plasticity research in a variety of fields. The next frontier in neuroscience is just waiting to be discovered, from technology advancements and precise therapies to a better comprehension of interconnections at the network level. These prospects have the potential to revolutionize our understanding of how the brain functions and present fresh approaches to treating neurological conditions, improving cognitive function, and creating cutting-edge neurotechnologies. Finding synaptic plasticity's full potential and the significant effects it will have on brain function and cognition are promises of future research.

ACKNOWLEDGMENTS

The authors are grateful to Director, Indian Institute of Information Technology Allahabad, India, for providing facilities for research and thankful to Ministry of Education, India, for providing research fellowship.

AUTHORS' CONTRIBUTION

All authors contributed to the study conception and design. Data collection, analysis, and manuscript writing were performed by Sneha Ranjan and Shrey Dwivedi. Conceptualization, supervision, manuscript editing, and review was done by Dr. Sangeeta Singh. The first draft of the manuscript was written by Sneha Ranjan and Shrey Dwivedi. All authors commented on previous versions of the manuscript. All authors read and approved the final manuscript.

COMPLIANCE WITH ETHICAL STANDARDS

This study does not involve any animals or humans as subjects for experimentation.

COMPETING INTERESTS

The authors have no competing interests to declare that are relevant to the content of this article. All authors certify that they have no affiliations with or involvement in any organization or entity with any financial interest or non-financial interest in the subject matter or materials discussed in this manuscript.

REFERENCES

1. Monday, H. R., Younts, T. J., & Castillo, P. E. (2018). Lifelong plasticity of neurotransmitter release: emerging mechanisms and contributions to brain function and disease. *Annual Review of Neuroscience*, *41*, 299–322.
2. Gjorgjieva, J., Drion, G., & Marder, E. (2016). Computational implications of biophysical diversity and multiple timescales in neurons and synapses for circuit performance. *Current Opinion in Neurobiology*, *37*, 44–52.
3. Nusser, Z. (2018). Creating diverse synapses from the same molecules. *Current Opinion in Neurobiology*, *51*, 8–15.
4. Grant, S. G. N., & Fransén, E. (2020). The synapse diversity dilemma: molecular heterogeneity confounds studies of synapse function. *Frontiers in Synaptic Neuroscience*, *12*, 590403.
5. Magee, J. C., & Grienberger, C. (2020). Synaptic plasticity forms and functions. *Annual Review of Neuroscience*, *43*, 95–117.
6. Triche, A., Maida, A. S., & Kumar, A. (2022). Exploration in neo-Hebbian reinforcement learning: computational approaches to the exploration–exploitation balance with bio-inspired neural networks. *Neural Networks*, *151*, 16–33.
7. Josselyn, S. A., & Tonegawa, S. (2020). Memory engrams: recalling the past and imagining the future. *Science*, *367*(6473), eaaw4325.
8. Stacho, M., & Manahan-Vaughan, D. (2022). The intriguing contribution of hippocampal *long-term depression* to spatial learning and lifelong memory. *Frontiers in Behavioral Neuroscience*, *16*, 806356.
9. Zhang, L., & Zhao, Z.-X. (2006). The impact of synapsins on synaptic plasticity and cognitive behaviors. *Neuroscience Bulletin*, *22*(1), 63–67.
10. Cull-Candy, S., Brickley, S., & Farrant, M. (2001). NMDA receptor subunits: diversity, development and disease. *Current Opinion in Neurobiology*, *11*(3), 327–335.
11. McBain, C. J., & Traynelis, S. F. (2006). Malevolent lurkers no more: NMDA receptors come of age. *The Journal of Physiology*, *575*(Pt 2), 317.
12. Lisek, M., Zylinska, L., & Boczek, T. (2020). Ketamine and calcium signaling—A crosstalk for neuronal physiology and pathology. *International Journal of Molecular Sciences*, *21*(21), 8410.
13. Nabavi, S., Fox, R., Proulx, C. D., Lin, J. Y., Tsien, R. Y., & Malinow, R. (2014). Engineering a memory with LTD and LTP. *Nature*, *511*(7509), 348–352.
14. Nicoll, R. A., & Malenka, R. C. (1999). Expression mechanisms underlying NMDA receptor-dependent long-term potentiation. *Annals of the New York Academy of Sciences*, *868*(1), 515–525.
15. Foster, T. C., Sharrow, K. M., Masse, J. R., Norris, C. M., & Kumar, A. (2001). Calcineurin links $Ca2^{+}$ dysregulation with brain aging. *Journal of Neuroscience*, *21*(11), 4066–4073.
16. Norris, C. M., Halpain, S., & Foster, T. C. (1998). Alterations in the balance of protein kinase/phosphatase activities parallel reduced synaptic strength during aging. *Journal of Neurophysiology*, *80*(3), 1567–1570.
17. Foster, T. C. (2007). Calcium homeostasis and modulation of synaptic plasticity in the aged brain. *Aging Cell*, *6*(3), 319–325.
18. Lee, H.-K., Kameyama, K., Huganir, R. L., & Bear, M. F. (1998). NMDA induces lifelong synaptic depression and dephosphorylation of the GluR1 subunit of AMPA receptors in hippocampus. *Neuron*, *21*(5), 1151–1162.
19. Kumar, A. (2010). Carbachol-induced lifelong synaptic depression is enhanced during senescence at hippocampal CA3–CA1 synapses. *Journal of Neurophysiology*, *104*(2), 607–616.

20. Martin, S. J., Grimwood, P. D., & Morris, R. G. M. (2000). Synaptic plasticity and memory: an evaluation of the hypothesis. *Annual Review of Neuroscience*, *23*(1), 649–711.
21. Lau, D., Bengtson, C. P., Buchthal, B., & Bading, H. (2015). BDNF reduces toxic extrasynaptic NMDA receptor signaling via synaptic NMDA receptors and nuclear-calcium-induced transcription of inhba/activin A. *Cell Reports*, *12*(8), 1353–1366.
22. Messaoudi, E., Ying, S.-W., Kanhema, T., Croll, S. D., & Bramham, C. R. (2002). Brain-derived neurotrophic factor triggers transcription-dependent, late phase *long-term potentiation in vivo*. *Journal of Neuroscience*, *22*(17), 7453–7461.
23. Anastasia, A., Deinhardt, K., Chao, M. V., Will, N. E., Irmady, K., Lee, F. S., Hempstead, B. L., & Bracken, C. (2013). Val66Met polymorphism of BDNF alters prodomain structure to induce neuronal growth cone retraction. *Nature Communications*, *4*(1), 2490.
24. Kowiański, P., Lietzau, G., Czuba, E., Waśkow, M., Steliga, A., & Moryś, J. (2018). BDNF: a key factor with multipotent impact on brain signaling and synaptic plasticity. *Cellular and Molecular Neurobiology*, *38*, 579–593.
25. Petanjek, Z., Judaš, M., Šimić, G., Rašin, M. R., Uylings, H. B. M., Rakic, P., & Kostović, I. (2011). Extraordinary neoteny of synaptic spines in the human prefrontal cortex. *Proceedings of the National Academy of Sciences*, *108*(32), 13281–13286.
26. Bourgeois, J. P., & Rakic, P. (1993). Changes of synaptic density in the primary visual cortex of the macaque monkey from fetal to adult stage. *The Journal of Neuroscience*, *13*(7), 2801. https://doi.org/10.1523/JNEUROSCI.13-07-02801.1993
27. Frey, U., & Morris, R. G. M. (1997). Synaptic tagging and *long-term potentiation*. *Nature*, *385*(6616), 533–536.
28. Paolicelli, R. C., Bolasco, G., Pagani, F., Maggi, L., Scianni, M., Panzanelli, P., Giustetto, M., Ferreira, T. A., Guiducci, E., & Dumas, L. (2011). Synaptic pruning by microglia is necessary for normal brain development. *Science*, *333*(6048), 1456–1458.
29. Ding, X., Wang, J., Huang, M., Chen, Z., Liu, J., Zhang, Q., Zhang, C., Xiang, Y., Zen, K., & Li, L. (2021). Loss of microglial SIRPα promotes synaptic pruning in preclinical models of neurodegeneration. *Nature Communications*, *12*(1), 2030.
30. Kano, M., Hashimoto, K., Chen, C., Abeliovich, A., Aiba, A., Kurihara, H., Watanabe, M., Inoue, Y., & Tonegawa, S. (1995). Impaired synapse elimination during cerebellar development in PKCγ mutant mice. *Cell*, *83*(7), 1223–1231. https://doi.org/10.1016/0092-8674(95)90147-7
31. Kano, M., Hashimoto, K., Kurihara, H., Watanabe, M., Inoue, Y., Aiba, A., & Tonegawa, S. (1997). Persistent multiple climbing fiber innervationof cerebellar purkinje cellsin mice lacking mGluR1. *Neuron*, *18*(1), 71–79. https://doi.org/10.1016/S0896-6273(01)80047-7
32. Ichikawa, R., Miyazaki, T., Kano, M., Hashikawa, T., Tatsumi, H., Sakimura, K., Mishina, M., Inoue, Y., & Watanabe, M. (2002). Distal extension of climbing fiber territory and multiple innervation caused by aberrant wiring to adjacent spiny branchlets in cerebellar Purkinje cells lacking glutamate receptor δ2. *Journal of Neuroscience*, *22*(19), 8487–8503.
33. Uesaka, N., Abe, M., Konno, K., Yamazaki, M., Sakoori, K., Watanabe, T., Kao, T.-H., Mikuni, T., Watanabe, M., & Sakimura, K. (2018). Retrograde signaling from progranulin to Sort1 counteracts synapse elimination in the developing cerebellum. *Neuron*, *97*(4), 796–805.
34. Uesaka, N., Uchigashima, M., Mikuni, T., Nakazawa, T., Nakao, H., Hirai, H., Aiba, A., Watanabe, M., & Kano, M. (2014). Retrograde semaphorin signaling regulates synapse elimination in the developing mouse brain. *Science*, *344*(6187), 1020–1023.
35. Kalil, K., & Dent, E. W. (2005). Touch and go: guidance cues signal to the growth cone cytoskeleton. *Current Opinion in Neurobiology*, *15*(5), 521–526.
36. Kolodkin, A. L., & Tessier-Lavigne, M. (2011). Mechanisms and molecules of neuronal wiring: a primer. *Cold Spring Harbor Perspectives in Biology*, *3*(6), a001727.
37. Nakamura, F., Ugajin, K., Yamashita, N., Okada, T., Uchida, Y., Taniguchi, M., Ohshima, T., & Goshima, Y. (2009). Increased proximal bifurcation of CA1 pyramidal apical dendrites in sema3A mutant mice. *Journal of Comparative Neurology*, *516*(5), 360–375.
38. Tran, T. S., Rubio, M. E., Clem, R. L., Johnson, D., Case, L., Tessier-Lavigne, M., Huganir, R. L., Ginty, D. D., & Kolodkin, A. L. (2009). Secreted semaphorins control spine distribution and morphogenesis in the postnatal CNS. *Nature*, *462*(7276), 1065–1069.
39. Zhang, Y., Cudmore, R. H., Lin, D.-T., Linden, D. J., & Huganir, R. L. (2015). Visualization of NMDA receptor–dependent AMPA receptor synaptic plasticity *in vivo*. *Nature Neuroscience*, *18*(3), 402–407.
40. Cembrowski, M. S., Wang, L., Sugino, K., Shields, B. C., & Spruston, N. (2016). Hipposeq: a comprehensive RNA-seq database of gene expression in hippocampal principal neurons. *Elife*, *5*, e14997.

41. Pasterkamp, R. J., & Giger, R. J. (2009). Semaphorin function in neural plasticity and disease. *Current Opinion in Neurobiology*, *19*(3), 263–274.
42. Uchida, Y., Ohshima, T., Sasaki, Y., Suzuki, H., Yanai, S., Yamashita, N., Nakamura, F., Takei, K., Ihara, Y., & Mikoshiba, K. (2005). Semaphorin3A signaling is mediated via sequential Cdk5 and GSK3β phosphorylation of CRMP2: implication of common phosphorylating mechanism underlying axon guidance and Alzheimer's disease. *Genes to Cells*, *10*(2), 165–179.
43. Cole, A. R., Noble, W., Aalten, L., van, Plattner, F., Meimaridou, R., Hogan, D., Taylor, M., LaFrancois, J., Gunn-Moore, F., & Verkhratsky, A. (2007). Collapsin response mediator protein-2 hyperphosphorylation is an early event in Alzheimer's disease progression. *Journal of Neurochemistry*, *103*(3), 1132–1144.
44. Schubert, V., Da Silva, J. S., & Dotti, C. G. (2006). Localized recruitment and activation of RhoA underlies dendritic spine morphology in a glutamate receptor–dependent manner. *The Journal of Cell Biology*, *172*(3), 453–467.
45. Chen, W. R., Lee, S., Kato, K., Spencer, D. D., Shepherd, G. M., & Williamson, A. (1996). Lifelong modifications of synaptic efficacy in the human inferior and middle temporal cortex. *Proceedings of the National Academy of Sciences of the United States of America*, *93*(15), 8011–8015. https://doi.org/10.1073/PNAS.93.15.8011
46. Malenka, R. C., & Bear, M. F. (2004). LTP and LTD: an embarrassment of riches. *Neuron*, *44*(1), 5–21.
47. Citri, A., & Malenka, R. C. (2008). Synaptic plasticity: multiple forms, functions, and mechanisms. *Neuropsychopharmacology*, *33*(1), 18–41.
48. Gaugler, J., James, B., Johnson, T., Reimer, J., Solis, M., Weuve, J., Buckley, R. F., & Hohman, T. J. (2022). 2022 Alzheimer's disease facts and figures. *Alzheimers & Dementia*, *18*(4), 700–789.
49. Roda, A. R., Serra-Mir, G., Montoliu-Gaya, L., Tiessler, L., & Villegas, S. (2022). Amyloid-beta peptide and tau protein crosstalk in Alzheimer's disease. *Neural Regeneration Research*, *17*(8), 1666.
50. Funamoto, S., Tagami, S., Okochi, M., & Morishima-Kawashima, M. (2020). Successive cleavage of β-amyloid precursor protein by γ-secretase. *Seminars in Cell & Developmental Biology*, *105*, 64–74.
51. Kumar, A., & Foster, T. C. (2019). Alteration in NMDA receptor mediated glutamatergic neurotransmission in the hippocampus during senescence. *Neurochemical Research*, *44*, 38–48.
52. Wang, X., & Zheng, W. (2019). Ca^{2+} homeostasis dysregulation in Alzheimer's disease: a focus on plasma membrane and cell organelles. *The FASEB Journal*, *33*(6), 6697–6712.
53. Parsons, M. P., & Raymond, L. A. (2014). Extrasynaptic NMDA receptor involvement in central nervous system disorders. *Neuron*, *82*(2), 279–293.
54. Jamjoom, A. A. B., Rhodes, J., Andrews, P. J. D., & Grant, S. G. N. (2021). The synapse in traumatic brain injury. *Brain*, *144*(1), 18–31.
55. Leroi, I., McDonald, K., Pantula, H., & Harbishettar, V. (2012). Cognitive impairment in Parkinson disease: impact on quality of life, disability, and caregiver burden. *Journal of Geriatric Psychiatry and Neurology*, *25*(4), 208–214.
56. Costa, C., Sgobio, C., Siliquini, S., Tozzi, A., Tantucci, M., Ghiglieri, V., Di Filippo, M., Pendolino, V., de Iure, A., & Marti, M. (2012). Mechanisms underlying the impairment of hippocampal *long-term potentiation* and memory in experimental Parkinson's disease. *Brain*, *135*(6), 1884–1899.
57. Esmaeili-Mahani, S., Haghparast, E., Nezhadi, A., Abbasnejad, M., & Sheibani, V. (2021). Apelin-13 prevents hippocampal synaptic plasticity impairment in Parkinsonism rats. *Journal of Chemical Neuroanatomy*, *111*, 101884.
58. Wang, Y., Feng, L., Liu, S., Zhou, X., Yin, T., Liu, Z., & Yang, Z. (2019). Transcranial magneto-acoustic stimulation improves neuroplasticity in hippocampus of Parkinson's disease model mice. *Neurotherapeutics*, *16*, 1210–1224.
59. Wu, Q., Takano, H., Riddle, D. M., Trojanowski, J. Q., Coulter, D. A., & Lee, V. M.-Y. (2019). α-Synuclein (αSyn) preformed fibrils induce endogenous αSyn aggregation, compromise synaptic activity and enhance synapse loss in cultured excitatory hippocampal neurons. *Journal of Neuroscience*, *39*(26), 5080–5094.
60. Hansen, N., & Manahan-Vaughan, D. (2014). Dopamine D1/D5 receptors mediate informational saliency that promotes persistent hippocampal lifelong plasticity. *Cerebral Cortex*, *24*(4), 845–858.
61. O'Carroll, C. M., Martin, S. J., Sandin, J., Frenguelli, B., & Morris, R. G. M. (2006). Dopaminergic modulation of the persistence of one-trial hippocampus-dependent memory. *Learning & Memory*, *13*(6), 760 769.
62. Calabresi, P., Ghiglieri, V., Mazzocchetti, P., Corbelli, I., & Picconi, B. (2015). Levodopa-induced plasticity: a double-edged sword in Parkinson's disease? *Philosophical Transactions of the Royal Society B: Biological Sciences*, *370*(1672), 20140184.
63. Calabresi, P., Castrioto, A., Di Filippo, M., & Picconi, B. (2013). New experimental and clinical links between the hippocampus and the dopaminergic system in Parkinson's disease. *The Lancet Neurology*, *12*(8), 811–821.

64. Kohl, Z., Abdallah, N., Ben, Vogelgsang, J., Tischer, L., Deusser, J., Amato, D., Anderson, S., Müller, C. P., Riess, O., & Masliah, E. (2016). Severely impaired hippocampal neurogenesis associates with an early serotonergic deficit in a BAC α-synuclein transgenic rat model of Parkinson's disease. *Neurobiology of Disease*, *85*, 206–217.
65. Pendolino, V., Bagetta, V., Ghiglieri, V., Sgobio, C., Morelli, E., Poggini, S., Branchi, I., Latagliata, E. C., Pascucci, T., & Puglisi-Allegra, S. (2014). l-DOPA reverses the impairment of Dentate Gyrus LTD in experimental parkinsonism via β-adrenergic receptors. *Experimental Neurology*, *261*, 377–385.
66. Nguyen, M., Wong, Y. C., Ysselstein, D., Severino, A., & Krainc, D. (2019). Synaptic, mitochondrial, and lysosomal dysfunction in Parkinson's disease. *Trends in Neurosciences*, *42*(2), 140–149.
67. Cubo, E., Shannon, K. M., Tracy, D., Jaglin, J. A., Bernard, B. A., Wuu, J., & Leurgans, S. E. (2006). Effect of donepezil on motor and cognitive function in Huntington disease. *Neurology*, *67*(7), 1268–1271.
68. Ransome, M. I., Renoir, T., & Hannan, A. J. (2012). Hippocampal neurogenesis, cognitive Deficits and affective disorDer in huntington's disease. *Neural Plasticity, 2012*. https://doi.org/10.1155/2012/874387
69. Wilkie, C. M., Barnes, J. R., Benson, C.-L. M., Brymer, K. J., Nafar, F., & Parsons, M. P. (2020). Hippocampal synaptic dysfunction in a mouse model of huntington disease is not alleviated by ceftriaxone treatment. *Eneuro*, *7*(3), ENEURO.0440-19.2020.
70. Zhang, H., Zhang, C., Vincent, J., Zala, D., Benstaali, C., Sainlos, M., Grillo-Bosch, D., Daburon, S., Coussen, F., & Cho, Y. (2018). Modulation of AMPA receptor surface diffusion restores hippocampal plasticity and memory in Huntington's disease models. *Nature Communications*, *9*(1), 4272.
71. Sepers, M. D., Smith-Dijak, A., LeDue, J., Kolodziejczyk, K., Mackie, K., & Raymond, L. A. (2018). Endocannabinoid-specific impairment in synaptic plasticity in striatum of Huntington's disease mouse model. *Journal of Neuroscience*, *38*(3), 544–554.
72. Quirion, J. G., & Parsons, M. P. (2019). The onset and progression of hippocampal synaptic plasticity deficits in the Q175FDN mouse model of Huntington disease. *Frontiers in Cellular Neuroscience*, *13*, 326.
73. Giralt, A., Brito, V., Chevy, Q., Simonnet, C., Otsu, Y., Cifuentes-Díaz, C., De Pins, B., Coura, R., Alberch, J., & Ginés, S. (2017). Pyk2 modulates hippocampal excitatory synapses and contributes to cognitive deficits in a Huntington's disease model. *Nature Communications*, *8*(1), 15592.
74. Kolodziejczyk, K., Parsons, M. P., Southwell, A. L., Hayden, M. R., & Raymond, L. A. (2014). Striatal synaptic dysfunction and hippocampal plasticity deficits in the Hu97/18 mouse model of Huntington disease. *PLoS One*, *9*(4), e94562.
75. Anglada-Huguet, M., Xifro, X., Giralt, A., Zamora-Moratalla, A., Martín, E. D., & Alberch, J. (2014). Prostaglandin E2 EP1 receptor antagonist improves motor deficits and rescues memory decline in R6/1 mouse model of Huntington's disease. *Molecular Neurobiology*, *49*, 784–795.
76. Murphy, K. P. S. J., Carter, R. J., Lione, L. A., Mangiarini, L., Mahal, A., Bates, G. P., Dunnett, S. B., & Morton, A. J. (2000). Abnormal synaptic plasticity and impaired spatial cognition in mice transgenic for exon 1 of the human Huntington's disease mutation. *Journal of Neuroscience*, *20*(13), 5115–5123.
77. Bruel-Jungerman, E., Davis, S., & Laroche, S. (2007). Brain plasticity mechanisms and memory: a party of four. *The Neuroscientist*, *13*(5), 492–505.
78. Cattaneo, E., Zuccato, C., & Tartari, M. (2005). Normal huntingtin function: an alternative approach to Huntington's disease. *Nature Reviews Neuroscience*, *6*(12), 919–930.
79. Di Filippo, M., Tozzi, A., Picconi, B., Ghiglieri, V., & Calabresi, P. (2007). Plastic abnormalities in experimental Huntington's disease. *Current Opinion in Pharmacology*, *7*(1), 106–111.
80. Karussis, D. (2014). The diagnosis of multiple sclerosis and the various related demyelinating syndromes: a critical review. *Journal of Autoimmunity*, *48*, 134–142.
81. Chiaravalloti, N. D., & DeLuca, J. (2008). Cognitive impairment in multiple sclerosis. *The Lancet Neurology*, *7*(12), 1139–1151.
82. Nistico, R., Mori, F., Feligioni, M., Nicoletti, F., & Centonze, D. (2014). Synaptic plasticity in multiple sclerosis and in experimental autoimmune encephalomyelitis. *Philosophical Transactions of the Royal Society B: Biological Sciences*, *369*(1633), 20130162.
83. Dutta, R., Chang, A., Doud, M. K., Kidd, G. J., Ribaudo, M. V., Young, E. A., Fox, R. J., Staugaitis, S. M., & Trapp, B. D. (2011). Demyelination causes synaptic alterations in hippocampi from multiple sclerosis patients. *Annals of Neurology*, *69*(3), 445–454.
84. Novkovic, T., Shchyglo, O., Gold, R., & Manahan-Vaughan, D. (2015). Hippocampal function is compromised in an animal model of multiple sclerosis. *Neuroscience*, *309*, 100–112.
85. Di Filippo, M., Chiasserini, D., Gardoni, F., Viviani, B., Tozzi, A., Giampà, C., Costa, C., Tantucci, M., Zianni, E., & Boraso, M. (2013). Effects of central and peripheral inflammation on hippocampal synaptic plasticity. *Neurobiology of Disease*, *52*, 229–236.

86. Nisticò, R., Mango, D., Mandolesi, G., Piccinin, S., Berretta, N., Pignatelli, M., Feligioni, M., Musella, A., Gentile, A., & Mori, F. (2013). Inflammation subverts hippocampal synaptic plasticity in experimental multiple sclerosis. *PloS One*, *8*(1), e54666.
87. Mori, F., Nistico, R., Mandolesi, G., Piccinin, S., Mango, D., Kusayanagi, H., Berretta, N., Bergami, A., Gentile, A., & Musella, A. (2014). Interleukin-1β promotes *long-term potentiation* in patients with multiple sclerosis. *Neuromolecular Medicine*, *16*, 38–51.
88. Viviani, B., Gardoni, F., Bartesaghi, S., Corsini, E., Facchi, A., Galli, C. L., Di Luca, M., & Marinovich, M. (2006). Interleukin-1β released by gp120 drives neural death through tyrosine phosphorylation and trafficking of NMDA receptors. *Journal of Biological Chemistry*, *281*(40), 30212–30222.
89. Lai, A. Y., Swayze, R. D., El-Husseini, A., & Song, C. (2006). Interleukin-1 beta modulates AMPA receptor expression and phosphorylation in hippocampal neurons. *Journal of Neuroimmunology*, *175*(1–2), 97–106.
90. Mancini, A., Gaetani, L., Di Gregorio, M., Tozzi, A., Ghiglieri, V., Calabresi, P., & Di Filippo, M. (2017). Hippocampal neuroplasticity and inflammation: relevance for multiple sclerosis. *Multiple Sclerosis and Demyelinating Disorders*, *2*(1), 1–12.
91. Volianskis, A., France, G., Jensen, M. S., Bortolotto, Z. A., Jane, D. E., & Collingridge, G. L. (2015). *Long-term potentiation* and the role of N-methyl-D-aspartate receptors. *Brain Research*, *1621*, 5–16.
92. Xu, W. (2011). PSD-95-like membrane associated guanylate kinases (PSD-MAGUKs) and synaptic plasticity. *Current Opinion in Neurobiology*, *21*(2), 306–312.
93. Lim, G. P., Calon, F., Morihara, T., Yang, F., Teter, B., Ubeda, O., Salem, N., Frautschy, S. A., & Cole, G. M. (2005). A diet enriched with the omega-3 fatty acid docosahexaenoic acid reduces amyloid burden in an aged Alzheimer mouse model. *Journal of Neuroscience*, *25*(12), 3032–3040.
94. Arundine, M., Chopra, G. K., Wrong, A., Lei, S., Aarts, M. M., MacDonald, J. F., & Tymianski, M. (2003). Enhanced vulnerability to NMDA toxicity in sublethal traumatic neuronal injury in vitro. *Journal of Neurotrauma*, *20*(12), 1377–1395.
95. Spaethling, J. M., Klein, D. M., Singh, P., & Meaney, D. F. (2008). Calcium-permeable AMPA receptors appear in cortical neurons after traumatic mechanical injury and contribute to neuronal fate. *Journal of Neurotrauma*, *25*(10), 1207–1216.
96. Ansari, M. A., Roberts, K. N., & Scheff, S. W. (2008). A time course of contusion-induced oxidative stress and synaptic proteins in cortex in a rat model of TBI. *Journal of Neurotrauma*, *25*(5), 513–526.
97. Pinheiro, P. S., & Mulle, C. (2008). Presynaptic glutamate receptors: physiological functions and mechanisms of action. *Nature Reviews Neuroscience*, *9*(6), 423–436.
98. Lee, D. J., Dallapiazza, R. F., De Vloo, P., & Lozano, A. M. (2018). Current surgical treatments for Parkinson's disease and potential therapeutic targets. *Neural Regeneration Research*, *13*(8), 1342.
99. Kalampokini, S., Lyros, E., Lochner, P., Fassbender, K., & Unger, M. M. (2020). Effects of subthalamic nucleus deep brain stimulation on facial emotion recognition in Parkinson's disease: a critical literature review. *Behavioural Neurology*, *2020*. https://doi.org/10.1155/2020/4329297
100. Girault, J.-A. (2012). Integrating neurotransmission in striatal medium spiny neurons. *Synaptic Plasticity: Dynamics, Development and Disease*, *970*, 407–429.
101. Giorgi, M., D'Angelo, V., Esposito, Z., Nuccetelli, V., Martorana, A., Stefani, A., Bernardi, G., & Sancesario, G. (2008). Lowered cAMP and cGMP signaling in the brain during levodopa-induced dyskinesias in hemiparkinsonian rats: new aspects in the pathogenetic mechanisms. *European Journal of Neuroscience*, *28*(5), 941–950.
102. Lee, Y.-Y., Park, J.-S., Leem, Y.-H., Park, J.-E., Kim, D.-Y., Choi, Y.-H., Park, E.-M., Kang, J. L., & Kim, H.-S. (2019). The phosphodiesterase 10 inhibitor papaverine exerts anti-inflammatory and neuroprotective effects via the PKA signaling pathway in neuroinflammation and Parkinson's disease mouse models. *Journal of Neuroinflammation*, *16*, 1–17.
103. Amin, L., Le, N., Mercer, R. C. C., Germann, U., Alam, J., & Harris, D. A. (2019). p4-506: role of p38α map kinase in amyloid-β derived diffusible ligand (ADDL) induced dendritic spine loss in hippocampal neurons. *Alzheimer's & Dementia*, *15*(7S_Part_29), 1507–1507. https://doi.org/10.1016/j.jalz.2019.08.052
104. Jiang, Y., Stavrides, P., Darji, S., Yang, D.-S., Bleiwas, C., Smiley, J., Germann, U., Alam, J., & Nixon, R. (2019). p4-692: effects of p38α map kinase inhibition on the neurodegenerative phenotype of the ts2 down syndrome mouse model. *Alzheimer's & Dementia*, *15*(7S_Part_31), 1597–1597. https://doi.org/10.1016/j.jalz.2019.09.057
105. Reidling, J. C., Relaño-Ginés, A., Holley, S. M., Ochaba, J., Moore, C., Fury, B., Lau, A., Tran, A. H., Yeung, S., & Salamati, D. (2018). Human neural stem cell transplantation rescues functional deficits in R6/2 and Q140 Huntington's disease mice. *Stem Cell Reports*, *10*(1), 58–72.

106. Schenk, D. (2002). Amyloid-β immunotherapy for Alzheimer's disease: the end of the beginning. *Nature Reviews Neuroscience*, *3*(10), 824–828.
107. Lin, C.-H., Huang, Y.-J., Lin, C.-J., Lane, H.-Y., & E Tsai, G. (2014). NMDA neurotransmission dysfunction in mild cognitive impairment and Alzheimer's disease. *Current Pharmaceutical Design*, *20*(32), 5169–5179.
108. Novak, P., Zilka, N., Zilkova, M., Kovacech, B., Skrabana, R., Ondrus, M., Fialova, L., Kontsekova, E., Otto, M., & Novak, M. (2019). AADvac1, an active immunotherapy for Alzheimer's disease and non Alzheimer tauopathies: an overview of preclinical and clinical development. *The Journal of Prevention of Alzheimer's Disease*, *6*, 63–69.
109. Ambili, T. R., Manimekalan, A., & Bhat, D. A. (2012). Sirna gene therapy for Alzheimer's disease targeting APP gene. *International Journal of Pharmacy and Pharmaceutical Sciences*, *4*(3), 341–346.
110. Zhang, Y., Chopp, M., Meng, Y., Katakowski, M., Xin, H., Mahmood, A., & Xiong, Y. (2015). Effect of exosomes derived from multipluripotent mesenchymal stromal cells on functional recovery and neurovascular plasticity in rats after traumatic brain injury. *Journal of Neurosurgery*, *122*(4), 856–867.
111. Stonesifer, C., Corey, S., Ghanekar, S., Diamandis, Z., Acosta, S. A., & Borlongan, C. V. (2017). Stem cell therapy for abrogating stroke-induced neuroinflammation and relevant secondary cell death mechanisms. *Progress in Neurobiology*, *158*, 94–131.
112. Fujishige, K., Kotera, J., & Omori, K. (1999). Striatum-and testis-specific phosphodiesterase PDE10A: isolation and characterization of a rat PDE10A. *European Journal of Biochemistry*, *266*(3), 1118–1127.
113. Xie, Y., Jackson, M. F., & MacDonald, J. F. (2013). Optogenetics and synaptic plasticity. *Acta Pharmacologica Sinica*, *34*(11), 1381–1385.
114. Pati, S., Salvi, S. S., Kallianpur, M., Vaidya, B., Banerjee, A., Maiti, S., Clement, J. P., & Vaidya, V. A. (2019). Chemogenetic activation of excitatory neurons alters hippocampal neurotransmission in a dose-dependent manner. *eNeuro*, *6*(6), ENEURO.0124-19.2019.
115. Murphy, T. H. (2015). Two-photon imaging of neuronal structural plasticity in mice during and after ischemia. *Cold Spring Harbor Protocols*, *2015*(6), pdb-prot087486.
116. Siegel, F., & Lohmann, C. (2013). Probing synaptic function in dendrites with calcium imaging. *Experimental Neurology*, *242*, 27–32.

3 Relevance of Synaptic Plasticity and Mitochondrial Dysfunction in Neurodegenerative Disorders

Sangita Paul
CSIR Institute of Genomics and Integrative Biology
New Delhi, India
and
Academy of Scientific and Innovative Research (AcSIR)
Ghaziabad, India

*Deepak Kumar and Anita Pal**
University of Allahabad
Prayagraj, Uttar Pradesh, India

3.1 INTRODUCTION

Neurodegenerative disorders are marked by a gradual deterioration of neurons, leading to a decline in various functions directed by the central nervous system (CNS) including motor and non-motor symptoms [1–3]. Neurons have a limited lifespan, but the gradual deterioration of neurons, their structure, and/or their activities, which is referred to as neurodegeneration, is a significant health concern and strongly associated with the emergence of various brain disorders [4,5]. Tragically, this neuronal degeneration results in behavioral and neurological impairments that range from mild memory loss to dementia [4].

Nowadays neurodegenerative disorders are a major global health burden in the western world as well as in India, particularly with the ageing population. As humans age, their cognitive capacity naturally declines to some extent. However, this decline occurs at a much faster rate in neurodegenerative diseases. These diseases not only affect an individual's well-being but also impose a substantial responsibility on their caregivers [6]. The prevalence of neurodegenerative diseases is increasing with ageing at a high pace, and the exact etiologies of these diseases are still unknown [7].

While PD is associated with progressively declining cognitive abilities [8], AD is associated with progressive decline in motor abilities [9], and their pathophysiology is largely determined by synaptic transmission and plasticity defects [10–12]. In the context of PD, there is a connection between the substantia nigra pars compacta and motor impairment. Furthermore, memory impairment in AD is associated with neuronal death in the hippocampus area. It is important to note that other neurodegenerative disorders, such as, amyotrophic lateral sclerosis (ALS) and Huntington's disease, are characterized by severe motor, neuropsychological, and cognitive impairments.

Given their importance in communication, neurons are essential to the human brain's healthy operation [13]. The brain is a sophisticated system for processing and storing information, where neurons serve as the fundamental components. Neurons possess specialized structures and functions, such as

* **The first two authors contributed equally.**

DOI: 10.1201/9781003464648-4

ion channel and ion pump proteins, that allow them to transmit nerve impulses and process various inputs. Synapses serve as the principal locations for the storing of information. Information storage is facilitated by neuronal ability to undergo permanent changes in response to activity, a process known as plasticity. This characteristic enables brain circuitry to undergo synaptic alterations in response to experience, hence, influencing future cognitive processes, behavioral patterns, and emotional states [14].

There are many pathologies associated with neurodegenerative diseases, but synaptic alteration and mitochondrial dysfunction also play a vital role [15]. The brain is crucial for cognitive processes since it facilitates the formation of new memories, procedural learning, and environmental adaptation. In recent years, mitochondrial abnormalities and oxidative stress have been linked to the process of neurodegeneration [16]. In most cases of neurodegenerative disease, misfolded proteins, oxidative stress, and inflammation result in the loss of neurons specific to the affected region as well as impairment of neuroplasticity [17,18].

A deficit in energy supply to the brain can result in synapse loss and neurodegeneration, making bioenergetic variables highly relevant [19]. It is estimated that the brain consumes a large amount of the body's total energy, despite its small size. Neurons require a significant amount of energy for their biological processes and are particularly vulnerable to disruptions. Neuronal energy failure arises due to changes in bioenergetics observed in neurodegenerative diseases, such as reduced ATP production and poor functioning of the electron transport chain (ETC). Insufficient ATP supply significantly affects the viability and performance of neurons [20]. The significance of mitochondrial malfunction in neuronal death as a consequence of loss of synaptic plasticity will be discussed in this chapter, along with the ways in which this phenomenon contributes to the development of a variety of neurological illnesses.

3.2 MITOCHONDRIAL FUNCTION AND SYNAPTIC PLASTICITY IN HEALTHY NEURONS

Mitochondria, considered to be the powerhouse of cells, are extremely active organelles within the axon and dendrites that can divide, fuse, and migrate with purpose [21,22]. The function of mitochondria in neurotransmission is to generate ATP and regulate Ca^{2+} levels at synapses. Most of the cellular demand for ATP is generated by mitochondria through oxidative phosphorylation [23].

These coins of energy are crucial for the neuronal physiology including maintenance of the ion gradients, vesicular trafficking, and intricate regulation of synaptic plasticity, allowing synapses to sustain their essential functions [24].

Given the substantial energy requirements for maintaining the balance of calcium during the transmission of nerve signals, the presence of mitochondria in both pre- and postsynaptic terminals is essential [23,25], but during the process of ATP biogenesis, a substantial quantity of reactive oxygen species is also produced as a secondary product. A significant amount of the oxygen is converted to hydrogen peroxide by manganese superoxide dismutase (MnSOD). Catalase and glutathione peroxidase convert hydrogen peroxide back into water. H_2O_2 produces reactive hydroxyl radical (OH) when it reacts with Fe^{2+} or Cu^{+}. This has the potential to cause DNA damage and membrane lipid peroxidation. To prevent oxidative damage even further, mitochondria include several well-known antioxidant compounds, such as nicotinamide, creatine, and coenzyme Q10 (ubiquinone) [21,26] (Figure 3.1).

Along with ATP production, another key aspect of mitochondria is the regulation of calcium signaling and redox signals. The magnitude and length of cytosolic Ca^{2+} responses are limited by the buffering of Ca^{2+} entering the cell during neuronal excitation along with the release of stored Ca^{2+} back into the cytosol gradually. Thus, mitochondria are crucial for several Ca^{2+}-dependent functions in neurons [27].

Synapses are specialized connections that allow neurons to interact with one another. Neuronal maintenance and proper brain functioning are impacted by synaptic transmission between individual neurons. Synaptic plasticity refers to the structural and functional alterations of neural circuits brought about by stimuli like learning and memory, environmental factors, and brain injury. The brain's capacity for behavioral adaptation to an ever-changing environment results from these dynamical processes occurring inside the neuronal system [28,29]. The axon or dendritic sprouting and development, the establishment of synapses, the building up of synapses following recurrent activation, and neurogenesis

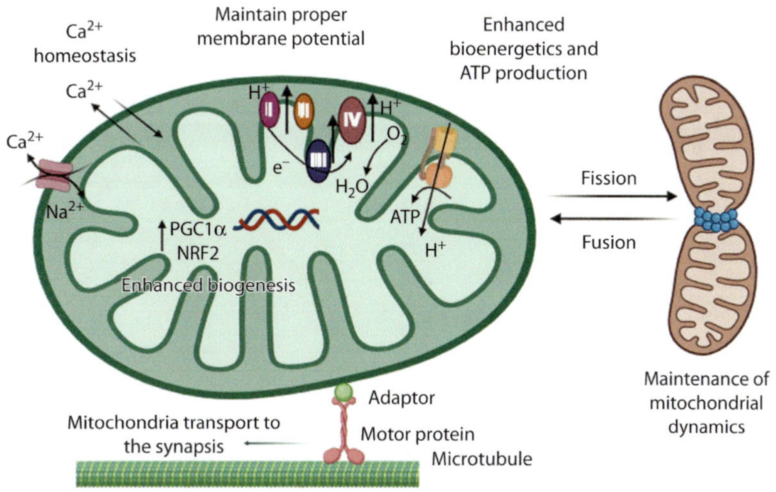

FIGURE 3.1 Diagram depicting various functions exhibited by mitochondria to maintain neuronal activity (in healthy conditions).

are a few examples of neuroplasticity [22]. This ability to adapt structurally and functionally is based on a variety of cellular and molecular processes including neurotransmission cytoskeletal remodeling, vesicular trafficking, and modulation of gene expression [21,30–34]. Long-term potentiation (LTP) and long-term depression (LTD) are two basic aspects of neuroplasticity (Figure 3.2).

The LTP is a process that involves continuously fortifying synapses, resulting in a sustained increase in signal transmission between neurons. The complementary phenomenon is known as LTD, which is the gradual weakening of synaptic connections between pre- and postsynaptic neural synapses. The synaptic plasticity is a crucial process for hippocampal learning [35]. Neural circuits are believed to be formed during the process of memory and learning through the involvement of NMDA receptors, which regulate LTP and LTD processes in the hippocampal pyramidal neurons. In a similar way, the AMPA receptors are crucial for facilitating rapid excitatory glutamatergic neurotransmission in the postsynaptic membrane of the mammalian brain.

The AMPAR channels are components of tetrameric subunits termed as GluA1-A4. The GluA1/A2 is a subtype of AMPAR observed in hippocampus pyramidal neurons, with the GluA2/A3 heterotetramer being the next most prevalent. It is commonly known that postsynaptic neurons' AMPAR trafficking is essential for the onset of LTD and LTP [36–39].

3.3 MOLECULAR AND CELLULAR MECHANISMS OF MITOCHONDRIAL DYSFUNCTION AND SYNAPTIC PLASTICITY LOSS

Mitochondria are crucial energy-generating organelles whose roles in neural synapses include calcium sequestration and ATP production for neurotransmitter release. The constant demand for rapid energy at synaptic terminals, the high concentration of mitochondria at synapses, and the preferential accumulation of toxic byproducts and aggregates of protein in and near these regions leaves synapses particularly vulnerable to damage. Any defect in the ability to generate efficient energy to fuel synaptic communication can lead to the emergence of various neurodegenerative disorders. In this case, the combined effect of multiple altered mechanisms results in and causes mitochondrial dysfunction (Figure 3.3).

3.3.1 Impaired Mitochondrial Dynamics

Mitochondrial dynamics refer to the alterations in the structure, number, and location of mitochondria in eukaryotic cells. These changes are crucial for the cells to carry out essential functions such as cell cycle regulation, energy production, movement, differentiation, apoptosis, and ageing [56].

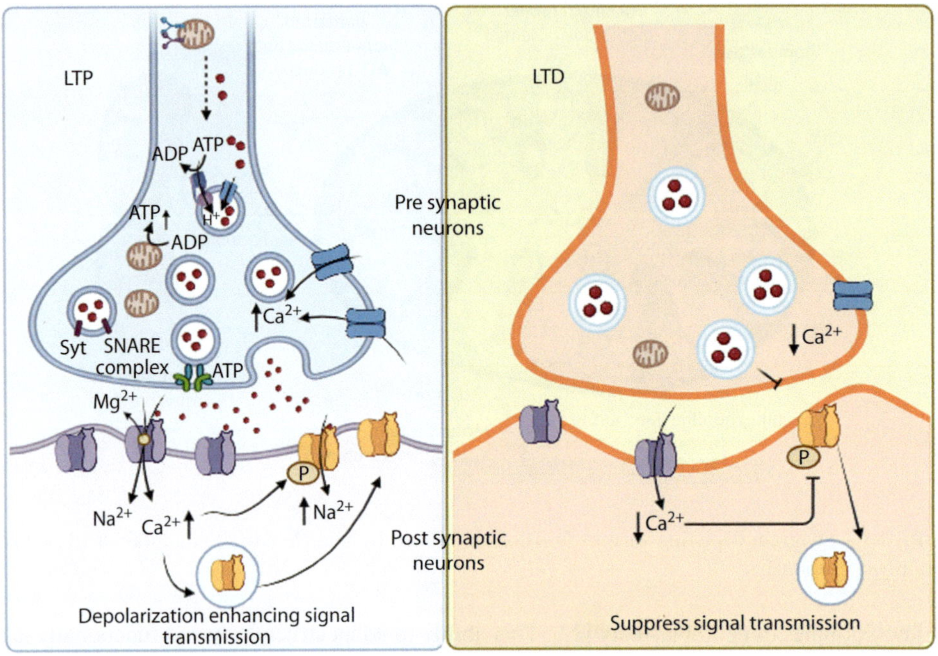

FIGURE 3.2 Schematic diagram illustrating the phenomena of long-term potentiation (LTP) and long-term depression (LTD). During LTP, the signal strength transmitted to the postsynaptic neuron is increased as a result of the heightened activation of presynaptic neuron results in enhanced neurotransmitters release. In LTD, the synaptic transmission is reduced due to insufficient activation and release of neurotransmitter from the presynaptic neuron. (Created by using biorender.com.)

It is essential to tightly regulate mitochondrial dynamics in neurons to maintain a healthy pool of mitochondria. Mitochondrial dynamics is the interplay between mitochondrial biogenesis, fusion, and fission. There are a number of proteins that play an important role in mitochondrial fusion, such as the mitofusins Mfn1/Mfn2 and OPA1, while Drp and Fis1 are critical for the mechanics of mitochondrial fission [40]. Individual mitochondria can undergo continual remodeling using such

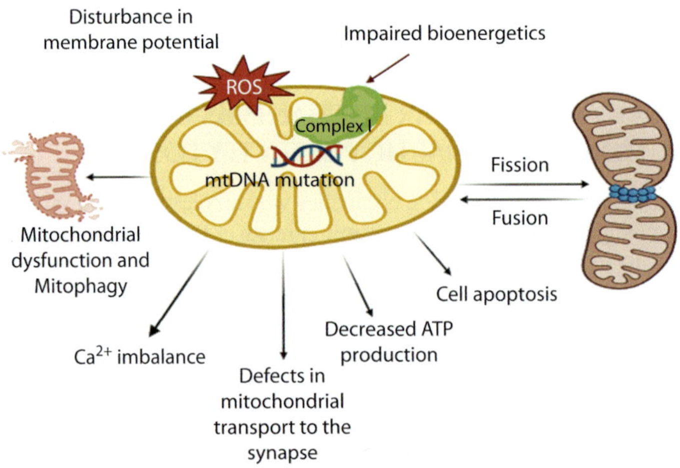

FIGURE 3.3 Diagram depicting various mechanisms of mitochondrial dysfunctions pertaining to neuronal loss (in disease condition).

dynamics; thus such events are crucial for neural activities [28,41,42]. Disruption of mitochondrial dynamics is a crucial pathological process in various diseases and conditions that are marked by impaired mitochondrial activity [43,44]. A growing body of research identifies impaired mitophagy and mitochondrial metabolism as key components of neurodegeneration and the process of aging [45]. When damaged mitochondria accumulate due to mitophagy dysregulation, an energy crisis is created. Several proteins involved in the processes of mitochondrial fusion and fission are altered in AD [46]. This crisis then mediates synaptic failure and neuronal degeneration, which ultimately leads to brain damage and cognitive deficiencies.

3.3.2 Imbalance of Calcium Ions

A delicate equilibrium needs to be struck for proper synaptic function and plasticity since ROS scavengers have been shown to prevent LTP, while ROS donors cause LTP [47]). Synaptic plasticity depends on Ca^{2+}-dependent protein kinases, which can be activated by elevated mitochondrial superoxide ROS. Apoptosis is frequently induced, and mitochondrial membrane permeabilization results from excessive Ca^{2+} absorption into the mitochondria. Degenerative neuropathology might also be influenced by Ca^{2+} dysregulation in other cell types that sustain neural activity [23]. Therefore, energy generation and the preservation of synaptic plasticity are significantly impacted by the intracellular balance of Ca^{2+} in neuronal mitochondria. The pathogenesis of AD is strongly influenced by the dysregulation of Ca^{2+} levels within neuronal mitochondria. The alteration of mitochondrial Ca^{2+} levels may be a promising pharmacological target for future AD treatment [23]. The ability of mitochondria to sequester calcium is reduced with age and neurological conditions. Both the ATP-dependent and glutamate-malate-ADP-driven pathways showed a decrease in mitochondrial Ca^{2+} uptake with age [47]. Moreover, the inhibition of mitochondrial respiration by rotenone leads to symptoms that closely resemble those observed in Parkinson's disease [48].

3.3.3 Defects in Mitochondrial Trafficking

Effective mitochondrial mobility within the synapses is crucial for this organelle's quick distribution to various locations to fulfill the high metabolic requirements of neurons. Microtubules transport mitochondria across axons to supply synapses with energy and maintain neuronal activity. Two families of motor proteins are crucial for the bidirectional transport of mitochondria. Kinesin proteins play an anterograde transport role, while dynein protein complexes are important for retrograde transport. Other motor proteins also involved in the process are trafficking kinesin binding proteins (TRAKs) and mitochondrial proteins Miro 1 and 2. Any defects in the motor proteins show defective mitochondrial trafficking to the synapses and loss of synaptic activity [49]. Furthermore, the anterograde transport of mitochondria is also associated with the KIF1B and KLC3 (kinesin superfamily members), which are also involved in the process. Furthermore, dyneins are also crucial for mitochondrial fission in addition to the axonal retrograde transport [50–52].

3.3.4 Role of Oxidative Stress

Oxidative stress is commonly recognized as a frequently mentioned mechanism leading to neuronal loss and degeneration. ROS is formed as a by-product of energy generation, particularly at the electron transport chain (ETC). The occurrence of oxidative stress takes place when the antioxidant defense mechanisms that are designed to prevent the production of ROS are either overwhelmed or weakened. The oxidative stress causes both the damage that occurs to mitochondrial DNA (mtDNA) and the damage that occurs to chromosomal DNA in the nucleus. Due to the fact that mtDNA is devoid of histones and is located in the vicinity of the ETC, it is especially vulnerable to oxidative degradation. As a matter of fact, mitochondria convert 0.2–2.0% of the molecular oxygen they utilize to superoxide anions into other ROS like OH and H_2O_2 [20,53,54].

The main location of mitochondrial ROS production is the ETC, a key component of oxidative phosphorylation. The vast majority of ROS found in mitochondria are superoxide anions. In mitochondria, superoxide and hydrogen peroxide are created when electrons in the ETC combine with O_2 and escape from complex I and complex III [54].

The enzymes Mn-SOD and Cu, Zn-SOD convert O_2^- at the ETC into H_2O_2, as O_2 does not easily pass through the biological membranes. The H_2O_2 diffuses into the cytosol and nucleus of the cell, whereas the enzymes glutathione peroxidase and catalase detoxify it, transforming it into H_2O. A state of oxidative stress is characterized by excessive accumulation of hydrogen peroxide (H_2O_2), which arises from an imbalance between the antioxidant defense system and the creation of H_2O_2 [54,55]. It is well-known that a rise in ROS formation and a disturbance of ATP synthesis are caused in large part by complex I damage and its suppression. Damage to ETC's complex I allows electrons to leak out, resulting in the production of ROS. For instance, ROS production damages complex I, which further inhibits complex I and increases the production of ROS [54].

3.3.5 Role of Bioenergetics

Mitochondria play a key role in facilitating the necessary energy for the production, release, and absorption of neurotransmitters. There is a hypothesis that degeneration of neurons is related to a reduction in mitochondrial number at synapses. The disruption of mitochondrial bioenergetics results in changes in neuro-metabolic coupling as well as changes in the processing of energy. Due to the absence of mitosis in neurons, certain neurons remain unaltered throughout an individual's lifespan. Research has demonstrated that certain brain regions, like the hippocampus, are capable of neurogenesis, which refers to the generation of new neurons. However, conflicting research have challenged these findings. Nevertheless, the replacement of degenerated neurons poses a significant challenge. Neurons require high levels of bioenergetics for synaptic function, neurotransmitter synthesis, release, absorption, and other related functions. In neurons, oxidative phosphorylation is the most common mechanism used to produce ATP. However, any disruption to this process might have detrimental consequences [20,53].

3.3.6 Mitophagy

Mitophagy, in which damaged mitochondria are removed by cells, is impaired when mitochondrial biomass is generated in pre-synaptic terminals. Controlling the quantity and quality of mitochondria depends critically on mitophagy, an evolutionarily conserved mechanism that uses autophagy to selectively eliminate unhealthy or unnecessary mitochondria [56,57].

A lack of ability to clear damaged mitochondria in combination with increased mitochondrial fission and altered PINK-1 and other mitophagy regulators leads to the accumulation of damaged mitochondria that worsen neurodegenerative processes. Some studies report increased mitophagy induction and downstream clearance of damaged mitochondria, which contribute to synaptic terminal energy depletion [58,59].

3.3.7 Mitochondrial Dysfunction and Neuronal Loss

Mitochondrial impairments are linked to a decrease in ATP production, the production of ROS, and the release of caspases. Excessive production of ROS can have detrimental effects on proteins, DNA, lipids, and proteins. It can impair the efficiency of cellular mechanisms, trigger inflammatory pathways, cause excitotoxicity, lead to protein aggregation, and induce apoptosis [60].

Further, the accumulation of toxic aggregates impairs axonal transport along microtubule networks, leading to compromised mitochondrial trafficking to the pre-synaptic terminal and insufficient energy to fuel neurotransmitter signaling, leading to loss of synaptic integrity. Dysfunctional microtubules, such as those resulting from hyperphosphorylated tau in AD, also impair retrograde transport of damaged mitochondria and mitophagosomes from the synaptic terminal, contributing to the persistence of defective mitochondria at the synapse. In conjunction with decreased mitochondrial transport to the axonal terminal, studies have shown that many neurodegenerative conditions are associated with hypometabolism in pre-synaptic neurons, indicative of dysregulated energy production and mitochondrial bioenergetics [61]. With impaired glucose import into the neuron and insufficient ATP production, mitochondrial function and intracellular processes slowly progress towards dysfunction until a threshold is surpassed, leaving neurons and larger neuronal networks vulnerable to degenerative collapse and cell death (Figure 3.4).

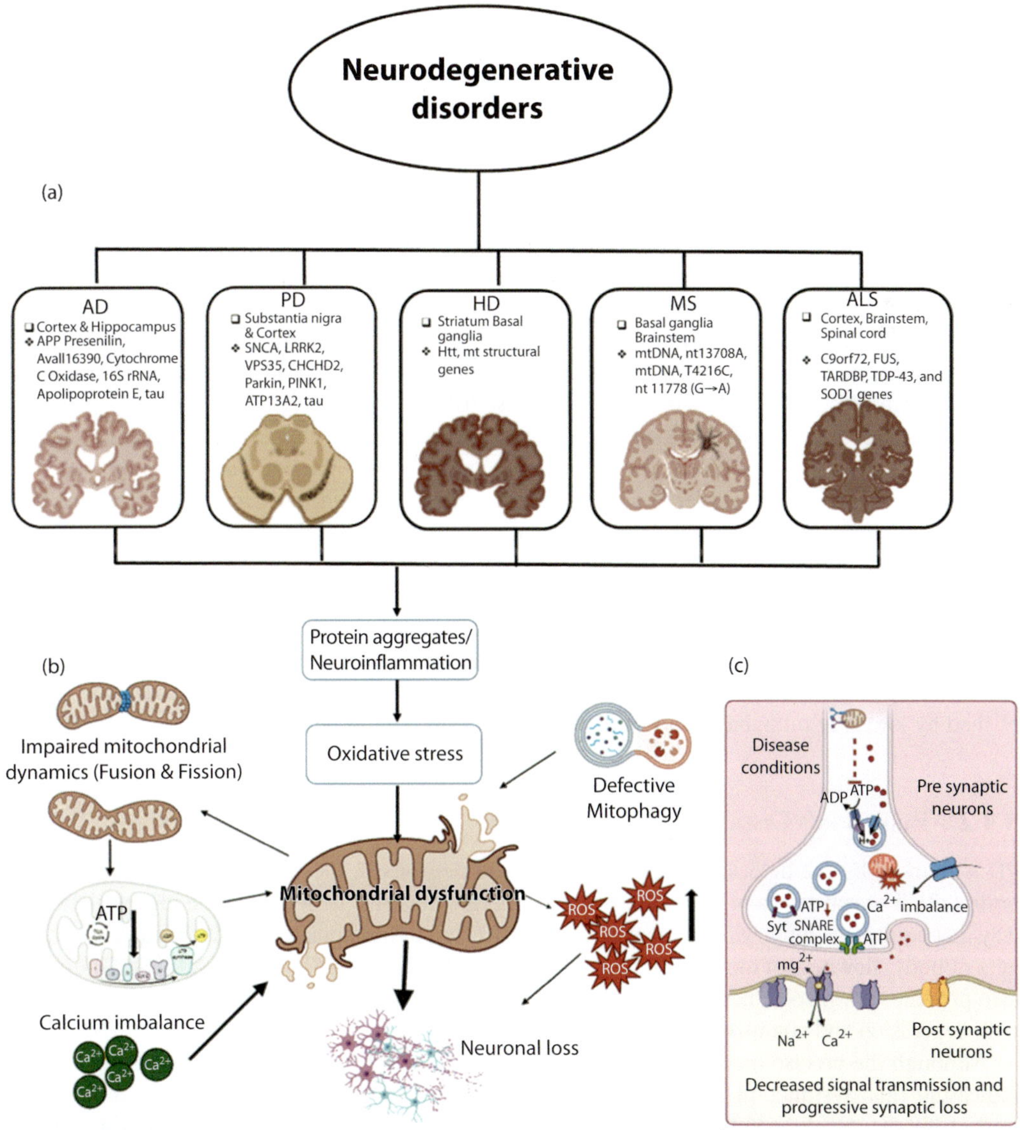

FIGURE 3.4 These schematic diagrams illustrate the process of neuronal death and loss of synaptic plasticity caused by mitochondrial dysfunction in neurodegenerative disorders such as AD, PD, HD, MS, and ALS. (a) Depiction of the precise brain region that is affected and the specific genes that experience mutations in various neurodegenerative disorders. (b) Diagram depicting the mechanistic processes linked to mitochondrial dysfunction caused by pathophysiological disruptions, including protein aggregates and neuroinflammation, leading to oxidative stress. (c) The impact of compromised mitochondrial function on the communication of signals among neurons.

3.4 MITOCHONDRIAL-SYNAPTIC ALTERATIONS IN NEURODEGENERATIVE DISORDERS

3.4.1 Alzheimer's Disease

In AD, neurofibrillary tangles and amyloid plaques build up in the brain, causing a gradual decline in memory and cognition. Protein aggregation disrupts the transmission of neurotransmitters, specifically acetylcholine, which is a distinctive pathological feature of AD [62,63]. There is a clear correlation between memory loss and cognitive deficiencies and synaptic injury, an early pathogenic event. The dysfunction of mitochondria is a prominent feature of AD pathology. Tau and amyloid-β (Aβ) both cause changes in the mitochondria. An increasing amount of data indicate that mitochondrial disruption is a substantial contributor to synapse degradation and failure in AD [49,64].

A transcription factor PGC-1 α and NRF1 and NRF2 play an important role in mitochondrial biogenesis. Aβ peptide can activate the NF-κB signaling in AD, which in turn suppresses the synthesis of PGC-1 α, NRF1, and NFR2 [65]. In the AD brain, the levels of mitochondrial fission and fusion proteins are reported to be significantly decreased. Several mitochondrial fission proteins such as Drpl interact with protein aggregates (Aβ monomers and oligomers), and such aberrant interactions become more pronounced as the AD advances. Therefore, increased synthesis of Aβ and its association with Drpl are important contributors to aberrant mitochondrial dynamics, synaptic injury, and mitochondrial fragmentation [46,66]. Postmortem AD brain samples reveal elevated levels of free radicals, lipid peroxidation, compromised ATP synthesis, lower amounts of pyruvate dehydrogenase, and cytochrome c oxidase [67,68].

Patients affected with AD exhibit axonal degeneration, which is characterized by enlarged areas in which an excessive number of mitochondria accumulate. It has been documented that mutations in presenilin 1 (PS1) hinder axonal transport mediated by kinesin-1 through elevated GSK3β activity, which in turn activate the phosphorylation of kinesin-1 light chains. This causes synaptic vesicles and mitochondrial impairments of hippocampal neurons [69,70]. Activation of parkin-mediated mitophagy in AD neurons is also quite prominent. As a consequence, the anterograde transport found in these neurons' axonal mitochondria is substantially reduced. The utilization of an antioxidant that specifically targets mitochondria was successful in rectifying these abnormalities and restoring mitochondrial mobility in Alzheimer's disease neurons. This was most likely accomplished by lowering mitophagy and Miro degradation [71,72] (Table 3.1).

3.4.2 Parkinson's Disease

PD is marked by the deterioration of dopaminergic neurons in the substantia nigra region of the brain [73]. The substantia nigra contains dopaminergic neurons, which form the nigrostriatal pathway that controls voluntary movements. The PD-affected individuals represent resting tremor, stiffness of body, slowness of movement, and postural instabilities [74–76]. Disease affected individuals also present with involvement of non-motor features and autonomic and sleep disturbances, along with depression and cognitive decline [77,78].

Although the precise mechanism behind PD is still undetermined, there is increasing evidence indicating that dysfunction in bioenergetics and impairment of mitochondria have a substantial impact on the development of the disease [79,80]. Mitochondria, critical in generating cellular energy and maintaining cellular homeostasis, are also linked with the progression of both sporadic and familial AD. Several cellular processes, such as reduced ATP synthesis, ROS generation, and impaired calcium regulation, are linked with PD pathogenesis. The energy shortage and oxidative stress ultimately lead to neuronal death [81,82].

In PD, the gradual loss of synapses and axons is linked with the dopamine neurotransmission deficit. The discovery that MPTP induces PD-like symptoms established the initial connection between mitochondrial malfunction and PD [83]. After absorption, MPTP passes across the

TABLE 3.1
Brain Areas Affected, Clinical Symptoms, Genetic Factors, and Pathways Associated with Mitochondrial Dysfunction in Different Neurodegenerative Disorders

Neurodegenerative Disorders	Brain Areas Affected	Clinical Symptoms	Genetic Factors	Mitochondrial Dysfunction (Affected Pathways)	References
Alzheimer's disease (AD)	✓ Neocortex ✓ Amygdala ✓ Hippocampus ✓ Brain stem monoaminergic neurons Basal forebrain cholinergic neurons	Cognitive decline Dementia Behavioral disturbances	APP Presenilin, AvaII16390 Apolipoprotein E tau Cytochrome C Oxidase, 16S rRNA Drp1 Fis 1 OPA1 Mfn1 Mfn2 Tomm 40 PGC-1α NRF1 NRF2	• Defects in mitochondrial dynamics • Synaptic injury • Enhanced mitochondrial fragmentation • Decreased ATP synthesis • Decreased anterograde transport	[46,66,130,131]
Amyotrophic lateral sclerosis (ALS)	✓ Brain and spinal cord ✓ Motor cortex	Weakness of muscles and paralysis	C9orf72 FUS TARDBP TDP-43 SOD1	• Poor mitochondrial trafficking • Decreased ATP synthesis • Defects in fusion/fission dynamics • Calcium buffering	[112,132–136]
Huntington's disease (HD)	✓ Striatum	Chorea Cognitive impairment Behavioral disturbances	HTT p^{53} PGC-1α Hip 1 Hip 14 HAP1 HAP 40 SH3GL3	• Decreased Ca^{2+} buffering • Reduced membrane potential (ΔΨ) • Caspase-mediated apoptotic pathways • Trafficking abnormalities	[107,108,137,138]

(Continued)

TABLE 3.1 *(Continued)*

Brain Areas Affected, Clinical Symptoms, Genetic Factors, and Pathways Associated with Mitochondrial Dysfunction in Different Neurodegenerative Disorders

Neurodegenerative Disorders	Brain Areas Affected	Clinical Symptoms	Genetic Factors	Mitochondrial Dysfunction (Affected Pathways)	References
Multiple sclerosis (MS)	✓ White matter	Neuroinflammation	HADC1 mtDNA Drp1	• Defects in mitochondrial dynamics • Defective mitophagy • Decreased ATP synthesis • Defects in ETC complex I and III	[125,139,140]
Parkinson's disease (PD)	✓ Substantia nigra Cortex Basal ganglia	Tremors Muscle rigidity Bradykinesia Shuffling gait Cognitive and neuropsychological deficits	SNCA LRRK2 VPS35 CHCHD2 Parkin PINK1 ATP13A2 tau	• Defective mitophagy • Decreased ATP synthesis • High oxidative stress • Calcium imbalance • Mitochondrial dynamics • Reduced membrane potential ($\Delta\Psi$) • Defects in mitochondrial complex I	[8,74,75,82,89,90,141,142]

blood-brain barrier and is processed by monoaminoxidase B in the astrocytes to become MPP+. The dopamine (DA) transporter actively absorbs extracellularly produced MPP+, which subsequently builds up inside the mitochondria of DA neurons and inhibits the ETC's mitochondrial complex I. MPTP-induced toxicity is a robust method for simulating neurodegeneration and the emergence of motor symptoms in animal models [84,85]. Transnationally, the MPTP investigations have shown that DA SNc neurons with mitochondrial CI inhibition can develop several features of idiopathic PD [86].

There are a number of genetic factors that can lead to early onset of PD, the most common of which are mutations in the PINK and PARK genes. These are essential for mitochondrial quality control processes and signaling cascades that react to the mitochondrial damage at the cellular level. In addition to initiating mitophagy, PINK1/PARK can also regulate the fission and fusion of mitochondria, vesicle production, and the process of mitochondrial biogenesis [87]. Loss of these protein functions causes ETC impairment in fibroblasts from PINK1 and Parkin familial Parkinson's disease patients, resulting in low ATP generation and elevated ROS levels [88]. Moreover, PINK1-KO animals showed deficiencies in mitochondrial membrane potential, decreased Ca^{2+} buffering capacity, and abnormalities in CI function [89,90].

The, alpha-synuclein, a very tiny and unstructured protein that builds up in Lewy bodies and Lewy neurites in PD, has several purposes at the synapse, including control of ion channels and dopamine release. Growing evidence demonstrate that alpha-synuclein can also interact with mitochondria and cause their malfunction in dopaminergic neurons. In addition, the removal of Pink1, Parkin, or DJ-1 from the striatum can cause dopamine reduction as well as a reduction in synaptic plasticity. Likewise, it has been demonstrated that LRRK2 interacts with several synaptic proteins and affects its transportation. In addition, the LRRK2 mutation reduces dopamine release in the striatum [91,92] (Table 3.1).

3.4.3 Huntington's Disease

HD is a progressive neurogenerative disorder characterized by an expansion of trinucleotide repeats (CAG) in the Huntington gene (*HTT*). As a result, misfolding proteins form aggregates of proteins. These aggregates of HTT cause the death of striatal neurons, affecting the striatum, cortical projection neurons, cortical white matter, and various subcortical areas of the brain. Symptoms appear in middle age, typically between 30s to 40s [93–95].

Huntington's disease manifested as cognitive deterioration, psychiatric manifestations, dementia, and chorea. The onset and advancement of the disease are influenced by the extent of polyglutamine repeat expansion leading to various cognitive and motor impairments [96–98]. The disease commonly manifests symptoms during adulthood, leading to a decline in an individual's motor skills, cognitive abilities, and verbal communication, ultimately necessitating full time care [93]. Gaining a comprehensive comprehension of synaptic plasticity is of utmost importance in the context of HD to devise therapeutic interventions that enhance neural connections and mitigate cognitive decline. HD is also characterized by a notable problem of impaired mitochondrial function, resulting in reduced glucose utilization and increased levels of lactate. The presence of the mutated huntingtin gene exacerbates the detrimental effects of mitDNA heteroplasmies, leading to an accelerated progression of the disease.

With the course of HD, striatal medium spiny neurons (MSNs) preferentially lose, resulting in motor dysfunction and cognitive deterioration. There is a growing body of evidence indicating that mitochondrial dysfunction is a major contributor to HD pathophysiology. Mutant HTT (mHTT)-mediated mitochondrial problems have been suggested to have a substantial impact on MSNs because of this neuronal subtype's high energy need [99]. Additionally, afflicted brain areas of HD patients show a substantial drop in glucose metabolism and a matching rise in lactate, suggesting a bioenergetics abnormality [100,101]. Furthermore, AD brain samples have reduced mitochondrial respiratory complex II, III, and IV activity [102] (Table 3.1).

Researchers have discovered that dysregulation of key transcription factors, such as p53 and PGC-1α, plays a significant role in HD by facilitating apoptosis, neurodegeneration, and mitochondrial dysfunction. Best known for being a tumor suppressor, p53 is a transcription factor that is activated and has increased expression in disease state. Strong mHTT-p53 interactions have been seen; these connections improve p53 stabilization by preventing the E3 ligase Mdm2 from recruiting. [103]. When p53 is overexpressed, many mitochondria-associated proteins (including Puma and Bax, which are connected to mitochondrial depolarization) are expressed, and apoptosis is triggered [104]. Also, p53 directly interacts with Drp1 to mediate mitochondrial-related necrosis and fragmentation in HD [105]. PGC-1α expression is significantly reduced in MSNs in transgenic mice models and individuals with HD. Since the CREB/TAF4 signaling pathway modulates PGC-1α expression, mHTT's interference with it contributes to the impairment of PGC-1α expression [106].

To regulate the passage of endocytic vesicles in both the anterograde and retrograde directions, the normal Htt protein engages in interactions with several trafficking mediators. These mediators include adaptin, endophilin, clathrin, and dynamin. By binding to HAP1, wild-type Htt promotes trafficking. This, in turn, causes interaction with kinesin and dynein/dynactin, two molecular motors that control mitochondrial and BDNF vesicle movement via microtubules. Phosphorylation of HTT plays a crucial role in determining whether vesicles should be transported in an anterograde or retrograde manner, acting as a molecular determinant. Trafficking abnormalities in HD are associated with abnormal protein interaction between mHTT and motor proteins; mHTT aggregates may also serve as physical barriers [107,108] (Table 3.1).

3.4.4 Amyotrophic Lateral Sclerosis

ALS is a progressive fatal neurodegenerative disorder characterized by deterioration of both upper and lower motor neurons (MNs) in the brain and spinal cord. Three to five years after symptoms first appear, patients expire from respiratory failure as a result of severe muscular weakness and eventual paralysis brought on by the increasing loss of MNs. Damage to the ETC, which impacts OXPHOS and ATP synthesis, is a major contributor to mitochondrial dysfunction. ALS patients also demonstrate impairments in the respiratory chain [109–111]. Thus far, ALS has been linked to mutations in more than 20 genes, such as C9orf72, FUS, TARDBP; TDP-43, and SOD1 [112]. It has also been documented that mitochondria are involved in the biology of ALS. The most common gene responsible for 20% of cases of familial ALS is the SOD1 gene [113]. The mitochondrial outer membrane, matrix, and inner membrane space have ALS-related mutant SOD1 proteins. SOD1 entrance and retention in mitochondria are facilitated by copper chaperone for SOD1 [114,115]. Animal model studies expressing copper chaperone for SOD1 lead to a higher concentration of mutant SOD1 in the mitochondria, resulting in a faster progression of the disease [51,116]. In differentiated NSC34 cells expressing SOD1 targeted at mitochondria, mutant SOD1 affected anterograde and retrograde mitochondrial transport. Furthermore, defective mitochondrial dynamics and fragmentation were brought on by mutant SOD1. SOD1 mutants cause abnormal mitochondrial fragmentation and clustering, altered mitochondrial transport, mitochondrial fragmentation, poor mitochondrial dynamics, and reduced mitochondrial density along neurites [116–118]. Motor neurons that were genetically modified to express the mutant ALS FUS protein exhibited a reduction in the size of their mitochondria. This reduction was further exacerbated by the presence of FUS protein in the cytoplasm [119] (Table 3.1).

3.4.5 Multiple Sclerosis

The hallmark of MS includes chronic inflammatory demyelination in the discrete regions of the white matter, along with subsequent neuroaxonal degeneration. Females usually develop the disease around 30, whereas males typically develop it around 33, with a 3:1 female-to-male ratio. The architecture and velocity of the mitochondria inside the myelinated axoplasm vary greatly, mostly depending on where they are located within the cell at the paranode, juxtanode, internode, or node of Ranvier where variations in energy demand and calcium levels occur. Ca^{2+} signaling is significantly elevated in peripheral nervous system (PNS) axons near the Ranvier node, where mitochondria are heavily concentrated. Significant alterations in mitochondrial dynamics result from the profound metabolic changes in axons during demyelination in MS conditions [120,121]. The combination of reduced energy production and loss of myelin in individuals with MS might worsen the process of neurodegeneration and lead to neuronal disability [122].

MS autopsy tissue showed abnormal levels of HDAC1, an enzyme that suppresses nuclear DNA transcription within the nucleus, in the cytoplasm and degenerated axons. This neuron-specific nuclear export of HDAC1 happened before neuritis injury and was linked to increased calcium entry into the neuron. Exposing neurons to glutamate and TNF-alpha impeded the forward movement of mitochondria in the axon by interfering with the interaction between cytoplasmic HDAC1 and kinesin [123,124].

Changes in fission and fusion can result in demyelination having a direct impact on mitochondrial dynamics. Such events alter both the distribution of mitochondria within the cell and their shape; they are essential to mitochondria. The role of Drp1 was investigated in both the experimental autoimmune encephalomyelitis- and cuprizone-induced animal models. Luo et al. reported an elevated expression of Drp1 in the damaged spinal cord during the peak of experimental autoimmune encephalomyelitis [125]. Aberrant expression of Drp1 in a laboratory setting revealed mitochondrial fragmentation. However, suppressing the activation of Drp1 had a beneficial effect on the protection of neurons in both EAE and CPZ [125]. In another study, it was shown that the dorsal root ganglia of patients with progressing MS were populated by respiratory defective neurons that have COXII but lack COXIV. Additionally, the number, size, and quantity of mitochondria were all higher in these respiratory defective neurons [126–128]. During the acute phase of relapsing-remitting multiple sclerosis, levels of PINK1 and Parkin were statistically significantly higher in cerebrospinal fluid and serum samples. Postmortem brain samples of AD exhibit mitochondrial abnormalities that are comparable to those seen in multiple sclerosis postmortem tissue [126,129]. This indicates that mitophagy plays a crucial role in the etiopathogenesis and development of MS (Table 3.1).

3.5 CONCLUSION

Although various neurodegenerative disorders have different spectrums of pathophysiology, there is compelling evidence suggesting that mitochondria play a crucial role in the underlying mechanisms of neurodegeneration and its linked mechanisms with synapse loss are prominent in several disorders. The mitochondria can be directly impacted by aberrant proteins, for instance beta-amyloid in AD and alpha-synuclein in PD. A progressive loss of synapses is well documented in the pathophysiology of AD, PD, HD, ALS, and MS. In such disorders, the cognitive impairment and other neurological symptoms are highly correlated with this loss of synapses. In this milieu, mitochondrial dysfunction is characterized primarily by defects in ATP production and increased oxidative stress. Energy production is based on the process of oxidative phosphorylation, which takes place inside the mitochondria.

A malfunction in these systems can result in energy deficiencies that impact cellular activities, particularly in tissues that require a lot of energy, such as the brain. The high energy requirements of synaptic terminals can be jeopardized by abnormalities in mitochondrial activity, which can ultimately result in synaptic dysfunction and loss. In addition, mitochondrial dynamics, which include transport, fission, and fusion, are among the many dynamic events that occur inside the mitochondria. Free radicals, which are mostly produced by mitochondrial metabolic processes that produce energy, also play an important role in neurodegeneration. Superoxide dismutase and glutathione are two examples of antioxidant scavenger enzymes that quickly neutralize any ROS byproducts under normal circumstances. However, in pathological conditions, the combination of dysfunctional antioxidant enzymes and dysregulated glucose metabolism leads to the uncontrolled production of

ROS, resulting in the accumulation of damaged mitochondrial proteins, global synaptic mitochondrial dysfunction, and a self-perpetuating state of oxidative stress. The mitochondria, on the other hand, are not only responsible for generating reactive oxygen species (ROS), but also for being targeted by them. In case of mitochondrial malfunction, ROS production becomes imbalanced, resulting in damage to DNA, lipids, and proteins.

Due to the growing body of research on the connection between mitochondrial malfunction and synaptic loss, novel treatments for neurodegenerative diseases that target the mitochondria are being investigated. Antioxidant techniques have been studied to lessen these processes. The potential for LTP and LTD to be used to treat neurodegenerative disorders is significant in present-day neuroscience. Many neurological disorders result from increased or lost synaptic drive brought on by diseases, brain injury, or childhood sensory deprivation. It may be possible to restore synaptic strength and, hence, improve CNS disorders associated with plasticity by manipulating synaptic strength using a variety of emerging technologies.

In summary, there is a reciprocal and intricate relationship between the loss of synapses and mitochondrial failure. Conversely, synaptic degeneration can result from disturbances in mitochondrial activity, and the loss of synapses can impact the health of mitochondria. Understanding and researching these pathways is essential to creating customized treatment plans for neurodegenerative diseases. Gaining insight into the mechanisms that cause changes in synaptic plasticity in these circumstances not only improves our understanding of how the diseases progress but also provides opportunities for creating specific therapeutic approaches that aim to protect cognitive function and decelerate neurodegenerative processes.

ACKNOWLEDGMENT

We would like to acknowledge University of Allahabad for its extensive support in completing this chapter.

CONFLICT OF INTEREST

The authors declare no conflict of interest.

ABBREVIATIONS

AD	Alzheimer's disease
ADP	Adenosine di-phosphate
ALS	Amyotrophic lateral sclerosis
AMPA	α-amino-3-hydroxy-5-methyl-4-isoxazolepropionic acid
AMPAR	α-amino-3-hydroxy-5-methyl-4-isoxazolepropionic acid receptor
ATP	Adenosine tri-phosphate
CAG	Cytosine–adenine–guanine
CCS	Copper chaperons for SOD1
CNS	Central nervous system
COX	Cytochrome c oxidase
CPZ	Cuprizone
CSF	Cerebrospinal fluid
DA	Dopamine
Drp	Dynamin-related protein
EAE	Encephalomyelitis
ETC	Electron transport chain
Fus	Fused in sarcoma
HD	Huntington's disease

HDAC	Histone deacetylase
KO	Knock out
LRRK	Leucine-rich repeat kinase
LTD	Long-term depression
LTP	Long -erm potentiation
Mfn	Mitofusin
mHTT	Mutant Huntington protein
MN	Motor neurons
MnSOD	Manganese superoxide dismutase
MPTP	1-methyl-4-phenyl-1, 2, 3, 6-tetrahydropyridine
MS	Multiple Sclerosis
MSN	Medium spiny neurons
NMDA	N-methyl-D-aspartate
NRF	Nuclear respiratory factor
OPA1	Optic atrophy-1
PARK	Parkin
PD	Parkinson's disease
PGC1	α Peroxisome proliferator-activated receptor gamma coactivator 1-alpha
PINK	PTEN-induced kinase
PNS	Peripheral nervous system
PS1	Presenilin
ROS	Reactive oxygen species
SN	Substantia nigra
SOD	Superoxide dismutase

REFERENCES

1. Gao, H.-M., and Hong, J.-S., 2008, "Why Neurodegenerative Diseases Are Progressive: Uncontrolled Inflammation Drives Disease Progression," Trends Immunol, **29**(8), pp. 357–365. https://doi.org/10.1016/j.it.2008.05.002.
2. Mayne, K., White, J.A., McMurran, C.E., Rivera, F.J., and de la Fuente, A.G., 2020, "Aging and Neurodegenerative Disease: Is the Adaptive Immune System a Friend or Foe?" Front Aging Neurosci, **12**, p. 572090. https://doi.org/10.3389/fnagi.2020.572090.
3. Verkhratsky, A., and Butt, A.M., 2023, "Neurodegenerative Diseases," *Neuroglia*, A. Verkhratsky, and A.M. Butt, eds., Academic Press, pp. 563–598. https://doi.org/10.1016/B978-0-12-821565-4.00012-2.
4. Lamptey, R.N.L., Chaulagain, B., Trivedi, R., Gothwal, A., Layek, B., and Singh, J., 2022, "A Review of the Common Neurodegenerative Disorders: Current Therapeutic Approaches and the Potential Role of Nanotherapeutics," Int J Mol Sci, **23**(3), p. 1851. https://doi.org/10.3390/ijms23031851.
5. Przedborski, S., Vila, M., and Jackson-Lewis, V., 2003, "Series Introduction: Neurodegeneration: What Is It and Where Are We?" J Clin Invest, **111**(1), pp. 3–10. https://doi.org/10.1172/JCI200317522.
6. Murman, D.L., 2015, "The Impact of Age on Cognition," Semin Hear, **36**(3), pp. 111–121. https://doi.org/10.1055/s-0035-1555115.
7. Feigin, V.L., Vos, T., Nichols, E., Owolabi, M.O., Carroll, W.M., Dichgans, M., Deuschl, G., Parmar, P., Brainin, M., and Murray, C., 2020, "The Global Burden of Neurological Disorders: Translating Evidence into Policy," Lancet Neurol, **19**(3), pp. 255–265. https://doi.org/10.1016/S1474-4422(19)30411-9.
8. Pal, A., Pegwal, N., Kaur, S., Mehta, N., Behari, M., and Sharma, R., 2018, "Deficit in Specific Cognitive Domains Associated with Dementia in Parkinson's Disease," J Clin Neurosci, **57**, pp. 116–120. https://doi.org/10.1016/j.jocn.2018.08.016.
9. Tan, S.H., Karri, V., Tay, N.W.R., Chang, K.H., Ah, H.Y., Ng, P.Q., Ho, H.S., Keh, H.W., and Candasamy, M., 2019, "Emerging Pathways to Neurodegeneration: Dissecting the Critical Molecular Mechanisms in Alzheimer's Disease, Parkinson's Disease," Biomed Pharmacother, **111**, pp. 765–777. https://doi.org/10.1016/j.biopha.2018.12.101.
10. Cramer, S.C., Sur, M., Dobkin, B.H., O'Brien, C., Sanger, T.D., Trojanowski, J.Q., Rumsey, J.M., Hicks, R., Cameron, J., Chen, D., Chen, W.G., Cohen, L.G., deCharms, C., Duffy, C.J., Eden, G.F., Fetz, E.E.,

Filart, R., Freund, M., Grant, S.J., Haber, S., Kalivas, P.W., Kolb, B., Kramer, A.F., Lynch, M., Mayberg, H.S., McQuillen, P.S., Nitkin, R., Pascual-Leone, A., Reuter-Lorenz, P., Schiff, N., Sharma, A., Shekim, L., Stryker, M., Sullivan, E.V., and Vinogradov, S., 2011, "Harnessing Neuroplasticity for Clinical Applications," Brain, **134**(6), pp. 1591–1609. https://doi.org/10.1093/brain/awr039.

11. Pisani, A., Centonze, D., Bernardi, G., and Calabresi, P., 2005, "Striatal Synaptic Plasticity: Implications for Motor Learning and Parkinson's Disease," Mov Disord, **20**(4), pp. 395–402. https://doi.org/10.1002/mds.20394.
12. Yuan, T.-F., Li, W.-G., Zhang, C., Wei, H., Sun, S., Xu, N.-J., Liu, J., and Xu, T.-L., 2020, "Targeting Neuroplasticity in Patients with Neurodegenerative Diseases Using Brain Stimulation Techniques," Transl Neurodegener, **9**(1), p. 44. https://doi.org/10.1186/s40035-020-00224-z.
13. van den Heuvel, M.P., and Sporns, O., 2013, "Network Hubs in the Human Brain," Trends Cogn Sci, **17**(12), pp. 683–696. https://doi.org/10.1016/j.tics.2013.09.012.
14. Citri, A., and Malenka, R.C., 2008, "Synaptic Plasticity: Multiple Forms, Functions, and Mechanisms," Neuropsychopharmacol, **33**(1), pp. 18–41. https://doi.org/10.1038/sj.npp.1301559.
15. Compagnoni, G.M., Di Fonzo, A., Corti, S., Comi, G.P., Bresolin, N., and Masliah, E., 2020, "The Role of Mitochondria in Neurodegenerative Diseases: The Lesson from Alzheimer's Disease and Parkinson's Disease," Mol Neurobiol, **57**(7), pp. 2959–2980. https://doi.org/10.1007/s12035-020-01926-1.
16. Magrané, J., Hervias, I., Henning, M.S., Damiano, M., Kawamata, H., and Manfredi, G., 2009, "Mutant SOD1 in Neuronal Mitochondria Causes Toxicity and Mitochondrial Dynamics Abnormalities," Hum Mol Genet, **18**(23), pp. 4552–4564. https://doi.org/10.1093/hmg/ddp421.
17. Farooqui, A.A., 2022, "Contribution of Neuroinflammation, Resolution, and Neuroprotection in Neurodegenerative Diseases," *Neuroinflammation, Resolution, and Neuroprotection in the Brain*, A.A. Farooqui, ed., Academic Press, pp. 121–160. https://doi.org/10.1016/B978-0-323-88460-0.00010-2.
18. Orth, M., and Schapira, A.H., 2001, "Mitochondria and Degenerative Disorders," Am J Med Genet, **106**(1), pp. 27–36. https://doi.org/10.1002/ajmg.1425.
19. Yassine, H.N., Self, W., Kerman, B.E., Santoni, G., Navalpur Shanmugam, N., Abdullah, L., Golden, L.R., Fonteh, A.N., Harrington, M.G., Gräff, J., Gibson, G.E., Kalaria, R., Luchsinger, J.A., Feldman, H.H., Swerdlow, R.H., Johnson, L.A., Albensi, B.C., Zlokovic, B.V., Tanzi, R., Cunnane, S., Samieri, C., Scarmeas, N., and Bowman, G.L., 2023, "Nutritional Metabolism and Cerebral Bioenergetics in Alzheimer's Disease and Related Dementias," Alzheimer's Dement, **19**(3), pp. 1041–1066. https://doi.org/10.1002/alz.12845.
20. Strope, T.A., Birky, C.J., and Wilkins, H.M., 2022, "The Role of Bioenergetics in Neurodegeneration," Int J Mol Sci, **23**(16), p. 9212. https://doi.org/10.3390/ijms23169212.
21. Mattson, M.P., Gleichmann, M., and Cheng, A., 2008, "Mitochondria in Neuroplasticity and Neurological Disorders," Neuron, **60**(5), pp. 748–766. https://doi.org/10.1016/j.neuron.2008.10.010.
22. Todorova, V., and Blokland, A., 2017, "Mitochondria and Synaptic Plasticity in the Mature and Aging Nervous System," Curr Neuropharmacol, **15**(1), pp. 166–173. https://doi.org/10.2174/1570159x14666160414111821.
23. Tang, J., Oliveros, A., and Jang, M.-H., 2019, "Dysfunctional Mitochondrial Bioenergetics and Synaptic Degeneration in Alzheimer Disease," Int Neurourol J, **23**(Suppl 1), pp. S5–10. https://doi.org/10.5213/inj.1938036.018.
24. Gazit, N., Vertkin, I., Shapira, I., Helm, M., Slomowitz, E., Sheiba, M., Mor, Y., Rizzoli, S., and Slutsky, I., 2016, "IGF-1 Receptor Differentially Regulates Spontaneous and Evoked Transmission via Mitochondria at Hippocampal Synapses," Neuron, **89**(3), pp. 583–597. https://doi.org/10.1016/j.neuron.2015.12.034.
25. Li, Z., Okamoto, K.-I., Hayashi, Y., and Sheng, M., 2004, "The Importance of Dendritic Mitochondria in the Morphogenesis and Plasticity of Spines and Synapses," Cell, **119**(6), pp. 873–887. https://doi.org/10.1016/j.cell.2004.11.003.
26. Beal, M.F., 2003, "Bioenergetic Approaches for Neuroprotection in Parkinson's Disease," Ann Neurol, **53**(Suppl 3), pp. S39–S47; discussion S47–48. https://doi.org/10.1002/ana.10479.
27. Walters, G.C., and Usachev, Y.M., 2023, "Mitochondrial Calcium Cycling in Neuronal Function and Neurodegeneration," Front Cell Dev Biol, **11**, p. 1094356. https://doi.org/10.3389/fcell.2023.1094356.
28. Cheng, A., Hou, Y., and Mattson, M.P., 2010, "Mitochondria and Neuroplasticity," ASN Neuro, **2**(5), p. e00045. https://doi.org/10.1042/AN20100019.
29. Malenka, R.C., and Nicoll, R.A., 1999, "Long-Term Potentiation–a Decade of Progress?" Science, **285**(5435), pp. 1870–1874. https://doi.org/10.1126/science.285.5435.1870.
30. Cheng, W.-T., Guo, Z.-X., Lin, C.-A., Lin, M.-Y., Tung, L.-C., and Fang, K., 2009, "Oxidative Stress Promotes Autophagic Cell Death in Human Neuroblastoma Cells with Ectopic Transfer of Mitochondrial PPP2R2B (Bβ2)," BMC Cell Biol, **10**, p. 91. https://doi.org/10.1186/1471-2121-10-91.

31. Greer, P.L., and Greenberg, M.E., 2008, "From Synapse to Nucleus: Calcium-Dependent Gene Transcription in the Control of Synapse Development and Function," Neuron, **59**(6), pp. 846–860. https://doi.org/10.1016/j.neuron.2008.09.002.
32. Mysore, S., Tai, C.-Y., and Schuman, E., 2008, "N-Cadherin, Spine Dynamics, and Synaptic Function," Front Neurosci, **2**, p. 2. https://doi.org//10.3389/neuro.01.035.2008.
33. Shah, S.A., and Schiff, N.D., 2010, "Central Thalamic Deep Brain Stimulation for Cognitive Neuromodulation: A Review of Proposed Mechanisms and Investigational Studies," Eur J Neurosci, **32**(7), pp. 1135–1144. https://doi.org/10.1111/j.1460-9568.2010.07420.x.
34. Shepherd, J.D., and Huganir, R.L., 2007, "The Cell Biology of Synaptic Plasticity: AMPA Receptor Trafficking," Annu Rev Cell Dev Biol, **23**, pp. 613–643. https://doi.org/10.1146/annurev.cellbio.23.090506.123516.
35. Bliss, T.V.P., and Cooke, S.F., 2011, "Long-Term Potentiation and Long-Term Depression: A Clinical Perspective," Clinics, **66**(Suppl 1), pp. 3–17. https://doi.org/10.1590/S1807-59322011001300002.
36. Chater, T.E., and Goda, Y., 2014, "The Role of AMPA Receptors in Postsynaptic Mechanisms of Synaptic Plasticity," Front Cell Neurosci, **8**, p. 401. https://doi.org/10.3389/fncel.2014.00401.
37. Ge, Y., Dong, Z., Bagot, R.C., Howland, J.G., Phillips, A.G., Wong, T.P., and Wang, Y.T., 2010, "Hippocampal Long-Term Depression Is Required for the Consolidation of Spatial Memory," Proc Natl Acad Sci USA, **107**(38), pp. 16697–16702. https://doi.org/10.1073/pnas.1008200107.
38. Lüscher, C., and Malenka, R.C., 2012, "NMDA Receptor-Dependent Long-Term Potentiation and Long-Term Depression (LTP/LTD)," Cold Spring Harb Perspect Biol, **4**(6), p. a005710. doi:10.1101/cshperspect.a005710.
39. Sumi, T., and Harada, K., 2020, "Mechanism Underlying Hippocampal Long-Term Potentiation and Depression Based on Competition between Endocytosis and Exocytosis of AMPA Receptors," Sci Rep, **10**(1), p. 14711. https://doi.org/10.1038/s41598-020-71528-3.
40. Hall, A.R., and Hausenloy, D.J., 2012, "The Shape of Things to Come: Mitochondrial Fusion and Fission in the Adult Heart," Cardiovasc Res, **94**(3), pp. 391–392. https://doi.org/10.1093/cvr/cvs150.
41. Detmer, S.A., and Chan, D.C., 2007, "Functions and Dysfunctions of Mitochondrial Dynamics," Nat Rev Mol Cell Biol, **8**(11), pp. 870–879. https://doi.org/10.1038/nrm2275.
42. Karbowski, M., and Youle, R.J., 2003, "Dynamics of Mitochondrial Morphology in Healthy Cells and during Apoptosis," Cell Death Differ, **10**(8), pp. 870–880. https://doi.org/10.1038/sj.cdd.4401260.
43. Rodrigues, T., and Ferraz, L.S., 2020, "Therapeutic Potential of Targeting Mitochondrial Dynamics in Cancer," Biochem Pharmacol, **182**, p. 114282. https://doi.org/10.1016/j.bcp.2020.114282.
44. Yapa, N.M.B., Lisnyak, V., Reljic, B., and Ryan, M.T., 2021, "Mitochondrial Dynamics in Health and Disease," FEBS Lett, **595**(8), pp. 1184–1204. https://doi.org/10.1002/1873-3468.14077.
45. Oliver, D., and Reddy, P.H., 2019, "Dynamics of Dynamin-Related Protein 1 in Alzheimer's Disease and Other Neurodegenerative Diseases," Cells, **8**(9), p. 961. https://doi.org/10.3390/cells8090961.
46. Wang, X., Su, B., Lee, H., Li, X., Perry, G., Smith, M.A., and Zhu, X., 2009, "Impaired Balance of Mitochondrial Fission and Fusion in Alzheimer's Disease," J Neurosci, **29**(28), pp. 9090–9103. https://doi.org/10.1523/JNEUROSCI.1357-09.2009.
47. Leslie, S.W., Chandler, L.J., Barr, E.M., and Farrar, R.P., 1985, "Reduced Calcium Uptake by Rat Brain Mitochondria and Synaptosomes in Response to Aging," Brain Res, **329**(1–2), pp. 177–183. https://doi.org/10.1016/0006-8993(85)90523-2.
48. Betarbet, R., Sherer, T.B., MacKenzie, G., Garcia-Osuna, M., Panov, A.V., and Greenamyre, J.T., 2000, "Chronic Systemic Pesticide Exposure Reproduces Features of Parkinson's Disease," Nat Neurosci, **3**(12), pp. 1301–1306. https://doi.org/10.1038/81834.
49. Cai, Q., and Tammineni, P., 2017, "Mitochondrial Aspects of Synaptic Dysfunction in Alzheimer's Disease," J Alzheimers Dis, **57**(4), pp. 1087–1103. https://doi.org/10.3233/JAD-160726.
50. Schwarz, T.L., 2013, "Mitochondrial Trafficking in Neurons," Cold Spring Harb Perspect Biol, **5**(6), p. a011304. https://doi.org/10.1101/cshperspect.a011304.
51. Shi, P., Wei, Y., Zhang, J., Gal, J., and Zhu, H., 2010, "Mitochondrial Dysfunction Is a Converging Point of Multiple Pathological Pathways in Amyotrophic Lateral Sclerosis," J Alzheimers Dis, **20**(Suppl 2), pp. S311–S324. https://doi.org/10.3233/JAD-2010-100366.
52. Varadi, A., Johnson-Cadwell, L.I., Cirulli, V., Yoon, Y., Allan, V.J., and Rutter, G.A., 2004, "Cytoplasmic Dynein Regulates the Subcellular Distribution of Mitochondria by Controlling the Recruitment of the Fission Factor Dynamin-Related Protein-1," J Cell Sci, **117**(Pt 19), pp. 4389–4400. https://doi.org/10.1242/jcs.01299.
53. Liu, X.-L., Wang, Y.-D., Yu, X.-M., Li, D.-W., and Li, G.-R., 2018, "Mitochondria-Mediated Damage to Dopaminergic Neurons in Parkinson's Disease (Review)," Int J Mol Med, **41**(2), pp. 615–623. doi:10.3892/ijmm.2017.3255.

54. Tirichen, H., Yaigoub, H., Xu, W., Wu, C., Li, R., and Li, Y., 2021, "Mitochondrial Reactive Oxygen Species and Their Contribution in Chronic Kidney Disease Progression Through Oxidative Stress," Front Physiol, **12**, p. 627837. https://doi.org//10.3389/fphys.2021.627837.
55. Chang, K.-H., and Chen, C.-M., 2020, "The Role of Oxidative Stress in Parkinson's Disease," Antioxidants, **9**(7), p. 597. https://doi.org/10.3390/antiox9070597.
56. Chen, W., Zhao, H., and Li, Y., 2023, "Mitochondrial Dynamics in Health and Disease: Mechanisms and Potential Targets," Sig Transduct Target Ther, **8**(1), pp. 1–25. https://doi.org/10.1038/s41392-023-01547-9.
57. Palikaras, K., Lionaki, E., and Tavernarakis, N., 2018, "Mechanisms of Mitophagy in Cellular Homeostasis, Physiology and Pathology," Nat Cell Biol, **20**(9), pp. 1013–1022. https://doi.org/10.1038/s41556-018-0176-2.
58. Gómez-Sánchez, R., Yakhine-Diop, S.M.S., Bravo-San Pedro, J.M., Pizarro-Estrella, E., Rodríguez-Arribas, M., Climent, V., Martin-Cano, F.E., González-Soltero, M.E., Tandon, A., Fuentes, J.M., and González-Polo, R.A., 2015, "PINK1 Deficiency Enhances Autophagy and Mitophagy Induction," Mol Cell Oncol, **3**(2), p. e1046579. https://doi.org/10.1080/23723556.2015.1046579.
59. Li, J., Yang, D., Li, Z., Zhao, M., Wang, D., Sun, Z., Wen, P., Dai, Y., Gou, F., Ji, Y., Zhao, D., and Yang, L., 2023, "PINK1/Parkin-Mediated Mitophagy in Neurodegenerative Diseases," Ageing Res Rev, **84**, p. 101817. https://doi.org/10.1016/j.arr.2022.101817.
60. Wu, Y., Chen, M., and Jiang, J., 2019, "Mitochondrial Dysfunction in Neurodegenerative Diseases and Drug Targets via Apoptotic Signaling," Mitochondrion, **49**, pp. 35–45. https://doi.org/10.1016/j.mito.2019.07.003.
61. Chakravorty, A., Jetto, C.T., and Manjithaya, R., 2019, "Dysfunctional Mitochondria and Mitophagy as Drivers of Alzheimer's Disease Pathogenesis," Front Aging Neurosci, **11**, p. 311. https://doi.org/10.3389/fnagi.2019.00311.
62. Francis, P., Palmer, A., Snape, M., and Wilcock, G., 1999, "The Cholinergic Hypothesis of Alzheimer's Disease: A Review of Progress," J Neurol Neurosurg Psychiatry, **66**(2), pp. 137–147.
63. Serrano-Pozo, A., Frosch, M.P., Masliah, E., and Hyman, B.T., 2011, "Neuropathological Alterations in Alzheimer Disease," Cold Spring Harb Perspect Med, **1**(1), p. a006189. https://doi.org/10.1101/cshperspect.a006189.
64. Swerdlow, R.H., Burns, J.M., and Khan, S.M., 2014, "The Alzheimer's Disease Mitochondrial Cascade Hypothesis: Progress and Perspectives," Biochim Biophys Acta, **1842**(8), pp. 1219–1231. https://doi.org/10.1016/j.bbadis.2013.09.010.
65. Bhatti, J.S., Bhatti, G.K., and Reddy, P.H., 2017, "Mitochondrial Dysfunction and Oxidative Stress in Metabolic Disorders — A Step towards Mitochondria Based Therapeutic Strategies," Biochim Biophys Acta Mol Basis Dis, **1863**(5), pp. 1066–1077. https://doi.org/10.1016/j.bbadis.2016.11.010.
66. Manczak, M., Calkins, M.J., and Reddy, P.H., 2011, "Impaired Mitochondrial Dynamics and Abnormal Interaction of Amyloid Beta with Mitochondrial Protein Drp1 in Neurons from Patients with Alzheimer's Disease: Implications for Neuronal Damage," Hum Mol Genet, **20**(13), pp. 2495–2509. https://doi.org/10.1093/hmg/ddr139.
67. Devi, L., Prabhu, B.M., Galati, D.F., Avadhani, N.G., and Anandatheerthavarada, H.K., 2006, "Accumulation of Amyloid Precursor Protein in the Mitochondrial Import Channels of Human Alzheimer's Disease Brain Is Associated with Mitochondrial Dysfunction," J Neurosci, **26**(35), pp. 9057–9068. https://doi.org/10.1523/JNEUROSCI.1469-06.2006.
68. Hirai, K., Aliev, G., Nunomura, A., Fujioka, H., Russell, R.L., Atwood, C.S., Johnson, A.B., Kress, Y., Vinters, H.V., Tabaton, M., Shimohama, S., Cash, A.D., Siedlak, S.L., Harris, P.L., Jones, P.K., Petersen, R.B., Perry, G., and Smith, M.A., 2001, "Mitochondrial Abnormalities in Alzheimer's Disease," J Neurosci, **21**(9), pp. 3017–3023. https://doi.org/10.1523/JNEUROSCI.21-09-03017.2001.
69. Pigino, G., Morfini, G., Pelsman, A., Mattson, M.P., Brady, S.T., and Busciglio, J., 2003, "Alzheimer's Presenilin 1 Mutations Impair Kinesin-Based Axonal Transport," J Neurosci, **23**(11), pp. 4499–4508. https://doi.org/10.1523/JNEUROSCI.23-11-04499.2003.
70. Stokin, G.B., Lillo, C., Falzone, T.L., Brusch, R.G., Rockenstein, E., Mount, S.L., Raman, R., Davies, P., Masliah, E., Williams, D.S., and Goldstein, L.S.B., 2005, "Axonopathy and Transport Deficits Early in the Pathogenesis of Alzheimer's Disease," Science, **307**(5713), pp. 1282–1288. https://doi.org/10.1126/science.1105681.
71. Calkins, M.J., Manczak, M., Mao, P., Shirendeb, U., and Reddy, P.H., 2011, "Impaired Mitochondrial Biogenesis, Defective Axonal Transport of Mitochondria, Abnormal Mitochondrial Dynamics and Synaptic Degeneration in a Mouse Model of Alzheimer's Disease," Hum Mol Genet, **20**(23), pp. 4515–4529. https://doi.org/10.1093/hmg/ddr381.
72. Ye, X., Sun, X., Starovoytov, V., and Cai, Q., 2015, "Parkin-Mediated Mitophagy in Mutant hAPP Neurons and Alzheimer's Disease Patient Brains," Hum Mol Genet, **24**(10), pp. 2938–2951. https://doi.org/10.1093/hmg/ddv056.

73. Dauer, W., and Przedborski, S., 2003, "Parkinson's Disease: Mechanisms and Models," Neuron, **39**(6), pp. 889–909. https://doi.org/10.1016/S0896-6273(03)00568-3.
74. Jankovic, J., 2008, "Parkinson's Disease: Clinical Features and Diagnosis," J Neurol Neurosurg Psychiatry, **79**(4), pp. 368–376. https://doi.org/10.1136/jnnp.2007.131045.
75. Maiti, P., Manna, J., and Dunbar, G.L., 2017, "Current Understanding of the Molecular Mechanisms in Parkinson's Disease: Targets for Potential Treatments," Transl Neurodegener, **6**(1), p. 28. https://doi.org/10.1186/s40035-017-0099-z.
76. Postuma, R.B., Berg, D., Stern, M., Poewe, W., Olanow, C.W., Oertel, W., Obeso, J., Marek, K., Litvan, I., Lang, A.E., Halliday, G., Goetz, C.G., Gasser, T., Dubois, B., Chan, P., Bloem, B.R., Adler, C.H., and Deuschl, G., 2015, "MDS Clinical Diagnostic Criteria for Parkinson's Disease," Mov Disord, **30**(12), pp. 1591–1601. https://doi.org/10.1002/mds.26424.
77. Chaudhuri, K.R., Healy, D.G., and Schapira, A.H., 2006, "Non-Motor Symptoms of Parkinson's Disease: Diagnosis and Management," Lancet Neurol, **5**(3), pp. 235–245. https://doi.org/10.1016/S1474-4422(06)70373-8.
78. Reichmann, H., Schneider, C., and Löhle, M., 2009, "Non-Motor Features of Parkinson's Disease: Depression and Dementia," Parkinsonism Relat Disord, **15**(Suppl 3), pp. S87–S92. https://doi.org/10.1016/S1353-8020(09)70789-8.
79. Abou-Sleiman, P.M., Muqit, M.M.K., and Wood, N.W., 2006, "Expanding Insights of Mitochondrial Dysfunction in Parkinson's Disease," Nat Rev Neurosci, **7**(3), pp. 207–219. https://doi.org/10.1038/nrn1868.
80. Moon, H.E., and Paek, S.H., 2015, "Mitochondrial Dysfunction in Parkinson's Disease," Exp Neurobiol, **24**(2), pp. 103–116. https://doi.org/10.5607/en.2015.24.2.103.
81. Exner, N., Lutz, A.K., Haass, C., and Winklhofer, K.F., 2012, "Mitochondrial Dysfunction in Parkinson's Disease: Molecular Mechanisms and Pathophysiological Consequences," EMBO J, **31**(14), pp. 3038–3062. https://doi.org/10.1038/emboj.2012.170.
82. Mounsey, R.B., and Teismann, P., 2010, "Mitochondrial Dysfunction in Parkinson's Disease: Pathogenesis and Neuroprotection," Parkinson's Disease, **2011**, p. e617472. https://doi.org/10.4061/2011/617472.
83. Chen, C., Turnbull, D.M., and Reeve, A.K., 2019, "Mitochondrial Dysfunction in Parkinson's Disease—Cause or Consequence?" Biology, **8**(2). https://doi.org/10.3390/biology8020038.
84. Burns, R.S., Chiueh, C.C., Markey, S.P., Ebert, M.H., Jacobowitz, D.M., and Kopin, I.J., 1983, "A Primate Model of Parkinsonism: Selective Destruction of Dopaminergic Neurons in the Pars Compacta of the Substantia Nigra by N-Methyl-4-Phenyl-1,2,3,6-Tetrahydropyridine," Proc Natl Acad Sci USA, **80**(14), pp. 4546–4550.
85. Meredith, G.E., and Rademacher, D.J., 2011, "MPTP Mouse Models of Parkinson's Disease: An Update," J Parkinsons Dis, **1**(1), pp. 19–33. https://doi.org/10.3233/JPD-2011-11023.
86. Henrich, M.T., Oertel, W.H., Surmeier, D.J., and Geibl, F.F., 2023, "Mitochondrial Dysfunction in Parkinson's Disease - a Key Disease Hallmark with Therapeutic Potential," Mol Neurodegener, **18**(1), p. 83. https://doi.org/10.1186/s13024-023-00676-7.
87. Kumar, M., Acevedo-Cintrón, J., Jhaldiyal, A., Wang, H., Andrabi, S.A., Eacker, S., Karuppagounder, S.S., Brahmachari, S., Chen, R., Kim, H., Ko, H.S., Dawson, V.L., and Dawson, T.M., 2020, "Defects in Mitochondrial Biogenesis Drive Mitochondrial Alterations in PARKIN-Deficient Human Dopamine Neurons," Stem Cell Rep, **15**(3), pp. 629–645. https://doi.org/10.1016/j.stemcr.2020.07.013.
88. Rakovic, A., Grünewald, A., Seibler, P., Ramirez, A., Kock, N., Orolicki, S., Lohmann, K., and Klein, C., 2010, "Effect of Endogenous Mutant and Wild-Type PINK1 on Parkin in Fibroblasts from Parkinson Disease Patients," Hum Mol Genet, **19**(16), pp. 3124–3137. https://doi.org/10.1093/hmg/ddq215.
89. Palacino, J.J., Sagi, D., Goldberg, M.S., Krauss, S., Motz, C., Wacker, M., Klose, J., and Shen, J., 2004, "Mitochondrial Dysfunction and Oxidative Damage in Parkin-Deficient Mice," J Biol Chem, **279**(18), pp. 18614–18622. https://doi.org/10.1074/jbc.M401135200.
90. Stichel, C.C., Zhu, X.-R., Bader, V., Linnartz, B., Schmidt, S., and Lübbert, H., 2007, "Mono- and Double-Mutant Mouse Models of Parkinson's Disease Display Severe Mitochondrial Damage," Hum Mol Genet, **16**(20), pp. 2377–2393. https://doi.org/10.1093/hmg/ddm083.
91. Piccoli, G., Condliffe, S.B., Bauer, M., Giesert, F., Boldt, K., De Astis, S., Meixner, A., Sarioglu, H., Vogt-Weisenhorn, D.M., Wurst, W., Gloeckner, C.J., Matteoli, M., Sala, C., and Ueffing, M., 2011, "LRRK2 Controls Synaptic Vesicle Storage and Mobilization within the Recycling Pool," J Neurosci, **31**(6), pp. 2225–2237. https://doi.org/10.1523/JNEUROSCI.3730-10.2011.
92. Reeve, A.K., Grady, J.P., Cosgrave, E.M., Bennison, E., Chen, C., Hepplewhite, P.D., and Morris, C.M., 2018, "Mitochondrial Dysfunction within the Synapses of Substantia Nigra Neurons in Parkinson's Disease," NPJ Parkinson's Dis, **4**(1), pp. 1–10. https://doi.org/10.1038/s41531-018-0044-6.

93. Ajitkumar, A., and De Jesus, O., 2023, "Huntington Disease," *StatPearls*, StatPearls Publishing. http://www.ncbi.nlm.nih.gov/books/NBK559166/.
94. Semaka, A., Kay, C., Belfroid, R.D.M., Bijlsma, E.K., Losekoot, M., van Langen, I.M., van Maarle, M.C., Oosterloo, M., Hayden, M.R., and van Belzen, M.J., 2015, "A New Mutation for Huntington Disease Following Maternal Transmission of an Intermediate Allele," Eur J Med Genet, **58**(1), pp. 28–30. https://doi.org/10.1016/j.ejmg.2014.11.005.
95. Telenius, H., Kremer, H.P., Theilmann, J., Andrew, S.E., Almqvist, E., Anvret, M., Greenberg, C., Greenberg, J., Lucotte, G., and Squitieri, F., 1993, "Molecular Analysis of Juvenile Huntington Disease: The Major Influence on (CAG)n Repeat Length Is the Sex of the Affected Parent," Hum Mol Genet, **2**(10), pp. 1535–1540. https://doi.org/10.1093/hmg/2.10.1535.
96. Dai, Y., Wang, H., Lian, A., Li, J., Zhao, G., Hu, S., and Li, B., 2023, "A Comprehensive Perspective of Huntington's Disease and Mitochondrial Dysfunction," Mitochondrion, **70**, pp. 8–19. https://doi.org/10.1016/j.mito.2023.03.001.
97. Intihar, T.A., Martinez, E.A., and Gomez-Pastor, R., 2019, "Mitochondrial Dysfunction in Huntington's Disease; Interplay Between HSF1, P53 and PGC-1α Transcription Factors," Front Cell Neurosci, **13**, p. 103. https://doi.org/10.3389/fncel.2019.00103.
98. MacDonald, M.E., Ambrose, C.M., Duyao, M.P., Myers, R.H., Lin, C., Srinidhi, L., Barnes, G., Taylor, S.A., James, M., Groot, N., MacFarlane, H., Jenkins, B., Anderson, M.A., Wexler, N.S., Gusella, J.F., Bates, G.P., Baxendale, S., Hummerich, H., Kirby, S., North, M., Youngman, S., Mott, R., Zehetner, G., Sedlacek, Z., Poustka, A., Frischauf, A.-M., Lehrach, H., Buckler, A.J., Church, D., Doucette-Stamm, L., O'Donovan, M.C., Riba-Ramirez, L., Shah, M., Stanton, V.P., Strobel, S.A., Draths, K.M., Wales, J.L., Dervan, P., Housman, D.E., Altherr, M., Shiang, R., Thompson, L., Fielder, T., Wasmuth, J.J., Tagle, D., Valdes, J., Elmer, L., Allard, M., Castilla, L., Swaroop, M., Blanchard, K., Collins, F.S., Snell, R., Holloway, T., Gillespie, K., Datson, N., Shaw, D., and Harper, P.S., 1993, "A Novel Gene Containing a Trinucleotide Repeat That Is Expanded and Unstable on Huntington's Disease Chromosomes," Cell, **72**(6), pp. 971–983. https://doi.org/10.1016/0092-8674(93)90585-E.
99. Pickrell, A.M., Fukui, H., Wang, X., Pinto, M., and Moraes, C.T., 2011, "The Striatum Is Highly Susceptible to Mitochondrial Oxidative Phosphorylation Dysfunctions," J Neurosci, **31**(27), pp. 9895–9904. https://doi.org/10.1523/JNEUROSCI.6223-10.2011.
100. Djoussé, L., Knowlton, B., Cupples, L.A., Marder, K., Shoulson, I., and Myers, R.H., 2002, "Weight Loss in Early Stage of Huntington's Disease," Neurology, **59**(9), pp. 1325–1330. https://doi.org/10.1212/01.wnl.0000031791.10922.cf.
101. Jenkins, B.G., Andreassen, O.A., Dedeoglu, A., Leavitt, B., Hayden, M., Borchelt, D., Ross, C.A., Ferrante, R.J., and Beal, M.F., 2005, "Effects of CAG Repeat Length, HTT Protein Length and Protein Context on Cerebral Metabolism Measured Using Magnetic Resonance Spectroscopy in Transgenic Mouse Models of Huntington's Disease," J Neurochem, **95**(2), pp. 553–562. https://doi.org/10.1111/j.1471-4159.2005.03411.x.
102. Gu, M., Gash, M.T., Mann, V.M., Javoy-Agid, F., Cooper, J.M., and Schapira, A.H., 1996, "Mitochondrial Defect in Huntington's Disease Caudate Nucleus," Ann Neurol, **39**(3), pp. 385–389. https://doi.org/10.1002/ana.410390317.
103. Bae, B.-I., Xu, H., Igarashi, S., Fujimuro, M., Agrawal, N., Taya, Y., Hayward, S.D., Moran, T.H., Montell, C., Ross, C.A., Snyder, S.H., and Sawa, A., 2005, "P53 Mediates Cellular Dysfunction and Behavioral Abnormalities in Huntington's Disease," Neuron, **47**(1), pp. 29–41. https://doi.org/10.1016/j.neuron.2005.06.005.
104. La Spada, A.R., and Morrison, R.S., 2005, "The Power of the Dark Side: Huntington's Disease Protein and P53 Form a Deadly Alliance," Neuron, **47**(1), pp. 1–3. https://doi.org/10.1016/j.neuron.2005.06.023.
105. Guo, X., Sesaki, H., and Qi, X., 2014, "Drp1 Stabilizes P53 on the Mitochondria to Trigger Necrosis under Oxidative Stress Conditions in Vitro and in Vivo," Biochem J, **461**(1), pp. 137–146. https://doi.org/10.1042/BJ20131438.
106. Cui, L., Jeong, H., Borovecki, F., Parkhurst, C.N., Tanese, N., and Krainc, D., 2006, "Transcriptional Repression of PGC-1alpha by Mutant Huntingtin Leads to Mitochondrial Dysfunction and Neurodegeneration," Cell, **127**(1), pp. 59–69. https://doi.org/10.1016/j.cell.2006.09.015.
107. Bossy-Wetzel, E., Petrilli, A., and Knott, A.B., 2008, "Mutant Huntingtin and Mitochondrial Dysfunction," Trends Neurosci, **31**(12), pp. 609–616. https://doi.org/10.1016/j.tins.2008.09.004.
108. Chang, D.T.W., Rintoul, G.L., Pandipati, S., and Reynolds, I.J., 2006, "Mutant Huntingtin Aggregates Impair Mitochondrial Movement and Trafficking in Cortical Neurons," Neurobiol Dis, **22**(2), pp. 388–400. https://doi.org/10.1016/j.nbd.2005.12.007.

109. van Es, M.A., Hardiman, O., Chio, A., Al-Chalabi, A., Pasterkamp, R.J., Veldink, J.H., and van den Berg, L.H., 2017, "Amyotrophic Lateral Sclerosis," Lancet, **390**(10107), pp. 2084–2098. https://doi.org/10.1016/S0140-6736(17)31287-4.
110. Genin, E.C., Abou-Ali, M., and Paquis-Flucklinger, V., 2023, "Mitochondria, a Key Target in Amyotrophic Lateral Sclerosis Pathogenesis," Genes, **14**(11), p. 1981. https://doi.org/10.3390/genes14111981.
111. Hardiman, O., Al-Chalabi, A., Chio, A., Corr, E.M., Logroscino, G., Robberecht, W., Shaw, P.J., Simmons, Z., and van den Berg, L.H., 2017, "Amyotrophic Lateral Sclerosis," Nat Rev Dis Primers, **3**(1), pp. 1–19. https://doi.org/10.1038/nrdp.2017.71.
112. Smith, E.F., Shaw, P.J., and De Vos, K.J., 2019, "The Role of Mitochondria in Amyotrophic Lateral Sclerosis," Neurosci Lett, **710**, p. 132933. https://doi.org/10.1016/j.neulet.2017.06.052.
113. Rosen, D.R., Siddique, T., Patterson, D., Figlewicz, D.A., Sapp, P., Hentati, A., Donaldson, D., Goto, J., O'Regan, J.P., Deng, H.-X., Rahmani, Z., Krizus, A., McKenna-Yasek, D., Cayabyab, A., Gaston, S.M., Berger, R., Tanzi, R.E., Halperin, J.J., Herzfeldt, B., Van den Bergh, R., Hung, W.-Y., Bird, T., Deng, G., Mulder, D.W., Smyth, C., Laing, N.G., Soriano, E., Pericak-Vance, M.A., Haines, J., Rouleau, G.A., Gusella, J.S., Horvitz, H.R., and Brown, R.H., 1993, "Mutations in Cu/Zn Superoxide Dismutase Gene Are Associated with Familial Amyotrophic Lateral Sclerosis," Nature, **362**(6415), pp. 59–62. https://doi.org/10.1038/362059a0.
114. Sturtz, L.A., Diekert, K., Jensen, L.T., Lill, R., and Culotta, V.C., 2001, "A Fraction of Yeast Cu, Zn-Superoxide Dismutase and Its Metallochaperone, CCS, Localize to the Intermembrane Space of Mitochondria. A Physiological Role for SOD1 in Guarding against Mitochondrial Oxidative Damage," J Biol Chem, **276**(41), pp. 38084–38089. https://doi.org/10.1074/jbc.M105296200.
115. Vande Velde, C., Miller, T.M., Cashman, N.R., and Cleveland, D.W., 2008, "Selective Association of Misfolded ALS-Linked Mutant SOD1 with the Cytoplasmic Face of Mitochondria," Proc Natl Acad Sci USA, **105**(10), pp. 4022–4027. https://doi.org/10.1073/pnas.0712209105.
116. Son, M., Puttaparthi, K., Kawamata, H., Rajendran, B., Boyer, P.J., Manfredi, G., and Elliott, J.L., 2007, "Overexpression of CCS in G93A-SOD1 Mice Leads to Accelerated Neurological Deficits with Severe Mitochondrial Pathology," Proc Natl Acad Sci USA, **104**(14), pp. 6072–6077. https://doi.org/10.1073/pnas.0610923104.
117. Magrané, J., Sahawneh, M.A., Przedborski, S., Estévez, ÁG., and Manfredi, G., 2012, "Mitochondrial Dynamics and Bioenergetic Dysfunction Is Associated with Synaptic Alterations in Mutant SOD1 Motor Neurons," J Neurosci, **32**(1), pp. 229–242. https://doi.org/10.1523/JNEUROSCI.1233-11.2012.
118. Sotelo-Silveira, J.R., Lepanto, P., Elizondo, V., Horjales, S., Palacios, F., Martinez-Palma, L., Marin, M., Beckman, J.S., and Barbeito, L., 2009, "Axonal Mitochondrial Clusters Containing Mutant SOD1 in Transgenic Models of ALS," Antioxid Redox Signal, **11**(7), pp. 1535–1545. https://doi.org/10.1089/ars.2009.2614.
119. Tradewell, M.L., Yu, Z., Tibshirani, M., Boulanger, M.-C., Durham, H.D., and Richard, S., 2012, "Arginine Methylation by PRMT1 Regulates Nuclear-Cytoplasmic Localization and Toxicity of FUS/TLS Harbouring ALS-Linked Mutations," Hum Mol Genet, **21**(1), pp. 136–149. https://doi.org/10.1093/hmg/ddr448.
120. Ohno, N., Kidd, G.J., Mahad, D., Kiryu-Seo, S., Avishai, A., Komuro, H., and Trapp, B.D., 2011, "Myelination and Axonal Electrical Activity Modulate the Distribution and Motility of Mitochondria at CNS Nodes of Ranvier," J Neurosci, **31**(20), pp. 7249–7258. https://doi.org/10.1523/JNEUROSCI.0095-11.2011.
121. Zhang, N., An, M.C., Montoro, D., and Ellerby, L.M., 2010, "Characterization of Human Huntington's Disease Cell Model from Induced Pluripotent Stem Cells," PLoS Curr, **2**. https://doi.org/10.1371/currents.RRN1193.
122. Campbell, G., and Mahad, D.J., 2018, "Mitochondrial Dysfunction and Axon Degeneration in Progressive Multiple Sclerosis," FEBS Lett, **592**(7), pp. 1113–1121. https://doi.org/10.1002/1873-3468.13013.
123. Kim, J.Y., Shen, S., Dietz, K., He, Y., Howell, O., Reynolds, R., and Casaccia, P., 2010, "HDAC1 Nuclear Export Induced by Pathological Conditions Is Essential for the Onset of Axonal Damage," Nat Neurosci, **13**(2), pp. 180–189. https://doi.org/10.1038/nn.2471.
124. Sorbara, C.D., Wagner, N.E., Ladwig, A., Nikić, I., Merkler, D., Kleele, T., Marinković, P., Naumann, R., Godinho, L., Bareyre, F.M., Bishop, D., Misgeld, T., and Kerschensteiner, M., 2014, "Pervasive Axonal Transport Deficits in Multiple Sclerosis Models," Neuron, **84**(6), pp. 1183–1190. https://doi.org/10.1016/j.neuron.2014.11.006.
125. Luo, F., Herrup, K., Qi, X., and Yang, Y., 2017, "Inhibition of Drp1 Hyper-Activation Is Protective in Animal Models of Experimental Multiple Sclerosis," Exp Neurol, **292**, pp. 21–34. https://doi.org/10.1016/j.expneurol.2017.02.015.

126. Atkinson, K.C., Osunde, M., and Tiwari-Woodruff, S.K., 2023, "The Complexities of Investigating Mitochondria Dynamics in Multiple Sclerosis and Mouse Models of MS," Front Neurosci, **17**, p. 1144896. https://doi.org/10.3389/fnins.2023.1144896.
127. Licht-Mayer, S., Campbell, G.R., Canizares, M., Mehta, A.R., Gane, A.B., McGill, K., Ghosh, A., Fullerton, A., Menezes, N., Dean, J., Dunham, J., Al-Azki, S., Pryce, G., Zandee, S., Zhao, C., Kipp, M., Smith, K.J., Baker, D., Altmann, D., Anderton, S.M., Kap, Y.S., Laman, J.D., Hart, B.A.'t, Rodriguez, M., Watzlawick, R., Schwab, J.M., Carter, R., Morton, N., Zagnoni, M., Franklin, R.J.M., Mitchell, R., Fleetwood-Walker, S., Lyons, D.A., Chandran, S., Lassmann, H., Trapp, B.D., and Mahad, D.J., 2020, "Enhanced Axonal Response of Mitochondria to Demyelination Offers Neuroprotection: Implications for Multiple Sclerosis," Acta Neuropathol, **140**(2), pp. 143–167. https://doi.org/10.1007/s00401-020-02179-x.
128. Zambonin, J.L., Zhao, C., Ohno, N., Campbell, G.R., Engeham, S., Ziabreva, I., Schwarz, N., Lee, S.E., Frischer, J.M., Turnbull, D.M., Trapp, B.D., Lassmann, H., Franklin, R.J.M., and Mahad, D.J., 2011, "Increased Mitochondrial Content in Remyelinated Axons: Implications for Multiple Sclerosis," Brain, **134**(Pt 7), pp. 1901–1913. https://doi.org/10.1093/brain/awr110.
129. Cossu, D., Yokoyama, K., Sechi, L.A., and Hattori, N., 2021, "Potential of PINK1 and PARKIN Proteins as Biomarkers for Active Multiple Sclerosis: A Japanese Cohort Study," Front Immunol, **12**, p. 681386. https://doi.org/10.3389/fimmu.2021.681386.
130. Martin, L.J., 2002, "Neurodegenerative Disorders," *Encyclopedia of the Human Brain*, V.S. Ramachandran, ed., Academic Press, pp. 441–463. https://doi.org/10.1016/B0-12-227210-2/00240-5.
131. Moreira, P.I., Carvalho, C., Zhu, X., Smith, M.A., and Perry, G., 2010, "Mitochondrial Dysfunction Is a Trigger of Alzheimer's Disease Pathophysiology," Biochim Biophys Acta Mol Basis Dis, **1802**(1), pp. 2–10. https://doi.org/10.1016/j.bbadis.2009.10.006.
132. Ferri, A., Cozzolino, M., Crosio, C., Nencini, M., Casciati, A., Gralla, E.B., Rotilio, G., Valentine, J.S., and Carrì, M.T., 2006, "Familial ALS-Superoxide Dismutases Associate with Mitochondria and Shift Their Redox Potentials," Proc Natl Acad Sci, **103**(37), pp. 13860–13865. https://doi.org/10.1073/pnas.0605814103.
133. Liu, W., Yamashita, T., Tian, F., Morimoto, N., Ikeda, Y., Deguchi, K., and Abe, K., 2013, "Mitochondrial Fusion and Fission Proteins Expression Dynamically Change in a Murine Model of Amyotrophic Lateral Sclerosis," Curr Neurovasc Res, **10**(3), pp. 222–230. https://doi.org/10.2174/15672026113109990060.
134. Muyderman, H., and Chen, T., 2014, "Mitochondrial Dysfunction in Amyotrophic Lateral Sclerosis – a Valid Pharmacological Target?" Br J Pharmacol, **171**(8), pp. 2191–2205. https://doi.org/10.1111/bph.12476.
135. Sasaki, S., Horie, Y., and Iwata, M., 2007, "Mitochondrial Alterations in Dorsal Root Ganglion Cells in Sporadic Amyotrophic Lateral Sclerosis," Acta Neuropathol, **114**(6), pp. 633–639. https://doi.org/10.1007/s00401-007-0299-1.
136. Zhao, W., Varghese, M., Yemul, S., Pan, Y., Cheng, A., Marano, P., Hassan, S., Vempati, P., Chen, F., Qian, X., and Pasinetti, G.M., 2011, "Peroxisome Proliferator Activator Receptor Gamma Coactivator-1alpha (PGC-1α) Improves Motor Performance and Survival in a Mouse Model of Amyotrophic Lateral Sclerosis," Mol Neurodegener, **6**(1), p. 51. https://doi.org/10.1186/1750-1326-6-51.
137. Damiano, M., Galvan, L., Déglon, N., and Brouillet, E., 2010, "Mitochondria in Huntington's Disease," Biochim Biophys Acta Mol Basis Dis, **1802**(1), pp. 52–61. https://doi.org/10.1016/j.bbadis.2009.07.012.
138. Okada, N., Yako, T., Nakamura, S., Shimazawa, M., and Hara, H., 2021, "Reduced Mitochondrial Complex II Activity Enhances Cell Death via Intracellular Reactive Oxygen Species in STHdhQ111 Striatal Neurons with Mutant Huntingtin," J Pharmacol Sci, **147**(4), pp. 367–375. https://doi.org/10.1016/j.jphs.2021.09.001.
139. Barcelos, I.P.D., Troxell, R.M., and Graves, J.S., 2019, "Mitochondrial Dysfunction and Multiple Sclerosis," Biology, **8**(2), p. 37. https://doi.org/10.3390/biology8020037.
140. Trapp, B.D., and Stys, P.K., 2009, "Virtual Hypoxia and Chronic Necrosis of Demyelinated Axons in Multiple Sclerosis," Lancet Neurol, **8**(3), pp. 280–291. https://doi.org/10.1016/S1474-4422(09)70043-2.
141. Alexander, G.E., 2004, "Biology of Parkinson's Disease: Pathogenesis and Pathophysiology of a Multisystem Neurodegenerative Disorder," Dialogues in Clinical Neuroscience, **6**, pp. 259–280.
142. Pal, A., Pegwal, N., Behari, M., and Sharma, R., 2019, "Is Dementia in Parkinson' Disease Related to Chronic Stress, Anxiety, and Depression?" Ann Indian Acad Neurol, **22**(4), pp. 409–413. https://doi.org/10.4103/aian.AIAN_341_18.

4 Cellular and Synaptic Plasticity in Parkinson's Disease, Alzheimer's Disease, and Huntington's Disease

*Nikhila Khola, Deepika, Somu Yadav, and Pawan Kumar Maurya**
Central University of Haryana
Mahendergarh, Haryana, India

4.1 INTRODUCTION

Neurodegenerative disorders (NDDs) are categorized by the loss of neurons. Common NDDs are Parkinson's disease Parkinson's disease (PD), Alzheimer's disease (AD), and Huntington's disease disease (HD). In addition to old age, environmental and genetic factors are also considered major contributors to NDDs (Lamptey et al., 2022). NDDs involve a gradual and slow degeneration of the neurons and axons in the brain and ultimately result in the abnormality of cellular function thus leading to cellular death. The symptomsof the disorder starts from the loss of memory and coordination and is followed by a complete loss of functional abilities. Patients with NDDs have a high amount of medical expensesand their normal life processes are disrupted (Jagaran & Singh, 2021). Synapses are specialized structures that occur as a connectivity between the neuronal extensions that form either a pre- or post-synaptic neuron. The formation of these synapses is not static, but they undergo a consistent change according to the behavioral needs of animals with respect to their changing environment. This results in the development of new connections or enhances the efficiency of the existing synapseswhich is referred as synaptic plasticity (Batool et al., 2019).

In recent times, an unpredicted relationship has been confirmed between synaptic transmission and neuronal pathology. Neuronal synapses are the sites of electrochemical signaling. Synapses are intercellular junctions that channel rapid point-to-point signals between neurons and between neurons circuits (Sudhof, 2021). Synaptic plasticity refers to neurophysiological intermingled connections that are involved in brain network development (Jaworski, Banach-Kasper, & Gralec, 2019). The main characteristic of neurons in the brain is how the data is encoded within a neuron and transferred through synapses in both digital and analogue patterns. The neurons communicate with each other at different volumes, and the setting upvolume at synapses is synaptic strength (Bello-Medina, Rodriguez-Martinez, Prado-Alcala, & Rivas-Arancibia, 2022). The change in synaptic strength is synaptic plasticity. Synaptic plasticity is associated with perusing and memory processes; it is also connected with clinical recovery after brain injury or trauma.

Neuroscientists mainly discuss two types of plasticity. Short-term plasticity refers to changes that occur in sub seconds, such as up or down adjustment of volume during normal conversation. Whereas long-term plasticity lasts for minutes, hours, and days. Long-term plasticity is the commanding one that controls how the brain stores information and creates memories (Mayford, Siegelbaum, & Kandel, 2012). Currently, new neurophysiological tools are accessible; for example, repetitive transcranial direct current stimulation and repetitive transcranial magnetic stimulation are initiated at clinical levels. These techniques are for analyzing the functioning of brain and modulate the cortical activity.

DOI: 10.1201/9781003464648-5

Plasticity basically indicates the capacity of the nervous system to make alterations in its connections functionally and anatomically in reaction to variations in the external environment (Loon et al., 2022). Synapses are not a static structure but dynamic connections underlying the adaptive development of neuronal circuits. Memory storage process is dependent on the synaptic transmission efficacy. Depending upon the external stimulus, the synaptic transmission can be strengthened or weakened. Numerous studies recommend that synaptic plasticity and its disability are one of the pathogenic factors in developing some NDDs such as AD, PD, and HD (Barnes et al., 2020). Synaptic plasticity at the cellular level plays an important role in conserving neuronal connections. It involves modification of the structure of dendritic spines and synapses and axonal modifications. Disrupted connections between neurons lead to dazzled changes and venerable diseases such as dementia and other diseases associated with dementia (Mango & Nistico, 2020).

These diseases are represented by diminished cognitive function along with emotional and personality changes. Along with cellular level plasticity, molecular factors are also responsible for synaptic transmission disorders such as β-amyloid deposition; α-synuclein accumulation; tau accumulation forming neurofibrillary tangles; impaired level of growth factors, for example, brain-derived neurotrophic factor (BDNF); nerve growth factor (NGF); glial-derived neurotrophic factor (GDNF); and disorganization of the immune system (Jiang, Li, Li, Xu, & Yang, 2021). Ultimately, with better interpretation and understanding of brain plasticity and neuronal plasticity mechanisms in neurodegeneration, more potent therapies and advancement in the quality of life for neurological patients can be accomplished.

4.2 SYNAPTIC TRANSMISSION—FROM PLASTICITY TO NEURODEGENERATION

The capacity of normal physiological movements and systems to respond to external stimuli and to resist agitations is very important for neurons to continue their normal functioning. It has been manifested that there are a variety of mechanisms responsible for maintaining the function of neuronal circuits and networks within a significant range (Cornell, Salinas, Huang, & Zhou, 2022). Synaptic plasticity is an important factor as it controls the homeostasis between excitatory and inhibitory activity, called the E/I ratio, within the networks. Synaptic transmission depends upon a temporally and spatially interrelated multistep process that allows neuronal communication (Aggarwal et al., 2023). These processes are interlinked by an ample number of specific proteins whose energetic interactions, regulations, and expressions define the potency of transmission and the mode of synaptic plasticity (Weir, Christiansen, Sandvig, & Sandvig, 2023).

At the cellular level, the synapse is an asymmetric organization composed of the presynaptic terminal with thousands of vesicles filled with neurotransmitters, the synaptic cleft, and the postsynaptic area having neurotransmitter receptors and signaling machinery (Cornell et al., 2022). One of the most attractive characteristics of the synapse is its plasticity. Synaptic plasticity is the hypothesized mechanism through which past experiences can affect future actions. Synaptic plasticity and brain network architecture play a vital role in brain impairment.(Yan & Rein, 2022). Particularly, excitability and static plasticity may play a role in driving a methodological neuronal network reformation after brain impairment. Alzheimer's disease and schizophrenia are mainly responsible for the decreased connectivity and impaired synaptic plasticity (Zhang et al., 2022). Damaged plasticity of synapsis could contribute damaged plasticity is one of the main neurophysiological conditions. Impaired plasticity and dysregulation of brain signaling can contribute to disease progression and clinical manifestations. The key difference between neurons and other cells of the CNS is that "neurons" are excitable (Malik, Eldaly, Chen, & Chan, 2022).

Nervous system activity within the normal stable range is very crucial for healthy and consistent brain function. However, this firmness must provide flexibility, which is important for memory development and the learning process (Xu, Wang, Lalchandani, & Ding, 2017). Therefore, homeostatic synaptic plasticity is required for maintaining the synaptic strength and intrinsic neuronal excitability to preserve the neural circuits functioning within the stable dynamic ranges throughout the developmental, environmental, and experimental challenges. Provoking links have recently been discovered that imply synaptic processes are necessary for the homeostatic control of synaptic function (Fakih et al., 2019).When homeostatic control becomes impaired, it may result in a diversity of neurological and neuropsychiatric disorders. Disruption of synaptic plasticity may lie at the heart of many neurological disorders; therefore, its study provides an encouraging route by which to develop treatment for brain disorders (Cast et al., 2021).

4.3 FACTORS AFFECTING SYNAPTIC PLASTICITY DURING NEURODEGENERATION

Synaptic plasticity is the internal mechanism important for the expansion and function of the brain. It is also important for deliberating Ωz which uses electical stimulation to measure adaptability, learning, and overall consciousness (Glasgow, McPhedrain, Madranges, Kennedy, & Ruthazer, 2019). However, the ecological causes such as trauma, psychological trauma, stress, exposure to particular elements or chemicals, disruption of synaptic plasticity and many other sociocultural impacts can have overpowering stimuli on brain excitability (Lopes, Pliassova, & Cunha, 2019). Therefore, all these factors can lead to psychological health diseases. In contrast to the brain ailments that occur due to stroke or physical trauma, psychological diseases are characterized by distributed pathology, predominately decreased cognitive function, as well as emotional personality changes (Paskus, Herring, & Roche, 2020).

Along with the environmental factors, some molecular factors also affect the synaptic communication such as β-amyloid deposition, α-synuclein accumulation, tau aggregation forming neurofibrillary tangles, and growth factors, for example, NGF, BDNF, and GDNF (Yan et al., 2022). Variations in the concentrations of synaptic proteins and these factors lead to adaptations in the structural plasticity of axons and dendritic structures that amend the communication between neurons. Damaged neuronal connections restrict the network of neurons, leading to disoriented variations and the development of dementia (Van Hook, 2020). Growth factors, such as NGF, BDNF, and GDNF, are critical for regulation of synaptic plasticity. Along with other factors, one of the factors that affects synapse plasticity and neurogenesis is astroglia cells. The method of formation of long-term storage of information and memories has been found to be mediated by both structural and functional plasticity (Battaglia, Di Fazio, Vicario, & Avenanti, 2023). Astrocytes are housekeeping cells. In recent times, it has been discovered that they are a vital part of the chemical synapse, as several studies have demonstrated that astrocytes exploit both ionotropic and metabotropic systems to regulate neuron-to-neuron communication (Ojala et al., 2021). Another recently reported factor responsible for synaptic dysfunctioning is tau protein phosphorylation. Tau protein is a protein involved in the cytoskeleton that supports microtubule stability and function. It is encoded by 16 exons on chromosome 17q21, which comprises the tau gene (Hill, Karikari, Moffat, Richardson, & Wall, 2019). Excessive phosphorylation of tau protein leads to dendritic spine loss. Mechanistically it can be explained that tau protein is involved in kinase protein activation that causes tau microtubule protein disorganization and cytotoxicity leading to ectopic cell cycle re-entry and resulting in neuronal cell death (Hunt et al., 2023).

Since microtubules are critical for the effective transport of synaptic components to axonal terminals and of post-synaptic components to dendritic terminals, the deterimental effect of tau proteins on the microtubules pose important hazards to synaptic functions (Kang et al., 2019). Cognitive capacity is dependent on a diversity of brain network activity patterns that characterize distinct operative modes, which are directly associated to fine changes in neuronal synchronicity (Lei & Wu, 2023). Therefore, any modifications in the synchronic events could help us to elaborate the cognitive deficits occurring during various neurodegenerative disorders. Current research suggests that early neuronal network dysfunctioning such as neuronal excitation, distorted fluctuating activity, and damage of neuronal symmetry are critical consequences that may lead to neurodegeneration.

4.3.1 Role of Synaptic Plasticity in Parkinson's Disease

Parkinson's is a neurodegenerative disease provoked by damage of dopaminergic neurons in the pars compacta of the substantia region in the brain. Primary symptoms include postural instability, tremors, bradykinesia, and rigidity (Chu, 2020). The damage of dopaminergic neurons in substantia regions induces several synaptic and molecular modifications in the ganglia of basal nuclei regions, which are assumed to cause the biochemical designs of action throughout the basal cortical ganglia grid, and ultimately results in the abnormal motor movements in PD patients (Ziaeemehr, Zarei, Valizadeh, & Mirasso, 2020).

The interference of an anomalous outline of movement within the grid is thought to be the basic mechanism triggering the healing consequences of suggestive treatment. Under healthy conditions, bidirectional synaptic plasticity at corticostriatial synapses is required for typical memory function and motor learning (Lerche et al., 2021). But when dopaminergic neurons are lost, then unidirectional plasticity performs a crucial function in disturbance of chemical balance of neurotransmitters. Deficiency of dopaminergic neurons in substantia regions leads to neuromodulation, which leads to various compatible changes at circuit, cellular, and molecular levels. It eventually leads to an indication of a immunological outline of activity in patients' brains (G. B. Marino et al., 2023). An altered corticostriatial plasticity relates to an impaired mesostriatal dopaminergic transmission and appears to represent a diagnostic mechanism for prompting the defective motor movement in PD. Anomalous plasticity in cortical regions develops a modified balance between dopaminergic and cholinergic neurons, resulting in cognitive impairment.

Although a bundle of research is being conducted around the globe related to PD, the defined mechanism leading to the onset of disease remains to be discovered (Speranza, di Porzio, Viggiano, de Donato, & Volpicelli, 2021). Diminishing ability of neurons in the basal ganglia circuit to endure synaptic plasticity is a significant component of several theories explaining neuronal deviations during neurodegeneration (Cole et al., 2020). Researchers using mice models have determined that the damage in structural and functional synaptic plasticity weakens memory retention, learning mechanisms, and motor performance (Imbriani, Martella, Bonsi, & Pisani, 2022). Postsynaptic and cellular variations occurring in the striatum in PD results in an immense dopaminergic loss that may lead to synaptic disability and corticostriatal transmission uncertainties (Zhang & Holscher, 2020). A defective form of synaptic plasticity therefore leads to the malfunctioning in the subunit conformation of glutamatergic ionotropic receptors, such as NMDA receptors, which enhances the clinical features of PD (Nagata et al., 2021). Stimulatingly, it has become gradually evident that the correct composition of NMDA receptor complex at the synaptic site is a major contributor in early phases of PD; therefore, it may represent an efficient target for therapeutic intervention (Tang et al., 2019).

As several studies suggest, dopamine depletion in neurons in PD induces many cellular and synaptic plasticity processes in striatum. It favors the excitation or cell autonomous increase in internal excitability in neurons. Loss of synaptic connections in striatum and alterations in cellular plasticity in the basal ganglia nuclei that are supposed to cause the immunological pattern of motion throughout the cortico-basal-ganglia thalamocortical network eventually impacts the abnormal movement symptoms of PD.

4.3.2 Role of Synaptic Plasticity in Alzheimer's Disease

Alzheimer's disease was first discovered in 1906 as a main cause of dementia by Alois Alzheimer. Since AD has a multifactorial etiology, definite diagnose of the disease till now is unknown. It is known that AD has interconnections with both environmental and genetic factors that leads to a diverse neurodegenerative disorder (Mango et al., 2019). It is an age-dependent neurodegenerative disorder characterized by residues of insoluble amyloid-β peptides in tau protein aggregation and extracellular matrix, which is located in neurofibrillary twists.

Research nowadays shows that cognitive impairment during early AD is due to synaptic abnormality caused by aggregation of non-fibrillar, oligomeric amyloid-β proteins which is responsible for neurodegeneration and synaptic loss (Qu et al., 2021). Even soluble oligomers of amyloid-β can negatively impact synaptic communication and plasticity at enormously low concentration, but the exact genetic mechanism is not fully understood Even though several regions of the brain are wounded during AD, the hippocampus, a network of symmetrical alignment and intellectual functions, is a considerable attention seeking area of research (Rezaei Asl, Sepehri, & Salami, 2019).

A recent AD hypothesis stated that amyloid-β leads to tau protein accumulation dissociated from neurofibrillary filaments and microtubules and forms neurofibrillary tangles, resulting in neuronal dysfunctioning and eventually neuronal death (M. Zhou et al., 2019). Using transgenic AD mice, Jackson et al. (2019) mentioned that, peculiarly, phosphorylation of tau protein is a significant moderator of Aβ-induced dendritic overactivity and synaptic dysfunctioning. Synaptic disorientation and loss is a crucial trait in many neurodegenerative disorders, mainly dementia and AD (Jackson et al., 2019).

Much research on transgenic animal models has proved that AD is primarily a disorder of synaptic functions, that is, synaptopathy (K. Zhou et al., 2019). The later stage of AD includes an inconvertible and significant loss of synapses and neurons. It has been proved that amyloid-β mechanically binds the synaptic proteins. Soluble amyloid-β also increases the level of DKK1, an endogenous factor of the wnt pathway, which activates the signaling that causes synaptic disassembly and localized cell loss. In recent times several studies have emphasized the role of hyperphosphorylation of tau protein and its relation with cognitive decline in AD patients, indicating that it has a more crucial role in synapse disorientation (Pei, Davies, Zhang, & Zhang, 2020). Neurofibrillary tangles, which are a characteristic feature in an AD brain, is not directly involved in synaptic plasticity, but soluble non-aggregated forms of tau are more involved. In a post-mortem of an AD brain, it was reported that pathological tau species are involved in impaired axonal transport. Disrupted axonal structure also impaired the conventional transport of different cellular parts such as mitochondria that play an imperative role in the normal functioning of synaptic communication.

This proved that a greater number of abnormal mitochondrial structures in synapses were reported in AD brains as compared to that in a normal healthy brain. An another study demonstrated thatthe presence of amyloid-β, and a large number of other synaptic proteins including tau protein, increase the probability of synaptic loss (Tu, Okamoto, Lipton, & Xu, 2014). Amyloid-β and tau proteins may have an unintended effect on synaptic function via instance inflammatory response or conflicts for glucose metabolism.

4.3.3 Role of Synaptic Plasticity in Huntington's Disease

Huntington's is a hereditary disease that leads to fatal neurodegenerative disease. It occurs due to a mutation in the huntingtin (HTT) gene; 35 or more repeats of polyglutamine causes HD. The disease is characterized by motor symptoms that include abnormal dance-like movements along with psychiatric disorders and cognitive impairment (Kim, Wallace, El-Rifai, Knudsen, & Sabatini, 2022). Later stages of the disease are also comprised of abnormal voluntary movements such as swallowing, speech, and body balancing, along with rigidity and bradykinesia. Histologically HD affects the basal ganglia, distinctively the striatum medium sized spiny neurons. In humans motor cortex regions are lost. The huntingtin protein is crucial for normal synaptic transmission (Jiang et al., 2022).

It is a scaffold protein and is involved in different processes of the brain, such as the formation of neurotransmitter vesicles and brain derived neurotrophic factor (BDNF), ciliogenesis, and cellular transport. BDNF signaling affects cell existence but also impacts the synaptic excitability (Dargaei, Liang, Serranilla, Santos, & Woodin, 2019). The disturbance of balance between these pathways leads to modification of corticostriatal signaling at the synapse, resulting in the development of the cognitive, movement, and psychiatric symptoms of HD (Wood et al., 2019). Disrupted

BDNF signaling also impacts multiple types of homeostatic synaptic plasticity, including new spine formation in the barrel cortex region. An insufficiency of BDNF signaling promotes HD's cognitive and psychiatric symptoms by disturbing the synaptic plasticity and interrupting the strength of cortical networks (Pupak et al., 2022). Reduced BDNF signaling results in a reduction in the striatal glutamatergic signaling crucial for spine formation, contributing to development of motor deficits in HD patients.

Huntingtin protein is a global protein in the brain, but it is remarkably abundant in the cerebral cortex and hippocampus regions. Most of the studies have suggested that a supplementary complicated portrait of synaptic dysfunction in the basal ganglia and in the cortex is an evolving concern (Braz et al., 2022). There also has been a developing curiosity in the function of glial cells, mainly astrocytes in HD. Cholesterol in the brain is synthesized in astrocytes. Its dysfunctioning can impact synaptic transport basically in two ways. First, cholesterol is the main factor of myelin. Reduced production of cholesterol can lead to impaired myelination, which ultimately causes weakening of electrical impulse transmission. Decreased myelin production is an early immunological condition in HD that results in neuronal loss and behavioral abnormalities (Keifer, 2022). Another role of cholesterol is in synaptic transmission per second. The concentration of cholesterol is a limiting factor for spontaneous synaptic vesicle fusion and neurotransmission of glutamate. It is also observed that BDNF promotes the production of cholesterol in neurons and helps in synthesizing lipid rafts (Espina et al., 2023). A decreased cholesterol centration leads to instability of synapses and loss of spines.

Several studies and research on animal models of HD infer parallel neurodevelopmental functional and morphological impairment in cortical and ganglial structures. It is evident that neuronal and non-neuronal elements both contribute to synapse dysfunction and loss as an active determinant to HD pathology (Miguez et al., 2015). It is precisely evident that rectifying synaptic dysfunctioning well before neuronal demise starts is possibly crucial for the progress of flourishing remedies.

4.4 PHARMACOLOGICAL INTERVENTION OF SYNAPTIC PLASTICITY

One hypothesis suggests that neuronal loss is caused by the concentration of calcium ions that enters inside the cells l. This hypothesis was proven by various studies, which showed a correlation between total cell calcium and neuronal cell death loading regulated by calcium accumulation measurements for isolated calcium influx pathways (Mahalakshmi, Maurya, Lee, & Bharath Kumar, 2020). Several studies suggest that most diseases of the central nervous system that are associated with neuronal loss and are correlated to glutamate excitotoxicity could be tempered using N-methyl-D-aspartate receptors (NMDA). NMDA receptors-mediated neuronal injury is involved in various neurological disorders such as epilepsy, dementia, brain trauma, and other neurodegenerative disorders (Gupta & Kaur, 2019). Non-selective antagonist compounds of NMDA receptors, have shown effective results in animal models, and some of the selected compounds also have undergone into clinical trials.

In Parkinson's disease pharmacological dopamine is replaced with 1-3,4-dihydroxylphenylalanine as an effective treatment. Recent advancement in neurotechnology for cell labeling and neural activity provides crucial information of changes in the Parkinsonian brain. The characterization and development of molecular and genetic α-synuclein represent an immense prospect for the field to analyze brain modifications at cellular and circuit levels as disease proceeds (Chu, 2020). L-DOPA treatment helps to cure irregular structural elasticity in the dendrites and the spines of striatal medium spiny neurons causing further short, long, functional modifications of glutamatergic and dopaminergic transmission (Okano, Takahata, Sugimoto, & Muraoka, 2019). Therefore, current clinical research has focused on non-dopaminergic neurotransmitter systems that may provide stability in motor movements as well as in cognitive impairment in PD. Cerebral modifications include deficits in inhibitory control (G. Marino et al., 2023). Preclinical studies suggest that impulsivity factors are linked to development of behavioral disorders with distinct anatomical, psychological, and pharmacological features. Therefore, not only dopamine, but also other neurotransmitters such as noradrenaline and serotonin, may be connected to the cognitive dysfunction in PD, so these might represent supplementary therapeutic targets.

A HD study shows that normal huntingtin protein attaches with a PSD-95 scaffold protein that inhibits the overexpression on NMDA receptors. It also suggests that PSD-95 protein is an indicator of neuronal toxicity caused by NMDA receptors and mutated huntingtin protein. As reported, NMDA receptors block Mg^{2+} ions in animal models of HD (Chi, Wang, Liu, Zhang, & Shen, 2023). Various studies using a BDNF based strategy along with other drugs, including antidepressants, that increased expression of neutrophins are expected to have extended clinical applications. Chemicals that elevated the BDNF levels synthesized by the compounds such as amphikines affect the development of HD-related neuropathology and is a desperate issue for future research (Kim et al., 2020). More current findings suggest that pathologically increased amyloid-β blocks neuronal uptake of glutamate at synapses and activation of NMDA signaling cascade is involved in synaptic loss (Jackson et al., 2019).

4.5 CONCLUSION

Synaptic plasticity has become an emerging theme of brain research as it is recognized that it contributes to a deeper understanding of nervous system functioning and composition. Synaptic plasticity is not only involved in memory formation, the learning process, and reactions to exterior stimuli, but it performs a crucial role in the recovery of the brain from injury and trauma. Intense research has proved in animal models that manipulating synaptic plasticity in neuronal signaling pathways and synaptic communications has a promising role in therapeutic and clinical interventions that will improve our health.

Plasticity modifications are alterations at the micro-scale level, but morphological and physiological changes occur throughout the entire nervous system and lifespan. Synaptic plasticity depends upon a number of factors, such as experience, trauma, injury, time, and individual factors such as genetics. Therefore, it is challenging because it is a highly complicated and variable process.

Irrespective of these challenges, there is remarkable development in this area providing new opportunities to link the multiple scales. For example, the relationship between precise markers of elasticity and individuals who can react to treatment can provide data for clinical trials. This information can be used to predict whether patients are expected to improve through changes in brain plasticity. One effective clinical biomarker is methylated exon 4 of BDNF, as it has a decreased response to anti-depressants. To understand the heterogeneity of disease and cause of manipulated signaling, it is necessary to have an improved biological understanding of wholesome neuroplasticity and how it can be disrupted. A future profitable direction is the development of individualization of treatment. Combining therapies could provide behavioral or pharmacological interventions as well. Improved therapeutic tools and the development of biomarkers have a probability to alter patient outcomes through personalized cure and measure of neural plasticity.

Prevailing data confirms that synaptic transmission is an extraordinarily multifaceted phenomenon, subject to improvements in signaling at both pre- and post-synaptic sites. Technical advancement in neuroscience research paves a new way to a more complete understanding of the functional consequences of synaptic plasticity and its inferences for health and disease. Synaptic plasticity includes synaptic responses, inhibitory synaptic responses, and the intrinsic excitability of neurons, all of which includes effects of one neural circuit function. Management of synaptic plasticity mechanisms is highly restricted to neural circuits and our understanding of how plasticity mediates pathological and adaptive experience. Applying the same approaches can provide new insights into molecular and cellular pathology of PD, AD, and HD, allowing for a greater opportunity to provide better understanding and efficacious treatment of these diseases. With more of the world's population living longer, individuals often develop age-related neurology disorders, which also affect lifestyle and well-being. Neuroplasticity can be influenced in both healthy and diseased brains.

Manipulating the neural circuits can provide new pathways that can play a huge role in rehabilitation and value-added quality of life. Recognition of the varied indications in the disease will aid the development of new therapeutic approaches. These aim to restore or replace a dysfunctional protein or pathway or restore synaptic strength. Synaptic plasticity is mainly the capability of neurons to modify the strength of their connections. It is a crucial neurophysiological process involved in the brain complex development and reform after damage.

REFERENCES

Aggarwal, A., Liu, R., Chen, Y., Ralowicz, A. J., Bergerson, S. J., Tomaska, F., … Podgorski, K. 2023. "Glutamate indicators with improved activation kinetics and localization for imaging synaptic transmission". *Nat Methods*, *20*(6), 925–934. doi:10.1038/s41592-023-01863-6

Barnes, J. R., Mukherjee, B., Rogers, B. C., Nafar, F., Gosse, M., & Parsons, M. P. 2020. "The relationship between glutamate dynamics and activity-dependent synaptic plasticity". *J Neurosci*, *40*(14), 2793–2807. doi:10.1523/JNEUROSCI.1655-19.2020

Batool, S., Raza, H., Zaidi, J., Riaz, S., Hasan, S., & Syed, N. I. 2019. "Synapse formation: From cellular and molecular mechanisms to neurodevelopmental and neurodegenerative disorders". *J Neurophysiol*, *121*(4), 1381–1397. doi:10.1152/jn.00833.2018

Battaglia, S., Di Fazio, C., Vicario, C. M., & Avenanti, A. 2023. "Neuropharmacological modulation of N-methyl-D-aspartate, noradrenaline and endocannabinoid receptors in fear extinction learning: Synaptic transmission and plasticity". *Int J Mol Sci, 24*(6). doi:10.3390/ijms24065926

Bello-Medina, P. C., Rodriguez-Martinez, E., Prado-Alcala, R. A., & Rivas-Arancibia, S. 2022. "Ozone pollution, oxidative stress, synaptic plasticity, and neurodegeneration". *Neurologia*, *37*(4), 277–286. doi:10.1016/j.nrleng.2018.10.025

Braz, B. Y., Wennagel, D., Ratie, L., de Souza, D. A. R., Deloulme, J. C., Barbier, E. L., … Humbert, S. 2022. "Treating early postnatal circuit defect delays Huntington's disease onset and pathology in mice". *Science*, *377*(6613), eabq5011. doi:10.1126/science.abq5011

Cast, T. P., Boesch, D. J., Smyth, K., Shaw, A. E., Ghebrial, M., & Chanda, S. 2021. "An autism-associated mutation impairs neuroligin-4 glycosylation and enhances excitatory synaptic transmission in human neurons". *J Neurosci*, *41*(3), 392–407. doi:10.1523/JNEUROSCI.0404-20.2020

Chi, X., Wang, L., Liu, H., Zhang, Y., & Shen, W. 2023. "Post-stroke cognitive impairment and synaptic plasticity: A review about the mechanisms and Chinese herbal drugs strategies". *Front Neurosci*, *17*, 1123817. doi:10.3389/fnins.2023.1123817

Chu, H. Y. 2020. "Synaptic and cellular plasticity in Parkinson's disease". *Acta Pharmacol Sin*, *41*(4), 447–452. doi:10.1038/s41401-020-0371-0

Cole, E. J., Stimpson, K. H., Bentzley, B. S., Gulser, M., Cherian, K., Tischler, C., … Williams, N. R. 2020. "Stanford accelerated intelligent neuromodulation therapy for treatment-resistant depression". *Am J Psychiatry*, *177*(8), 716–726. doi:10.1176/appi.ajp.2019.19070720

Cornell, J., Salinas, S., Huang, H. Y., & Zhou, M. 2022. "Microglia regulation of synaptic plasticity and learning and memory". *Neural Regen Res*, *17*(4), 705–716. doi:10.4103/1673-5374.322423

Dargaei, Z., Liang, X., Serranilla, M., Santos, J., & Woodin, M. A. 2019. "Alterations in hippocampal inhibitory synaptic transmission in the R6/2 mouse model of Huntington's disease". *Neuroscience*, *404*, 130–140. doi:10.1016/j.neuroscience.2019.02.007

Espina, M., Di Franco, N., Branas-Navarro, M., Navarro, I. R., Brito, V., Lopez-Molina, L., … Gines, S. 2023. "The GRP78-PERK axis contributes to memory and synaptic impairments in Huntington's disease R6/1 mice". *Neurobiol Dis*, *184*, 106225. doi:10.1016/j.nbd.2023.106225

Fakih, D., Zhao, Z., Nicolle, P., Reboussin, E., Joubert, F., Luzu, J., … Reaux-Le Goazigo, A. 2019. "Chronic dry eye induced corneal hypersensitivity, neuroinflammatory responses, and synaptic plasticity in the mouse trigeminal brainstem". *J Neuroinflammation*, *16*(1), 268. doi:10.1186/s12974-019-1656-4

Glasgow, S. D., McPhedrain, R., Madranges, J. F., Kennedy, T. E., & Ruthazer, E. S. 2019. "Approaches and limitations in the investigation of synaptic transmission and plasticity". *Front Synaptic Neurosci*, *11*, 20. doi:10.3389/fnsyn.2019.00020

Gupta, M., & Kaur, G. 2019. "Withania somnifera (L.) Dunal ameliorates neurodegeneration and cognitive impairments associated with systemic inflammation". *BMC Complement Altern Med*, *19*(1), 217. doi:10.1186/s12906-019-2635-0

Hill, E., Karikari, T. K., Moffat, K. G., Richardson, M. J. E., & Wall, M. J. 2019. "Introduction of tau oligomers into cortical neurons alters action potential dynamics and disrupts synaptic transmission and plasticity". *eNeuro*, *6*(5). doi:10.1523/ENEURO.0166-19.2019

Hunt, S., Leibner, Y., Mertens, E. J., Barros-Zulaica, N., Kanari, L., Heistek, T. S., … de Kock, C. P. J. 2023. "Strong and reliable synaptic communication between pyramidal neurons in adult human cerebral cortex". *Cereb Cortex*, *33*(6), 2857–2878. doi:10.1093/cercor/bhac246

Imbriani, P., Martella, G., Bonsi, P., & Pisani, A. 2022. "Oxidative stress and synaptic dysfunction in rodent models of Parkinson's disease". *Neurobiol Dis*, *173*, 105851. doi:10.1016/j.nbd.2022.105851

Jackson, J., Jambrina, E., Li, J., Marston, H., Menzies, F., Phillips, K., & Gilmour, G. 2019. "Targeting the synapse in Alzheimer's disease". *Front Neurosci*, *13*, 735. doi:10.3389/fnins.2019.00735

Jagaran, K., & Singh, M. 2021. "Nanomedicine for neurodegenerative disorders: Focus on Alzheimer's and Parkinson's diseases". *Int J Mol Sci*, *22*(16). doi:10.3390/ijms22169082

Jaworski, T., Banach-Kasper, E., & Gralec, K. 2019. "GSK-3beta at the intersection of neuronal plasticity and neurodegeneration". *Neural Plast*, *2019*, 4209475. doi:10.1155/2019/4209475

Jiang, Q., Tong, J., Xian, Y., Kerner, R. A., Dunfield, S. P., Xiao, C., … Zhu, K. 2022. "Surface reaction for efficient and stable inverted perovskite solar cells". *Nature*, *611*(7935), 278–283. doi:10.1038/s41586-022-05268-x

Jiang, Y., Li, K., Li, X., Xu, L., & Yang, Z. 2021. "Sodium butyrate ameliorates the impairment of synaptic plasticity by inhibiting the neuroinflammation in 5XFAD mice". *Chem Biol Interact*, *341*, 109452. doi:10.1016/j.cbi.2021.109452

Kang, H. J., Kim, K. O., Kim, J. W., Kim, S. W., Park, M. S., Kim, H. R., … Kim, J. M. 2019. "A longitudinal study of the associations of BDNF genotype and methylation with poststroke anxiety". *Int J Geriatr Psychiatry*, *34*(11), 1706–1714. doi:10.1002/gps.5185

Keifer, J. 2022. "Regulation of AMPAR trafficking in synaptic plasticity by BDNF and the impact of neurodegenerative disease". *J Neurosci Res*, *100*(4), 979–991. doi:10.1002/jnr.25022

Kim, A., Garcia-Garcia, E., Straccia, M., Comella-Bolla, A., Miguez, A., Masana, M., … Rodriguez, M. J. 2020. "Reduced fractalkine levels lead to striatal synaptic plasticity deficits in Huntington's disease". *Front Cell Neurosci*, *14*, 163. doi:10.3389/fncel.2020.00163

Kim, S., Wallace, M. L., El-Rifai, M., Knudsen, A. R., & Sabatini, B. L. 2022. "Co-packaging of opposing neurotransmitters in individual synaptic vesicles in the central nervous system". *Neuron*, *110*(8), 1371–1384.e1377. doi:10.1016/j.neuron.2022.01.007

Lamptey, R. N. L., Chaulagain, B., Trivedi, R., Gothwal, A., Layek, B., & Singh, J. 2022. "A review of the common neurodegenerative disorders: Current therapeutic approaches and the potential role of nanotherapeutics". *Int J Mol Sci*, *23*(3). doi:10.3390/ijms23031851

Lei, Z., & Wu, P. 2023. "Short-term plasticity, multimodal memory, and logical responses mimicked in stretchable hydrogels". *Matter*, *6*(2), 429–444.

Lerche, S., Schulte, C., Wurster, I., Machetanz, G., Roeben, B., Zimmermann, M., … Brockmann, K. 2021. "The mutation matters: CSF profiles of GCase, sphingolipids, alpha-synuclein in PD(GBA)". *Mov Disord*, *36*(5), 1216–1228. doi:10.1002/mds.28472

Loon, A., Zamudio, F., Sanneh, A., Brown, B., Smeltzer, S., Brownlow, M. L., … Selenica, M. B. 2022. "Accumulation of C-terminal cleaved tau is distinctly associated with cognitive deficits, synaptic plasticity impairment, and neurodegeneration in aged mice". *Geroscience*, *44*(1), 173–194. doi:10.1007/s11357-021-00408-z

Lopes, J. P., Pliassova, A., & Cunha, R. A. 2019. "The physiological effects of caffeine on synaptic transmission and plasticity in the mouse hippocampus selectively depend on adenosine A(1) and A(2A) receptors". *Biochem Pharmacol*, *166*, 313–321. doi:10.1016/j.bcp.2019.06.008

Mahalakshmi, B., Maurya, N., Lee, S. D., & Bharath Kumar, V. 2020. "Possible neuroprotective mechanisms of physical exercise in neurodegeneration". *Int J Mol Sci*, *21*(16). doi:10.3390/ijms21165895

Malik, A., Eldaly, A. B. M., Chen, K., & Chan, L. L. 2022. "Neuronal oscillatory signatures in the developing mouse visual cortex after short-term monocular deprivation". *Cereb Cortex*, *32*(12), 2657–2667. doi:10.1093/cercor/bhab372

Mango, D., & Nistico, R. 2020. "Role of ASIC1a in normal and pathological synaptic plasticity". *Rev Physiol Biochem Pharmacol*, *177*, 83–100. doi:10.1007/112_2020_45

Mango, D., Saidi, A., Cisale, G. Y., Feligioni, M., Corbo, M., & Nistico, R. 2019. "Targeting synaptic plasticity in experimental models of Alzheimer's disease". *Front Pharmacol*, *10*, 778. doi:10.3389/fphar.2019.00778

Marino, G., Campanelli, F., Natale, G., De Carluccio, M., Servillo, F., Ferrari, E., … Calabresi, P. 2023. "Intensive exercise ameliorates motor and cognitive symptoms in experimental Parkinson's disease restoring striatal synaptic plasticity". *Sci Adv*, *9*(28), eadh1403. doi:10.1126/sciadv.adh1403

Zhou, M., Wang, H., Zeng, X., Yin, P., Zhu, J., Chen, W., … Liang, X. 2019. "Mortality, morbidity, and risk factors in China and its provinces, 1990-2017: A systematic analysis for the Global Burden of Disease Study 2017". *Lancet, 394*(10204), 1145–1158. doi:10.1016/S0140-6736(19)30427-1

Marino, G. B., Ngai, M., Clarke, D. J. B., Fleishman, R. H., Deng, E. Z., Xie, Z., … Ma'ayan, A. 2023. "geneRanger and TargetRanger: Processed gene and protein expression levels across cells and tissues for target discovery". *Nucleic Acids Res, 51*(W1), W213–W224. doi:10.1093/nar/gkad399

Mayford, M., Siegelbaum, S. A., & Kandel, E. R. 2012. "Synapses and memory storage". *Cold Spring Harb Perspect Biol, 4*(6). doi:10.1101/cshperspect.a005751

Miguez, A., Garcia-Diaz Barriga, G., Brito, V., Straccia, M., Giralt, A., Gines, S., … Alberch, J. 2015. "Fingolimod (FTY720) enhances hippocampal synaptic plasticity and memory in Huntington's disease by preventing p75NTR up-regulation and astrocyte-mediated inflammation". *Hum Mol Genet, 24*(17), 4958–4970. doi:10.1093/hmg/ddv218

Nagata, T., Dwyer, C. A., Yoshida-Tanaka, K., Ihara, K., Ohyagi, M., Kaburagi, H., … Yokota, T. 2021. "Cholesterol-functionalized DNA/RNA heteroduplexes cross the blood-brain barrier and knock down genes in the rodent CNS". *Nat Biotechnol, 39*(12), 1529–1536. doi:10.1038/s41587-021-00972-x

Ojala, K. S., Ginebaugh, S. P., Wu, M., Miller, E. W., Ortiz, G., Covarrubias, M., & Meriney, S. D. 2021. "A high-affinity, partial antagonist effect of 3,4-diaminopyridine mediates action potential broadening and enhancement of transmitter release at NMJs". *J Biol Chem, 296*, 100302. doi:10.1016/j.jbc.2021.100302

Okano, M., Takahata, K., Sugimoto, J., & Muraoka, S. 2019. "Selegiline recovers synaptic plasticity in the medial prefrontal cortex and improves corresponding depression-like behavior in a mouse model of Parkinson's disease". *Front Behav Neurosci, 13*, 176. doi:10.3389/fnbeh.2019.00176

Paskus, J. D., Herring, B. E., & Roche, K. W. 2020. "Kalirin and Trio: RhoGEFs in synaptic transmission, plasticity, and complex brain disorders". *Trends Neurosci, 43*(7), 505–518. doi:10.1016/j.tins.2020.05.002

Pei, Y. A., Davies, J., Zhang, M., & Zhang, H. T. 2020. "The role of synaptic dysfunction in Alzheimer's disease". *J Alzheimers Dis, 76*(1), 49–62. doi:10.3233/JAD-191334

Pupak, A., Singh, A., Sancho-Balsells, A., Alcala-Vida, R., Espina, M., Giralt, A., … Brito, V. 2022. "Altered m6A RNA methylation contributes to hippocampal memory deficits in Huntington's disease mice". *Cell Mol Life Sci, 79*(8), 416. doi:10.1007/s00018-022-04444-6

Qu, W., Yuan, B., Liu, J., Liu, Q., Zhang, X., Cui, R., … Li, B. 2021. "Emerging role of AMPA receptor subunit GluA1 in synaptic plasticity: Implications for Alzheimer's disease". *Cell Prolif, 54*(1), e12959. doi:10.1111/cpr.12959

Rezaei Asl, Z., Sepehri, G., & Salami, M. 2019. "Probiotic treatment improves the impaired spatial cognitive performance and restores synaptic plasticity in an animal model of Alzheimer's disease". *Behav Brain Res, 376*, 112183. doi:10.1016/j.bbr.2019.112183

Speranza, L., di Porzio, U., Viggiano, D., de Donato, A., & Volpicelli, F. 2021. "Dopamine: The neuromodulator of long-term synaptic plasticity, Reward and Movement Control". *Cells, 10*(4). doi:10.3390/cells10040735

Sudhof, T. C. 2021. "The cell biology of synapse formation". *J Cell Biol, 220*(7). doi:10.1083/jcb.202103052

Tang, Y., Yang, X. K., Zhang, X. W., Wu, W. T., Zhang, F. L., Jiang, H., … Huang, W. H. 2019. "Harpagide, a natural product, promotes synaptic vesicle release as measured by nanoelectrode amperometry". *Chem Sci, 11*(3), 778–785. doi:10.1039/c9sc05538j

Tu, S., Okamoto, S., Lipton, S. A., & Xu, H. 2014. "Oligomeric Abeta-induced synaptic dysfunction in Alzheimer's disease". *Mol Neurodegener, 9*, 48. doi:10.1186/1750-1326-9-48

Van Hook, M. J. 2020. "Temperature effects on synaptic transmission and neuronal function in the visual thalamus". *PLoS One, 15*(4), e0232451. doi:10.1371/journal.pone.0232451

Weir, J. S., Christiansen, N., Sandvig, A., & Sandvig, I. 2023. "Selective inhibition of excitatory synaptic transmission alters the emergent bursting dynamics of in vitro neural networks". *Front Neural Circuits, 17*, 1020487. doi:10.3389/fncir.2023.1020487

Wood, T. E., Barry, J., Yang, Z., Cepeda, C., Levine, M. S., & Gray, M. 2019. "Mutant huntingtin reduction in astrocytes slows disease progression in the BACHD conditional Huntington's disease mouse model". *Hum Mol Genet, 28*(3), 487–500. doi:10.1093/hmg/ddy363

Xu, T., Wang, S., Lalchandani, R. R., & Ding, J. B. 2017. "Motor learning in animal models of Parkinson's disease: Aberrant synaptic plasticity in the motor cortex". *Mov Disord, 32*(4), 487–497. doi:10.1002/mds.26938

Yan, Z., & Rein, B. 2022. "Mechanisms of synaptic transmission dysregulation in the prefrontal cortex: Pathophysiological implications". *Mol Psychiatry, 27*(1), 445–465. doi:10.1038/s41380-021-01092-3

Yan, Z. P., Yuan, L., Zhang, Y., Mao, M. X., Liao, X. J., Ni, H. X., … Zuo, J. L. 2022. "A chiral dual-core organoboron structure realizes dual-channel enhanced ultrapure blue emission and highly efficient circularly polarized electroluminescence". *Adv Mater, 34*(36), e2204253. doi:10.1002/adma.202204253

Zhang, Y., Fan, Q., Hou, Y., Zhang, X., Yin, Z., Cai, X., … Zheng, X. 2022. “Bacteroides species differentially modulate depression-like behavior via gut-brain metabolic signaling”. *Brain Behav Immun*, *102*, 11–22. doi:10.1016/j.bbi.2022.02.007

Zhang, Z. Q., & Holscher, C. 2020. “GIP has neuroprotective effects in Alzheimer and Parkinson’s disease models”. *Peptides*, *125*, 170184. doi:10.1016/j.peptides.2019.170184

Zhou, K., Yuan, C., Dai, B., Wang, K., Chen, Y., Ma, D., … Cui, M. 2019. “Environment-sensitive near-infrared probe for fluorescent discrimination of abeta and tau fibrils in AD brain”. *J Med Chem*, *62*(14), 6694–6704. doi:10.1021/acs.jmedchem.9b00672

Zhou, M., Wang, H., Zeng, X., Yin, P., Zhu, J., Chen, W., … Liang, X. 2019. “Mortality, morbidity, and risk factors in China and its provinces, 1990-2017: A systematic analysis for the Global Burden of Disease Study 2017”. *Lancet*, *394*(10204), 1145–1158. doi:10.1016/S0140-6736(19)30427-1

Ziaeemehr, A., Zarei, M., Valizadeh, A., & Mirasso, C. R. 2020. “Frequency-dependent organization of the brain’s functional network through delayed-interactions”. *Neural Netw*, *132*, 155–165. doi:10.1016/j.neunet.2020.08.003

5 Role of Epigenetics on Synaptic Plasticity in Neurodegenerative Disorders

*Vineeta Singh and Vijaya Nath Mishra**
Banaras Hindu University
Varanasi, India

5.1 INTRODUCTION

The brain is made up of many kinds of glial cells and neurons. Highly polarized cells, neurons are distinguished by their dendritic tree, which has many spines, and axons. The brain is dynamic and constantly changing according to environment, experiences, and way of life. These modifications show up in the connections, structures, and functions that neurons perceive. The central nervous system is made up of neurons, which are the fundamental anatomical and functional components. Synapses, or connections between neurons, allow for the transfer of impulses and communication. This creates the neural networks that support brain activity. Maintaining synaptic resilience or fragility is crucial for preserving neural plasticity. The brain's remarkable capacity to rearrange its connections allows it to create new neurons and connections. Because of this adaptive feature, the brain may create new neural networks that are impacted by both internal and external inputs. Neuronal plasticity is the ability of the brain to modify both its structure and function in response to changes in the environment. Moreover, synaptic plasticity refers to the capacity of individual synapses to alter the intensity and effectiveness of synaptic transmission at the cellular level.

Mammalian brain plasticity is a basic feature that demonstrates the brain's capacity to change neural circuits and, in turn, influence feelings, ideas, and actions. For learning, memory formation, brain growth, and general cognitive functioning, synaptic plasticity is crucial. The dynamic generation of new brain connections is dependent on crucial processes such synapse formation, maturation, and spine morphogenesis. Changes in neuronal connectivity can lead to specific gene transcription programs related to neuronal plasticity [1].

The field of neuroscience is vast and complex, with its explorations delving deep into the intricacies of the human brain and its complex functions. Within this vast landscape, the intersection of two fundamental processes, synaptic plasticity and epigenetics, has emerged as a captivating and enlightening realm [2]. The convergence of research in neuroscience, genetics, and epigenetics has paved the way for a deeper understanding of neurodegenerative disorders (NDs), shedding light on the complex mechanisms that drive their onset and progression. These discoveries have emphasized the constant reorganization of synaptic connections, which plays a crucial part in memory formation. Genetic and epigenetic factors are linked to changes in neural plasticity, highlighting the significant impact of various external influences on the brain's structure and function. The brain, among all organs, is the most susceptible to the influence of epigenetic factors. Food, exercise, environmental changes, and stressors can impact neural activity across different stages of life, including embryonic development, adolescence, adulthood, and ageing. These external factors affect the brain through a range of epigenetic mechanisms such as DNA methylation, protein acetylation, and the regulation of non-coding RNAs and microRNAs [3].

DOI: 10.1201/9781003464648-6

These epigenetic processes play a central regulatory role in neural plasticity, which is essential for brain development and cognitive function. However, the exact ways in which neural plasticity shapes the brain's physiology and morphology remain somewhat unclear, despite extensive research. Neural plasticity is known to influence the remodeling of brain structures and functions, yet there is still much to explore and understand in this area. The study of neural and synaptic plasticity is critical for understanding the cellular and molecular mechanisms that regulate the nervous system's plasticity. Epigenetics has emerged as a promising field for modulating neuroplasticity through targeted interventions. By gaining insights into neural plasticity and its underlying dynamics, scientists can better understand its role in neurodegenerative diseases and other conditions [4].

Neuroepigenetics, an emerging field of neuroscience, is poised to provide valuable insights into how neurological remodeling occurs in response to both spontaneous and context-dependent learning. While there is a complex interplay between neurons, neuronal plasticity, and epigenomic mechanisms, research into specific epigenetic drivers and their underlying processes is still in its early stages. The understanding of the dynamic nature of epigenetic alterations and their association through neural plasticity is essential for future progress [5].

At the heart of synaptic plasticity is the dynamic adaptability of synapses, which are the connections between neurons. These changes, crucial for learning and memory, involve consolidation or weakening of synapses through processes such as long-term potentiation (LTP) and long-term depression (LTD). Without altering the DNA sequence itself, epigenetics—which includes alterations to DNA and associated proteins—is essential for controlling gene expression [6]. The intersection of synaptic plasticity and epigenetics is a key area of exploration, offering insights into the dynamic interplay between these processes. NDs such as Alzheimer's disease (AD), Huntington's disease (HD), and Parkinson's disease (PD) pose significant challenges for researchers and clinicians alike. By delving into the connections between synaptic plasticity and epigenetic mechanisms, scientists can advance our understanding of these conditions and work towards developing more effective treatments [7]. Understanding synaptic plasticity and epigenetic mechanisms has become crucial in deciphering the complex dynamics of these disorders. This chapter navigates through the intricate dance of these processes, shedding light on how disruptions in synaptic plasticity and aberrant epigenetic patterns contribute to the pathogenesis of neurodegenerative conditions. The dynamic interplay between synaptic plasticity and epigenetics involves a sophisticated molecular tango.

5.2 HISTORY OF SYNAPTIC PLASTICITY

Plasticity, a concept that emerged in the realm of science over a century ago, is credited to Santiago Ramón y Cajal, a pioneer of modern neuroscience and a renowned Spanish scientist. In 1894, he proposed that the brain can accumulate information by altering synaptic connections. Further, Ramón Cajal 3 year before predates Charles Sherrington's introduction of synapses as connections between nerve cells. Ramón y Cajal further established that neurons are distinct entities and identified synapses as the points of communication between them, contributing to neuron theory. He also emphasized the significance of dendritic spines, asserting that they are real structures playing a vital role in facilitating synaptic connectivity [8].

Further, in the late 1940s the synaptic plasticity concept gained prominence through the groundbreaking work of Polish neurophysiologist Jerzy Konorski and Canadian psychologist Donald Hebb. Konorski defined plasticity as "permanent functional transformations," while Hebb associated testable physiological characteristics with synaptic plasticity [9]. Synaptic plasticity refers to the dynamic nature of connections among nerve cells in the brain, emphasizing that these connections can change over time. The mammalian brain's remarkable plasticity allows for modifications in neural circuits, leading to changes in behavior, emotions, and responses to sensory input.

Synaptic plasticity encompasses activity-dependent alterations in synaptic communication, playing a critical role in the brain's ability to convert fleeting experiences into long-lasting memories. Researchers Timothy Bliss and Terry Lomo demonstrated an increase in synaptic strength in the mammalian hippocampus through electrical stimulation [10]. This increase in postsynaptic responses, known as long-term potentiation (LTP), was further complemented by findings from Serena Dudek and Mark Bear, who uncovered that synapses can change in both directions—either increasing (LTP) or decreasing [long-term depression (LTD)] in strength. LTP and LTD have been observed in various brain regions, including the hippocampus, cerebellum, cerebral cortex, and amygdala, where sensory input is linked to motor output, such as in fear conditioning paradigms [11].

5.3 SYNAPTIC PLASTICITY

Essentially, in a particular neural circuit, synaptic plasticity refers to strengthening or weakening of synaptic contacts due to increased or decreased stimuli levels of neurons involved. Synaptic plasticity directs the regulation of both presynaptic and postsynaptic neurons by making changes in the neurons of synaptic machinery [12]. As is well known, it regulates the amount of neurotransmitter released from the presynaptic neuron into the synapse, changes the number of neurotransmitter receptors in the postsynaptic membrane, or alters the sensitivity of the receptor to released neurotransmitters. Synaptic plasticity is generally observed at synapses involved in the transmission of glutamate-mediated excitability and, also, at other synapses that result in GABAergic inhibition. On quantitative analysis it was found that synaptic plasticity occurs on timescales ranging from tens of milliseconds to lifelong changes in synaptic transmission. Thus, synaptic plasticity is classified under two categories: short-term and long-term synaptic plasticity.

5.3.1 Short-term Synaptic Plasticity

Short-term synaptic plasticity happens at periods from sub-seconds to minutes [13].

5.3.1.1 Post-tetanic Potentiation

Post-tetanic potentiation (PTP) is a transient form of plasticity. When neurons subjected to high-frequency stimulation (typically between 200 milliseconds and 5 seconds) at rates of 10–200 Hz, a long-lasting form of plasticity can be observed. This results from an increased influx of Ca^{2+} into the presynaptic terminal, driven by the elevated number of tetanizing stimuli influenced by action potential activity. In some cases, repeated continuous activation of synapses can lead to synaptic depression, which may persist for several seconds or minutes.

5.3.1.2 Synaptic Depression

Synaptic depression is a kind of short-term plasticity that arises from the activation of presynaptic receptors or postsynaptic processes in a feedback loop. Repetitive stimulation of a synapse leads to a reduction in postsynaptic potentials, causing short-term synaptic depression (STD), which may have varying effects on network calculations. In 2012, Benita JM and her colleagues suggested that cerebral activation had an impact on STD. This kind of synaptic transmission is strongly associated with the probability of transmitter release and is a shared feature of both excitatory and inhibitory synapses.

5.3.1.3 Habituation

The behavior that makes memory research the simplest is habituation. It explains how repeated sensory stimulus causes physiological and behavioral responses to gradually decline. After extended exposure to any activity or circumstance, this leads to a decrease in response to an environmental stimulus.

5.3.1.4 Sensitization

The process by which repeated delivery of a stimulus increases a response's gradual amplification is known as sensitization. Fear can facilitate a quick response to a potentially dangerous stimulus. Synaptic transmission can also be triggered by sensitization when an irritable stimulus precedes a benign environmental stimulation.

5.3.2 Long-term Synaptic Plasticity

Long-term synaptic plasticity refers to enduring, activity-dependent changes in synaptic strength. This can either enhance synaptic strength through long-term potentiation (LTP) or diminish it through long-term depression (LTD) [14]. Additionally, these forms of plasticity can be classified further as follows.

5.3.2.1 Long-Term Potentiation

Long-lasting strengthening of synapses, known as LTP, occurs when post-tetanic potentiation persists for days as a result of repeated synaptic activity. This process is linked to an increased influx of Ca^{2+} in the postsynaptic terminal, which enhances synaptic strength. LTP develops quickly and is most frequently observed in the hippocampus.

5.3.2.2 Long-Term Depression

LTD is marked by a reduction in synaptic strength, making synapses less effective at transmitting neuronal signals. It plays a key role in adapting synaptic networks for various physiological activities in response to prolonged patterns of stimulation. Glutamate-mediated LTD can be triggered by the activation of N-methyl-D-aspartate receptors (NMDAR). This type of long-term synaptic plasticity can alter the efficiency of synapses for extended periods, from hours to years, and is believed to contribute to the formation of lasting memories stored in brain circuits.

5.4 FUNDAMENTAL ELEMENTS OF SYNAPTIC PLASTICITY AND ITS DIVERSE MECHANISMS

The transmission of neuronal impulses in the brain occurs through synaptic junctions, consisting of presynaptic and postsynaptic terminals separated by a synaptic cleft. A single neuron can get various stimuli by presynaptic cells, making the stimuli to integrate information from different neurons its transmission in the form of electrochemical code. Specialized synapses facilitate the flow of information through neurons in the brain. Neurotransmitters, stored in small vesicles at the presynaptic terminal, are key players in this process [15]. The postsynaptic terminal contains specific receptors for various neurotransmitters. Neurons generate electrical impulses, known as action potentials, that originate at the cell body and travel down the axon. The release of neurotransmitter-filled vesicles is voltage-dependent and triggered by an action potential, transforming the electrical impulse into a chemical signal at the synapse. The neurotransmitter then diffuses across the synaptic cleft and binds to receptors, generating an electrical impulse in the postsynaptic neuron. When the postsynaptic neuron reaches a minimum electrical threshold needed for firing, it initiates an action potential [16]. Different types of neural plasticity, defined by molecular mechanisms and involved molecules, have been proposed and are described below.

5.4.1 Postsynaptic Density Proteins Mediated Synaptic Plasticity

Glutamate receptors [such as α-amino-3-hydroxy-5-methyl-4-isoxazolepropionic acid (AMPA) and NMDA receptors] are anchored and stabilized by the postsynaptic density (PSD) protein complex, which is made up of many interwoven proteins within the postsynaptic neuron. PSD-95, synapse-associated protein, and other multi-domain proteins such as glutamate receptor interaction protein interact with these glutamate receptors. An important aspect of neuronal plasticity is this relationship that regulates both activity-dependent and activity-independent receptor targeting and trafficking. Studies indicate that PSD-95 has an impact on the functioning of AMPA-type glutamate receptors, which in turn enhances synaptic strength and promotes long-term potentiation (LTP). It has been discovered that decreasing the levels of PSD-95 results in a reduction of AMPA receptor expression and synaptic strength [17].

5.4.2 Neuroligin and Neurexin Dependent Neuronal Plasticity

Neuronal plasticity, which relies on the interaction between neuroligin and neurexin, plays a crucial role in maintaining accurate synaptic coordination in the brain. Synaptic transmission and plasticity appear to be dependent on the functional interaction of presynaptic and postsynaptic neurons across the synapse. The extracellular connections that these transmembrane proteins create improve the adhesion between dendrites and axons. While neurexin can cause postsynaptic differentiation and receptor clustering in dendrites, neuroligin can independently result in the creation of fully functioning presynaptic terminals. The development of synaptic function is triggered by the reciprocal interaction between neurexin and neuroligin [18].

5.4.3 Dendritic Spine Enlargement

Dendritic spine structures, found on dendrites, play crucial roles in brain connectivity and serve as sites for long-term synaptic plasticity related to memory. Actin filaments are a basic unit of the dendritic spine, which is widely distributed throughout the nervous system, and further, they indicate synaptic locations and are particularly associated with convergent neurons. In the central nervous system, over 90% of excitatory synapses occur exclusively on dendritic spines. Larger dendritic spines exhibit stronger synaptic transmission due to having more neurotransmitter receptors and a larger PSD protein compared to smaller ones. Dendritic spine plasticity is crucial for memory formation, particularly in the medial prefrontal cortex (PFC) and the hippocampus. Estrogen levels impact memory changes across a female's lifespan, with increased estrogen enhancing dendritic spine density and memory performance in the PFC and hippocampus [19].

5.4.4 Brain-Derived Neurotrophic Factor

Brain-derived neurotrophic factor (BDNF), a member of the neurotrophin family, facilitates dendritic spine formation and improves memory performance. It is essential for LTP-induced alterations in synaptic structure and function and believed to modify the synaptic proteome. BDNF serves as a marker for cognitive progression and influences excitatory synapses in the central nervous system during activity-dependent plasticity. Estrogen also elevates PFC and hippocampal BDNF concentrations, and in activity-dependent plasticity, BDNF acts as a vital regulator [20]. BDNF plays a role in promoting the growth, maturation, survivability, and maintenance of neuronal cells. Both estrogen receptors (ERs) and tyrosine kinase B (TrkB) receptors, associated with differentiation and plasticity of neurons in the hippocampus, have common targets, effects, and mechanisms of action. Their co-expression triggers various neurobiological functions related to learning and memory, controlling synaptic development and activity-dependent changes in synapse structure and function [21].

5.4.5 NMDAR-dependent Long-Term Potentiation and Long-Term Depression

The neural plasticity attributed to the onset of LTP and LTD has been extensively investigated in glutamate neurons, where glutamate serves as the primary excitatory neurotransmitter in the central nervous system (approximately 5–15 mmol glutamate per kg brain tissue). Playing a pivotal metabolic role in the brain, glutamate acts as a nexus between various biochemical pathways, exerting excitatory effects on nerve cells and controlling signal transmission within these cells. Two main types of glutamate receptors, NMDARs and AMPARs, play crucial roles in this process. NMDAR, an essential ionotropic glutamate receptor and ion channel protein found in neurons, is activated during LTP when glutamate molecules attach to its receptors due to a low firing rate of the presynaptic neuron. AMPARs, predominant ionotropic glutamate receptors, mediate basal synaptic transmission [22]. Simultaneous activation of both receptors results in the opening of the NMDA receptor, allowing the influx of calcium into the postsynaptic neuron and initiating LTP. Calcium further stimulates the production of more AMPA receptors, facilitating additional influx of calcium and sodium ions into the postsynaptic neuron. During strong and high-frequency firing from the first neuron, the depolarization through AMPA receptors is potent enough to displace magnesium from the blocked NMDAR. Consequently, the opened NMDAR permits increased entry of sodium and calcium into the postsynaptic terminal. The positive charge of sodium ions induces sufficient depolarization, triggering a cascade effect that fires the second neuron. The influx of calcium enhances the deposition of more AMPA receptors by the second neuron at the synapse, contributing to the strengthening of the synapse through a series of sequential events. NMDARs also play a crucial role in triggering LTD and regulating presynaptic and postsynaptic firing. The induction of LTP or LTD depends on the concentration of calcium ions present in the postsynaptic neuron. A moderate concentration of Ca^{2+} below its threshold level leads to LTD. NMDAR-dependent LTD increases the endocytosis of AMPAR, responsible for synaptic depression. Moreover, NMDAR-dependent LTD enhances short-term potentiation (STP), strengthening the neuronal responsiveness of previously depressed synapses [23].

5.5 EPIGENETICS

The term "epigenetics" was devised by Conrad Waddington in 1942 to address one of the fundamental problems of developmental biology. Epigenetic factors mediate gene-environment interaction. Epigenetic mechanisms play a pivotal role in both activity-dependent and long-term regulation of gene expression during development and throughout life. Epigenetics underlines stable alteration in gene levels without changing the underlying DNA sequence. Although epigenetics originally described non-genetic mitotically heritable changes, this view is being reconsidered since epigenetic mechanisms also strongly influence the genome plasticity of non-dividing cells. Here, we discuss the epigenetic mechanisms that include DNA methylation, chromatin remodeling by histone modifications, and alteration of genomic landscape by transposable elements and non-coding RNAs [24].

5.5.1 DNA Methylation

Chromatin is a highly organized structure that can be locally regulated by epigenetic mechanisms to dynamically regulate or stabilize the expression of the genes it carries. One of the best-known epigenetic mechanisms is DNA methylation. It stands out among epigenetic modifications because it modifies DNA directly and can be stable. In mammals, DNA methylation consists of transferring a methyl group to the fifth position of the pyrimidine ring of cytosines (5mC), generally in dinucleotide CpG sequences. About 1% of cytosines and 75% of CpG dinucleotides are methylated in mammalian genomes. Because methylated cytosines are 10–50 times more likely to undergo mutations (by deamination of C to T), the representation of CpGs in the genome has decreased during evolution. However, clusters of CpG-dense regions called CpG islands are commonly found in the promoter region of genes, and promoter-associated CpG islands are typically minimally methylated [25]. Increased 5mC generally results in transcriptional repression and is often associated with gene silencing. 5mC is thought to interfere with gene expression via two general, not mutually exclusive, mechanisms. The universal methyl donor S-adenosyl-L-methionine (SAM) necessary for DNA methylation is synthesized from methyl donors (methionine and choline) and cofactors [26]. The methyl group from SAM is transferred to cytosine by DNA methyltransferases (DNMTs), which exist in four primary forms in mammals: DNMT1, 2, 3a, and 3b. DNMT1 is involved in the maintenance of DNA methylation, in particular across cell division. It methylates the newly synthesized DNA strand during replication from the parent strand template. Thus, although reversible, 5mC can stably mark DNA and be maintained across mitosis and, in some cases, meiosis. In contrast to maintenance DNMTs, DNMT3a and DNMT3b are de novo DNMTs. Their mechanisms of action have yet to be fully understood. However, they are known to involve domain-specific recognition sites and might be involved in recruitment to specific DNA sequences via protein–protein interaction and small RNAs. The methylation of DNA is an epigenetic mark of particular interest due to its strong influence on gene transcription. Specifically, it was found that the inhibition of gene transcription correlates with upregulated DNA methylation levels at the 5' gene regions, promoter, and first exon. DNA methylation is a ubiquitous mechanism for controlling gene levels in both eukaryotes and prokaryotes. Contrary to prokaryotes, where cytosine and adenine could be modified, eukaryotic DNA methylation is generally restricted to cytosine. In the brain, cytosine methylation primarily occurs at CpG (C-phosphate-G) dinucleotides in two primary forms, 5-methylcytosine (5mC) and 5-hydroxymethylcytosine (5hmC) [26].

5.5.2 DNA Hydroxymethylation (5hmC)

It was long thought that 5mC was an irreversible epigenetic mark, in part because no DNA demethylation mechanism or enzyme could be identified. However, the fact that DNA in the zygote is first demethylated and then remethylated during development has always argued for the existence of an active demethylation process. The recent discovery of an intermediate epigenetic mark between DNA methylation and demethylation in the form of 5hmC strongly suggests that DNA demethylation does occur in mammalian cells [27]. 5hmC is generated by hydroxylation of 5mC by ten–eleven translocation (TET) proteins (TET1–3). 5-Hydroxylation is thought to be the first step of a cascade of chemical reactions leading to the removal of 5mC. Genome-wide analyses have revealed that 5hmC and 5mC are differently distributed on the genome. While 5mC occurs mostly in inter- and intragenic regions (CpG-islands surrounding promoter regions being largely unmethylated) and silences genes in most cases, 5hmC is primarily confined to the 50 end and correlates with gene transcription. Interestingly, the level of 5hmC in the body is the highest in the brain, suggesting an important role for this modification in neural functions. Although not much is known about the biological functions of 5hmC, the view that it is an epigenetic mark on its own that is associated with gene transcription is gaining momentum [28].

5.5.3 Nucleosome and Histones

In the nucleus, DNA molecules are packed into a complex structure called chromatin, the basic unit of which is a nucleosome. The main protein component of a nucleosome are the histones. In the nucleosome, 146 base pairs of DNA are wrapped around eight core histone proteins (2x H2A, H2B, H3, H4); the adjacent nucleosomes are linked by a histone H1 [29]. Hydrophilic N-terminal tails of different histones possess residue-specific modifications: lysine (K) methylation, acetylation (both modifications are described below in more detail), and ubiquitination, serine (S) phosphorylation, SUMOylation, and arginine (R) methylation [30]. Different histone modifications modulate the affinity of histones for the DNA and the chromatin remodeling complex, which includes transcriptional factors, resulting in altered expression of the genes critical for neuronal plasticity. In addition to DNA methylation, histone posttranslational modifications (PTMs) play a critical role in chromatin remodeling. Some of these modifications are transient, while others are stable and potentially heritable. Together with DNA methylation, histone PTMs induce local and global structural changes in the chromatin. They can partition the genome into distinct domains of transcriptionally active chromatin (euchromatin) or transcriptionally inactive chromatin (heterochromatin). They also alter the net electrical charge of nucleosomes and control the loosening or tightening of inter- and intra-nucleosomal DNA–histone interactions. For instance, acetylated histones (which are associated with transcriptionally active states) are more likely to be displaced from DNA, thus inducing a loosening of the chromatin [31]. Consistently, genome-wide studies have demonstrated that nucleosome density is typically lower at promoter regions that carry acetylated histones than in the coding region. Histone PTMs can also help recruit binding partners that can be positive or negative. Methylation of lysine four on histone 3 (H3K4me) can prevent the binding of histone deacetylases (HDACs), thus favoring histone acetylation. In contrast, H3K18 acetylation can favor the binding of histone acetyltransferases (HATs) and transcription factors such as cAMP-response element binding protein (CREB)-binding protein (CBP). However, since PTMs are multiple and occur in combinations, it is not easy to define their individual impact. Further, they have a different role depending on their location within a histone tail. For instance, the displacement of PTMs was shown to lead to the repression of usually transcribed genes. In addition to carrying complex combinations of PTMs, histones are also expressed as multiple sequence variants encoded by different genes that are associated with distinct transcriptional profiles [32].

5.5.4 Transposable Elements

Transposons are mobile genetic elements that comprise about 50% of the genome in mammals, including humans. The transposons have long been assumed to be parasitic and capable of moving themselves around the genome via their RNA-intermediate (retrotransposons) or transposing their DNA (DNA transposons) directly into new genomic locations, potentially causing mutations that could be either lethal, tumorigenic, neurotoxic, or conversely, contribute to genomic diversity and evolution. However, recently growing evidence implicates transposon-mediated regulation of gene expression and neuronal diversity that contribute to neuronal plasticity [33].

5.5.5 Non-coding RNAs (ncRNAs)

Similar to transposons, ncRNAs have long been considered junk genomes. Immense development in understanding ncRNA's role in cell plasticity was promoted by the discoveries of the vital importance of ncRNAs in mediating RNA interference-dependent gene silencing in the nucleus [small interfering RNAs (siRNAs), PIWI-interacting RNAs (piRNAs), and long-non-coding RNAs (lncRNAs)] and spatial-temporal regulation of gene transcription in the cytoplasm (lncRNAs, microRNAs, or miRNAs) [34]. In recent years, non-coding RNAs (ncRNAs) and small RNAs (sncRNAs) have emerged as pivotal regulators of gene expression, playing essential roles in both transcriptional and post-transcriptional regulation, thereby contributing to non-genetic control mechanisms. SncRNAs work by getting integrated into RNA-induced silencing complexes and further instruct the silencing machinery through binding to specific sequences on the target RNA(s). They display binary function; first, they target chromatin-modifying enzymes and modify their function, and second they mobilize the silencing complex to move and silence specific regions of the genome.

Moreover, several miRNAs have been identified as targets of chromatin-modifying enzymes. For example, miR-1 and miR-140 show regulatory action on HDAC4. In addition, they modulate the HDAC expression during development, while miR-290 targets transcriptional repressors of de novo DNMTs, thereby maintaining optimal DNA methylation levels in embryonic stem cells. Additionally, the miR-29 family of miRNAs targets different DNMTs (DNMT3a, DNMT3b) and TET1-3. Further they maintain the balance between DNA methylation and hydroxy-methylation in the genome [35].

5.6 BRAIN-SPECIFIC EPIGENETICS

Numerous neuronal and non-neuronal cell types, including oligodendrocytes, astrocytes, and microglia, require sophisticated regulatory systems for activity-dependent gene expression due to the complexity of the nervous system and the variety of environmental inputs. While many epigenetic alterations are incorporated into the epigenome during cell division, neurons have a special and complex epigenetic machinery that is still active in post-replicative cells since they are non-dividing cells. Replication-independent DNA methylation and demethylation, which are brought about by neural activity, are essential for memory, learning, and synaptic plasticity [36].

The intricate nature of the nervous system and the wide range of environmental stimuli necessitate complex regulatory mechanisms for activity-dependent gene expression in different types of neuronal and non-neuronal cells such as oligodendrocytes, astrocytes, and microglia.

A recent study suggests that the process of modifying nucleosomes, which is crucial for the growth of dendritic branches, the ability of synapses to change over time, and the formation of long-term memories, is controlled by a particular complex called neuronal Brg1/hBrm associated factor (nBAF), which is found only in neurons. Furthermore, the process of replacing histones in nucleosomes that contain H3.3 plays a crucial role in regulating the patterns of gene expression that are dependent on neuronal and glial cell activity [37].

Furthermore, the gene expression pattern of each neuron is distinct as a result of the impact of transposable elements located in close proximity to the gene. The brain contains a significant amount of non-coding RNAs (ncRNAs), both small and long. These ncRNAs are important for various functions such as modifying the structure of chromatin, silencing target genes, and transporting mRNA transcripts to distant dendritic spines in neurons. Once there, the transcripts can be translated locally in response to neuronal activity. This process is dependent on the activity of the neurons [38].

5.7 EPIGENETIC MECHANISMS SHAPING BRAIN PLASTICITY

The remarkable capacity of the brain, in contrast to other organs, to incorporate environmental information and modify its function accordingly is one of its distinguishing features. This ability is referred to as plasticity. The subsequent subsections present two instances that demonstrate the role of epigenetic pathways in shaping brain plasticity in response to life experiences. Drug addiction is a situation in which a person's behavior is fundamentally and permanently changed by a single contact with an addictive substance. The impact of early-life stress on psychological development and long-term health is the subject of the second case [39]. Environmental influences, including addiction and early-life stress, can cause quick and long-lasting alterations through intricate epigenetic mechanisms.

5.7.1 Drug Addiction

A chronic illness known as drug addiction is characterized by obsessive drug seeking and usage, even in the face of negative outcomes. This condition leads to long-lasting changes in behavior, such as intense desire and a return to previous habits, which are connected to physical changes in the brain's reward pathways. These pathways include the ventral tegmental area, nucleus accumbens (NAc), amygdala, hippocampus, and prefrontal cortex (PFC). Certain alterations result in the deterioration of the structure and operation of neurons in certain areas. Several illicit substances also impact crucial physiological signaling pathways linked to transcription factors including delta FosB and CREB. These pathways have the potential to influence many genes in different regions of the brain, and their effects may persist long after drug usage has stopped. Research has providing understanding of epigenetic processes that operates these complex types of transcriptional regulation. By the perspective of epigenetics, the enduring susceptibility to recurrence is especially noteworthy, since it infers that drug exposure might induce biological mechanisms that seize and maintain long-term modifications in synaptic plasticity. Following long-term (but not short-term) cocaine use, HDAC5 undergoes phosphorylation by CaMKII, resulting in its transportation out of the nucleus and a broad elevation in histone acetylation. When HDAC5 is overexpressed in the NAc of adult mice using a virus-mediated technique, it reduces the pleasurable effects of cocaine. On the other hand, when there is a lack of HDAC5, it leads to heightened sensitivity to the rewarding effects of cocaine after long-term use [40].

The increased sensitivity to rewards in mice missing HDAC5 is normalized when HDAC5 is delivered by a virus to the nucleus accumbens (NAc), suggesting that HDAC5 plays a direct role in this process. The use of cocaine leads to changes in histone methylation, namely involving the enzyme G9a, which is continuously reduced in the NAc after drug exposure. Elevating G9a levels in mice mitigates the overall decline in H3K9me2 induced by cocaine, hence diminishing their propensity for the drug. The first induction of deltaFosB, an immediate early gene essential for the beneficial effects of the medication, is required for the decrease in G9a and H3K9me2. Cocaine cannot lower H3K9me2 levels when deltaFosB activity is inhibited in the NAc; however, artificially elevating deltaFosB may mimic the effects of cocaine by lowering G9a and H3K9me2. New research suggests that histone alterations occur not just in the brain, but also in the sperm of cocaine-exposed mice and in the brains of their offspring, leading to increased acetylation at the BDNF gene. This implies that cocaine may have the ability to cause alterations in the genetic material that can be passed down to future generations. Moreover, the exposure to cocaine causes changes in DNA methylation, indicating the participation of the methylated DNA-binding protein MeCP2 may be facilitated by short RNAs [41]. Subsequent epigenomic research is expected to reveal the full range of epigenetic modifications brought on by cocaine, providing information on how to design epigenetic medications to lessen the long-term behavioral effects of drug addiction and prevent cravings and relapses.

5.7.2 Maternal Care

A prevalent kind of early life stress in the natural world is inadequate mother care and neglect. Rodent mothers, like their human counterparts, display notable variances in the quantity and value of maternal attention they deliver to their babies. Nevertheless, the level of care provided by each mother stays essentially constant for all her kids. Motherly care in rats and mice is considered as duration of time that mothers devote engaging in licking and nursing activities with their offspring during the first week after birth [42]. The consequences of early maternal care may be seen in the way animals react to medications. The decrease in G9a and H3K9me2, which are consequences that occur after cocaine exposure, depends on the activation of deltaFosB, an instant early gene that is vital for the pleasurable effects of the drug. When deltaFosB is deactivated in the NAc, cocaine does not lower H3K9me2 levels. However, if deltaFosB is conditionally overexpressed, it reduces G9a and H3K9me2, mimicking the effects of cocaine. Moreover, recent studies have shown that histone changes are not limited to the brain. There is evidence of heightened acetylation at the BDNF gene in the sperm of mice exposed to cocaine, which is then passed on to their offspring's brains. There is a chance that cocaine may have transgenerational consequences by causing epigenetic alterations in the germline. Following exposure to cocaine, DNA methylation, an epigenetic modification, undergoes changes. The protein MeCP2, which binds to methylated DNA, is likely implicated in this process and may operate via short RNAs [43]. Future epigenomic investigations seek to determine the whole spectrum of epigenetic alterations caused by cocaine use. It is believed that these discoveries may open the door to the creation of epigenetic medications that can lessen cravings and relapses while also reducing the long-term behavioral effects of drug addiction.

5.7.3 Early Life Stress

Variations in maternal care may have long-lasting impacts that persist throughout life, which implies that more intense experiences may have even more profound and long-lasting repercussions. According to recent studies, early life stress brought on by frequent mother dissociation in the first two weeks of lifespan modifies the stress response and includes epigenetic alterations. Early childhood stress results in hyperactivity of the HPA axis, elevated corticosterone production, and heightened sensitivity to severe stress in adulthood [44]. The observed properties are connected to a rise in the production of the hormone arginine vasopressin (AVP) in a specific region of the hypothalamus. This increase is caused by the reduction in methylation of a crucial AVP enhancer region located at a MeCP2 DNA-binding site with strong affinity. Typically, when the binding site of MeCP2 is methylated, it inhibits the production of AVP. However, when DNA methylation is decreased, this inhibition is lifted, leading to increased levels of AVP. The results indicate that early life stress has the ability to continuously and permanently change the epigenome. Inadequate maternal care is another circumstance that might result in enduring epigenetic alterations. Offspring of rats that were subjected to 30 minutes of daily stress during the first week after giving birth had notably reduced levels of BDNF mRNA in the prefrontal cortex when they reached adulthood. The differential DNA methylation patterns in a crucial regulatory region of the BDNF gene (exon IV) are associated with this. Whereas rats reared in normal settings had little to no DNA methylation, stressed-mother rats showed greater levels of methylation across 12 CpG sites in this area. Furthermore, the progeny of rats subjected to maltreatment exhibited heightened DNA methylation at the BDNF promoter region, a consequence that was not totally counteracted by cross-fostering [45]. These results indicate that there are mechanisms that enable long-lasting alterations in DNA methylation to be inherited across generations, independent of the postnatal experiences of the afflicted person. Specifically, early life events that are very distressing may lead to the exact transfer of their effects throughout numerous generations, indicating that one generation's exposure to stress can impact future generations. A severe disorder that impacts behavior throughout life is persistent and unexpected mother separation in mice, compounded by unpredictable maternal stress throughout the early postnatal stage of life. In adults, it results in depressive-like behaviors, social disengagement, cognitive deficits, and decreased behavioral control. Surprisingly, it also has a detrimental effect on their descendants across many generations. There seems to be germline participation in the transmission process, which happens via both male and female parents and is unrelated to maternal behavior. DNA methylation is thought to be a key mechanism in this process since stressed adult animals' brains show altered methylation patterns in various genes, including hypermethylation of some genes near their promoter regions [46].

5.8 EPIGENETIC MODULATOR IN SYNAPTIC PLASTICITY

Epigenetic mechanisms are constantly evolving and reversible processes regulated by different epigenetic modulators. These modulators have shown promising potential in regulating synaptic plasticity in animal models. Signals are regulated by epigenetic modulators from inflammation, environmental factors, and ageing. They can also alter chromosomal structure. Chromatin modification plays a crucial part in the progression of neurological disorders. Epigenetic research is generally focused on DNA methylation and patterns of histone post-translational modifications (such as acetylation and deacetylation) [47]. Various enzymes, such as histone acetyltransferase (HAT), HDAC, histone methyltransferase (HMT), and kinases, contribute the addition or removal of molecular marks from the histone-DNA complex. Hence, they regulate "epigenetic" modulation in the brain. Synapse synthesis is essential for learning and memory formation; however, the biochemical mechanisms responsible for synaptogenesis are complex and not yet fully understood. The term "epigenetic remodeling" is initiated by molecular interplay of various epigenetic modulators associated with synaptic plasticity. In this context, special attention is paid to significant epigenetic signature molecules such as MEF2, MECP2, HDAC, and DNMTs.

5.8.1 RE1 (Repressor-Element-1)-Silencing Transcription Factor (REST)

Acting as a transcriptional regulator for over 2000 genes, neuron-specific proteins encoding proteins crucial for neuronal plasticity, REST-mediated transcription emerges as a distinctive process governing cellular gene expression in neurons, crucial for maintaining neuronal homeostasis. Minimal levels of REST enable the upregulation of specific REST-related target genes [47]. Throughout neuronal differentiation and cell specification, REST assumes a significant role by repressing genes involved in neuronal differentiation. They are also involved in the repression of genes of neuronal maturation. Comprising three efficient domains, including a DNA binding domain (DBD) and two inhibitory domains at the N and C termini, REST employs eight zinc fingers (ZF) in the DBD to bind to a 21-pair consensus element, known as RE1. Gene suppression is initiated through the employment of two corepressor complexes, mSin3 and mammalian Co-REST, which then bind to chromatin-modifying enzymes. REST and its cofactor complexes collectively control neural gene expression in target cells during neuronal development and plasticity. Crucially, REST plays a vital role in maintaining synaptic plasticity and regulating glutamate receptors, with reduced or absent REST expression, which is linked to altered synaptic plasticity [48].

5.8.2 BDNF-dependent Synaptic Plasticity

The covalent alteration of cytosine residues in CpG sites, known as DNA methylation, is catalyzed through a highly conserved family of enzymes called DNMTs. The DNMT enzyme catalyzes the allocation of a methyl group from SAM to the carbon-5 position of a cytosine residue, leading to the creation of 5-methylcytosine (5mC). Subsequently, this 5mC undergoes oxidation via a sequence of reactions, resulting in the formation of three separate chemical compounds: 5-hydroxymethylcytosine, 5-formylcytosine, and 5-carboxycytosine [49]. The oxidation of 5mC into 5hmC by TET marks the start of the active DNA demethylation process. The base-excision DNA repair mechanism has a regulatory role in neurons and other cell types. The hippocampus's co-localization of TET1 and NeuN, a neuronal marker, points to their critical function in DNA demethylation-dependent brain plasticity. When TET1 is overexpressed, the central nervous system converts more 5mC to 5hmC. Moreover, the process is crucial for the demethylation and activation of fibroblast growth factor 1 (FGF1) and BDNF promoters. Furthermore, active DNA demethylation, which aids in controlling impulse transmission across synapses, is facilitated by TET3, the enzyme that is most highly expressed in the brain [50].

5.8.3 Myocyte Enhancer Factor 2 (MEF2) in Synaptic Plasticity

MEF2 is a transcription factor that controls the expression of a diverse set of genes, which have a role in the progress and expansion of many tissues, including the brain. The structure has three functional domains: the N-terminal domain, which includes the MCM1, Agamous, and MADS box, as well as the MEF2 domain. These components are responsible for binding to DNA, forming dimers, and interacting with cofactors. Additionally, they contain a transcriptional activation domain at the C-terminus [51]. MEF2 has been shown to have a vital function in controlling the development of neurites and impacting the creation of synapses. It is a flexible protein that is controlled by chemical changes such as phosphorylation, acetylation, and SUMOylation. When a cell is active, activated MEF2 increases the survival of neurons and their aptitude to change and activates the entrance of calcium via NMDAR, which depends on NMDAR. Through a variety of chromatin-modifying enzymes, MEF2 modifies target gene expression at the promoter area in a calcium-dependent way. In addition, it is well-established that mitogen-activated protein kinase (MAPK), a key player in several growth signaling pathways, strongly activates MEF2 activity [52].

5.8.4 Neuronal Activation and Synaptogenesis through Epigenetic Marker (MECP2)

The process of modifying neuronal activation and synaptogenesis via epigenetic mechanisms is primarily influenced by the epigenetic marker MECP2. A member of the methyl-binding domain (MBD) protein family, MECP2 plays a crucial role as a mediator in the intricate methylome landscape of the brain. This is the main epigenetic marker in the adult brain, interacting with chromatin modifiers to establish a connection between DNA methylation and chromatin structure [53]. MECP2 is a crucial factor in the process of transmitting signals between nerve cells in the central nervous system (CNS). It has a significant impact on both the spontaneous transmission of chemical messengers in the brain and the ability of synapses to change their strength temporarily. MECP2 utilizes post-translational changes, including phosphorylation, acetylation, and SUMOylation, to control synaptic transmission, which is crucial for its functioning. MECP2 is closely regulated due to its several functionally defined domains, including the MBD and transcriptional repression domain (TRD). Recent investigations have shown that the MBD, rather than the nuclear localization signal (NLS), is responsible for determining the nuclear localization of MECP2. MECP2 is subject to phosphorylation at two specific sites, serine 80 (S80) and serine 421 (S421). Phosphorylation at residue S421 leads to a decrease in phosphorylation at residue S80. The phosphorylation pattern at these sites is associated with the transition from the resting to the depolarized states of the neuron. A higher number of synapses boosts the strength of synaptic connections and results in greater amounts of MECP2 in neurons [53]. In addition, it was shown that the phosphorylation of serine 421 (S421) in MECP2, which is reliant on activity, promotes the transcriptional induction of BDNF. N-methyl-D-aspartate receptors (NMDAR) and L-type potential-sensitive calcium channels (L-VSCC) allow calcium to enter neurons, and BDNF's interaction with its receptor TrkB initiates this process. During this interaction, calcium-dependent protein kinase II (CamK II) becomes active and subsequently phosphorylates MECP2 at S421. Because neurotransmission is bidirectional, it is evident that neurotransmitter modulation is influenced by MECP2 levels, which in turn affect neuronal plasticity and the preservation of the excitation/inhibition balance [54].

5.8.5 Synaptic Regulation by Histone Modification

Histone acetyltransferase (HAT) is crucial for the process of acetylating lysine residues located at the N-terminus of histones. Because of this acetylation, positively charged histone tails and negatively charged DNA have less electrostatic contact, which strengthens the binding of transcription factors and activators to DNA and increases the expression of genes [55]. HDAC can undo this process of acetylation, resulting in the suppression of gene activity. The reversible post-translational modification known as acetylation is mediated by HDACs, which catalyze the opposite process, and HATs, which transfer acetyl groups from acetyl-CoA to lysine residues.

Histone methylation and demethylation, performed by histone methyltransferase (HMT) and histone demethylase (HDM), respectively, have distinct functions in controlling gene transcription based on the changed amino acid residue. Histone acetylation, a well-recognized alteration in chromatin structure linked to alterations in memory, learning, and neural plasticity, is closely related to active transcription. HDAC inhibitors increase histone acetylation, leading to an elevation in long-term potentiation (LTP) and thereby improving memory performance. In contrast, when HDAC2 is overexpressed, it leads to a reduction in the number of synapses, which in turn reduces synaptic plasticity and impairs memory [56].

5.8.6 Regulatory Role of BDNF in Synaptic Plasticity

As a crucial epigenetic regulator in the brain, BDNF is essential in controlling DNA methylation-mediated neural plasticity. It influences neuronal survival, development, growth, and plasticity by functioning as a broadly distributed regulator of neurotransmitter activity. BDNF is a protein that is generated by a gene with 11 exons, each of which produces transcripts in response to different stimuli. It contributes to long-term potentiation (LTP) in an activity-dependent way [57]. The endoplasmic reticulum and Golgi apparatus process the precursor form, pre-pro-BDNF, to create mature BDNF.

Epigenetic modifications at the BDNF regulatory region, such as CpG methylation, play an essential role in controlling gene transcription. Elevated BDNF levels in neurons lead to a reduction in CpG methylation, thereby enhancing transcriptional activity. Synaptic activity also contributes to the activation of BDNF transcription.

The receptor for both BDNF and NT4 ligands is the TrkB receptor, which is present in different parts of the brain. TrkB and BDNF bind to each other, causing ligand-dependent dimerization that phosphorylates certain tyrosine residues in the cytoplasmic domain of the receptor [58]. Intracellular signaling processes, including the recruitment of adaptor proteins like growth factor receptor-associated protein 2 (Grb2) and Shc, are triggered by this interaction. The MAPK/ERK pathway is activated as a result of these events, and CREB is then phosphorylated. CREB attaches to the BDNF promoter after it has been phosphorylated, which increases BDNF expression, improves synaptic plasticity, and aids in neuronal survival [58].

5.8.7 Transposons

Transposons exert regulatory control on adjacent genes via epigenetic processes. The retrotransposon LINE1 is extensively researched due to its significant representation, accounting for about 20% of the human genome. The retrotransposition and expression of the LINE1 gene are partially controlled by traditional epigenetic processes [59]. Alternatively, epigenetic processes might lead to the suppression of some gene expressions by increasing the methylation of transposons, such as LINE1. This can cause the establishment of heterochromatin in nearby promoter areas, ultimately preventing the transcription of Genome. On the other hand, when LINE1 is located close to genes, it can attract transcription factors and initiate abnormal transcription. Hence, the epigenetic condition of transposons may significantly impact the phenotypic of individual cells, including the distinct response of each neuron to environmental stimuli. Single-neuron sequencing was employed by Evrony et al. to determine that each individual neuron may have a different genotype due to LINE1 retrotransposition [60]. Neural plasticity may be made possible by the unique epigenetic state of every transposon and the existence of somatic mosaicism, which highlight the extraordinary complexity of brain function.

Mounting evidence indicates that the transcription of transposons in the human brain may be regulated in a way that depends on the specific cell and tissue and is responsive to stimuli like stress. Importantly, this regulation does not have any noticeable impact on the pace of transposition.

5.8.8 Post-transcriptional Regulation of Synaptic Plasticity Genes

Epigenetic processes control genes related to brain plasticity through regulation that goes beyond transcriptional control at the promoter level. In comparison to all other organs, the brain has the most intricate method for controlling gene expression following transcription. Due to the advantages of high synaptic plasticity during evolution, the brain has evolved complex post-transcriptional systems. The primary mechanisms behind the epigenetic regulation of gene expression after transcription include alternative splicing, mRNA stability, and the spatial organization of transcripts inside cells resulting from activity-dependent mRNA trafficking.

The prevalence of alternatively spliced transcripts, which have the potential to produce distinct protein isoforms, is greatest in brain tissues. Neuronal and synaptic plasticity also depends on the movement of mRNA transcripts, namely those encoding neurotrophic factors like BDNF or synaptic proteins, to distant cellular compartments like dendrites where they may engage in active translation [61]. Efficient regulation of gene expression after transcription relies on the integration of many epigenetic changes.

5.8.9 Alternative Splicing

DNA methylation and non-coding RNAs, namely miRNAs like miR-124, contribute to the maturation of pre-mRNAs by facilitating the splicing process. Several genes have higher amounts of DNA methylation in exons as opposed to introns. In addition, the areas located at the junctions between exons and introns often exhibit low levels of methylation. This enables the spliceosome to efficiently identify these regions during transcription and aid in the process of splicing. These findings indicate that alterations in the methylation patterns of exons and introns may impact alternative splicing and function as an epigenetic mechanism to control gene expression [62].

Intronic copies of LINE1 retrotransposons have functional splicing sites that can be detected by the spliceosome. This suggests that LINE1 may play a role in alternative splicing, leading to the production of abnormal mRNA transcripts. The spliceosome is formed by the proteins and small nuclear RNAs (snRNAs) (non-coding in nature), which interact to form the small nuclear ribonucleoprotein (snRNP) complex. The survival motor neuron (SMN) protein is a vital component of the snRNP complex. Spinal muscular atrophy (SMA) is a severe neurological condition that arises due to insufficient amounts of SMN protein. Potential treatment approaches seek to augment the expression of SMN [63].

Additional research is essential to completely comprehend the procedures that control the alternative splicing of genes specific to neurons, despite the presence of different transcripts commonly observed throughout the brain. Recent improvements in RNA-seq technology is providing a novel understanding to the role of the neural-specific SR-related protein of 100 kDa (nSR100) found at neural microexon alternative splicing process. Neural microexons are brief RNA segments consisting of 3–15 nucleotide pairs. The cationic amino acids that are readily available on the protein surface of these neuronal microexons play a vital role in actively modifying protein-protein connections that are critical for synapse development and plasticity [64]. Post-transcriptional regulation of gene expression extends beyond alternative splicing and encompasses the interaction between protein-coding mRNA transcripts and antisense lncRNA (AS-lncRNA), which are also referred to as natural antisense transcripts. The lncRNA gene is commonly situated in close proximity to the target gene, but it is transcribed from the complementary DNA strand utilizing a separate promoter. The expression of AS-lncRNA can be modulated, either augmented or diminished, reliant on the specific gene. This can result in compromised brain plasticity and potentially give rise to severe neurodegenerative disorders. Elevated expression of the long non-coding RNA HTTAS_v1 leads to reduced expression of the HTT protein. This is achieved by connecting and separating the Huntingtin transcript. Raised expression of the long non-coding RNA HTTAS_v1 proposes a conceivable mechanism for the advancement of HD [65]. In contrast, elevated levels of lncRNA BACE1-AS enhance the stability of its mRNA target, β-secretase-1, resulting in an augmentation of BACE1 protein synthesis and amyloid-β formation. Consequently, this contributes to the advancement of AD [66].

Utilizing AntagoNATs, a specific type of inhibitory oligonucleotides, for the purpose of targeting NAT expression has been proposed as a novel therapeutic strategy to enhance mRNA levels of specific brain plasticity genes such as BDNF, GDNF, and EPHB2. Additionally, scientists have suggested that the BDNF-AS transcript could decrease the synthesis of BDNF mRNA by binding to the BDNF genomic region and inducing inhibitory histone modifications. However, it remains unclear whether the rise in BDNF mRNA induced by AntagoNAT is limited to specific exons and dependent on activity. Both types of BDNF expression are essential for synaptic plasticity [67].

5.8.10 MicroRNAs and 3′-UTR

The brain-specific microRNA miR-137 has a critical function in controlling glutamatergic transmission and reducing the harmful effects of glutamate on the brain [68]. Overexpression of miR-137 decreases AMPA receptor-dependent transmission, which in turn suppresses active synapses. This ultimately impairs long-term potentiation (LTP) and causes deficiencies in hippocampus-dependent learning and memory. In addition, research on the whole genome has shown specific variations in the human has-miR-137 gene that are linked to schizophrenia. These variations are also related to an increased number of miR-137 gene targets.

The structure and function of the mouse brain are significantly affected by conditional knockouts of the Dicer gene, which is crucial for microRNA processing. The impact of these effects is most noticeable in the developing neocortex, resulting in significant reduction in cortical size and increased mortality quickly after the cessation of breastfeeding. Conversely, in adult forebrain-specific knockouts, the loss of brain-specific microRNAs is linked to increases in spine morphogenesis and memory function [69].

5.9 EPIGENETICS BEHIND THE IMPAIRED NEURONAL PLASTICITY: NEURODEGENERATIVE DISORDERS

Neurons are essential for the operation of the human brain since they enable communication. Although the majority of neurons are generated in the brain, they are found throughout the whole body. The majority of neurons are produced by neural stem cells throughout infancy, and as people become older, they create fewer neurons overall [70]. While neurons do not have eternal life, their gradual decline, accompanied by the deterioration of their structure and function (referred to as neurodegeneration), is a crucial characteristic of several brain illnesses and presents a significant health hazard. Neurodegeneration encompasses synapse dysfunction, disruptions in neural networks, and the buildup of chemically modified protein variations in the brain. NDs encompass a group of diseases characterized by the progressive degeneration of the nervous system. Examples of NDs include AD, PD, prion disease, amyotrophic lateral sclerosis, motor neuron disease, HD, spinal muscular atrophy, and spinocerebellar ataxia [71]. The occurrence of these serious degenerative brain illnesses may be caused by the dysfunction in the regulation of certain genes via epigenetic mechanisms. However, our knowledge about the role of epigenetics in the development of NDs is still restricted. Previous discussions have explored the possible contributions of microRNAs (miRNAs) and long non-coding RNAs (lncRNAs) to the development of NDs. This study will examine supplementary epigenetic pathways that might potentially contribute to the development of AD, HD, and PD.

AD is a serious condition characterized by the progressive degeneration of neurons and synapses, resulting in cognitive deterioration. It is also associated with the accumulation of amyloid-β plaques [72]. The correlation between the buildup of amyloid plaques in the frontal cortex and hippocampus neurons in deceased individuals with AD and the decrease in overall methylation levels has been shown. AD brains have shown alterations in DNA methylation, namely at the promoters of genes linked to AD. More precisely, the genes responsible for amyloid precursor protein (APP), presenilin 1 (PS1), and beta-site APP cleaving enzyme show reduced DNA methylation in individuals with AD. On the other hand, genes involved in the breakdown of Aβ and NEP are shown to have increased methylation in AD patients.

New research suggests that aberrant distribution of 5hmC in the brain may be responsible for the development of neurodegenerative disorders and brain tumors. AD patients had a significantly higher expression of TET1 and 5hmC in their cerebellum, para-hippocampus gyrus, and hippocampal areas compared to individuals with normal neurological health [73]. The impact of medications that manipulate histones and modify DNA methylation on the build-up of Aβ and the activation of genes associated with the disease is not thoroughly comprehended. However, one specific medication, GW3965, which acts as an agonist of the liver X receptor, has been connected to alterations in DNA methylation at different genes linked to plasticity in the hippocampus. Studies have demonstrated that GW3965 enhances cognitive function in mouse models of AD. The identified genes play a role in the development of synapses and the formation of new neurons. These findings indicate that alterations in DNA methylation in the hippocampus may contribute to the cognitive deterioration associated with AD [74].

Various research studies have also revealed irregular DNA methylation type in various brain areas, affecting both CpG and non-CpG sites. The results specify modified DNA methylation may contribute to the advancement of neurodegenerative disorders. A comprehensive investigation on human individuals has extensively analyzed the association between DNA methylation and genes associated within postmortem brain tissues. Studies have shown that people with AD have lower levels of methylation in the genes PSEN1 and PSEN2. These genes have a vital function in the synthesis of amyloid-β peptides, which are a critical element in the progression of AD. This hypomethylation occurs at both CpG and non-CpG sites. Nevertheless, there is a lack of consensus in the scientific literature regarding the notable changes in methylation of the amyloid precursor protein gene (APP) in the brains of individuals with AD. However, studies have shown that people with AD had varying amounts of methylation in several genes linked to AD, such as APOE, PLD3, and EPHA1 [75]. Several long non-coding RNAs (lncRNAs) have been shown to exhibit differential expression in the hippocampus of mice utilized in AD research. A study revealed that the quantities of long non-coding RNA (lncRNA) LoNA were increased in the brains of mice with the APP/PS1 genetic mutation. The study observed that reducing levels of LoNA in the hippocampus of these mice led to improved cognitive functions, namely in learning and memory, compared to a control group [76].

HD is an inherited neurological condition resulting from an increase in the number of CAG triplet repeats in the huntingtin gene (HTT). This mutation is linked to disturbances in the intracellular transportation of BDNF mRNA. HD is characterized by involuntary and excessive movements and currently does not have a definitive pharmaceutical cure [77]. In addition, individuals with HD experience a reduction in methylation at the D4S95 gene, which is recognized for its high variability, as well as a notable decrease in the amounts of 5hmC in the 5' UTR region of the adenosine A2A receptor (A2AR) [78].

PD is predominantly credited to the buildup of the protein α-synuclein, resulting in the development of Lewy bodies. The gradual accumulation of these compounds leads to the gradual decline of dopaminergic neurons of substantia nigra. PD has motor symptoms that become evident after the loss of over 70% of dopaminergic neurons. Current pharmaceutical treatments generally focus on mitigating motor symptoms but do not have any impact on the course of the disease [79]. PD is regulated by a combination of genes, such as α-synuclein and parkin. An epigenetic investigation has demonstrated that the development of PD is impacted by a reduction in DNA methylation in the promoter region of the α-synuclein gene (SNCA) in the brains of individuals with PD. Matsumoto et al. discovered a substantial decrease in methylation levels inside a particular genomic region called a CpG island, located within the initial intron of the SNCA gene. Furthermore, scientists observed an increase in the activity of the SNCA gene in the substantia nigra of individuals who have been diagnosed with PD [80]. In addition, studies have shown that individuals suffering from PD display abnormal DNA methylation patterns in both their brain and blood cells. The magnitude of these patterns can be augmented or diminished, contingent upon the particular gene under investigation. Exposing PC12 cells to DNMT inhibitor, 5-Aza-2'-Deoxycytidine, intensifies neurotoxicity and increases the expression of genes associated with PD in dopamine nerve cells. To decrease the toxicity caused by α-synuclein in dopaminergic neurons, there are several approaches that can be taken. One option is to use histone deacetylase (HDAC) inhibitors, which can reduce the activity of the HDAC silent information regulator 2 (SIRT2) through a process called epigenetic inhibition. Another option is to enhance histone acetylation by using the HDAC inhibitor valproic acid. Additionally, the toxicity can also be reduced by using RNA interference to inhibit the agent HDAC SIRT2 [81].

5.10 CONCLUSION

Synaptic plasticity is crucial for the functional organization, recovery, and maturation of synaptic connections, which have a profound impact on both the structure and function of the brain. Neural circuits may be reshaped by it, especially when it comes to how many synaptic connections neurons get. Neuroepigenetics research is a broad area that connects fundamental neurobiology with clinical epigenetics. Both in healthy and sick conditions, the intricacy of brain regions and cells results in structural and temporal genetic alterations. Understanding the epigenetic control of genes particular to the brain is essential for controlling the functioning and adaptability of neurons in response to environmental stimuli, both in the short and long term. Adverse childhood events, such as prenatal impacts and abuse or neglect, have the potential to induce lasting changes in gene activity and have a role in the onset of mental illnesses. The initial difficulties may also have an effect on future generations by causing changes in the brain's structure and influencing social and emotional behaviors across generations. Epigenetic reprogramming plays a crucial role in regulating synaptic strength, intrinsic excitability, and synaptic density, all of which are essential for optimal brain function. Multiple biochemical, genetic, and epigenetic pathways contribute to either improving or hindering the ability of neurons to change and adapt, with several epigenetic regulators implicated. This chapter describes the many forms of synaptic plasticity, their functional significance, and the epigenetic processes that modify synaptic genes and epigenetic markers. The text also explores the possible function of natural chemicals as epigenetic modulators in enhancing redox equilibrium and safeguarding the nervous system. Gaining a comprehension of how various epigenetic pathways and modulators interact might provide valuable understanding of the epigenetic processes and consequences associated with neurodegeneration. Furthermore, early difficulties may rewire social and emotional behavior as well as cause transgenerational alterations in brain structure that affect subsequent generations.

REFERENCES

1. Mateos-Aparicio P, Rodríguez-Moreno A. The impact of studying brain plasticity. Front Cell Neurosci. 2019;13:66. doi: 10.3389/fncel.2019.00066
2. Abel T, Poplawski S. Epigenetic advances in clinical neuroscience. Dialogues Clin Neurosci. 2014;16(3):27375. doi: 10.31887/DCNS.2014.16.3/tabel
3. Toraño EG, García MG, Fernández-Morera JL, Niño-García P, Fernández AF. The impact of external factors on the epigenome: In Utero and over lifetime. Biomed Res Int. 2016;2016:2568635. doi: 10.1155/2016/2568635
4. Wang J, Hodes GE, Zhang H, et al. Epigenetic modulation of inflammation and synaptic plasticity promotes resilience against stress in mice. Nat Commun. 2018;9(1):477. doi: 10.1038/s41467-017-02794-5
5. Sweatt JD. The emerging field of neuroepigenetics. Neuron. 2013;80(3):624–32. doi: 10.1016/j.neuron.2013.10.023
6. Citri A, Malenka RC. Synaptic plasticity: Multiple forms, functions, and mechanisms. Neuropsychopharmacology. 2008;33(1):18–41. doi: 10.1038/sj.npp.1301559
7. Yuan TF, Li WG, Zhang C, Wei H, Sun S, Xu NJ, Liu J, Xu TL. Targeting neuroplasticity in patients with neurodegenerative diseases using brain stimulation techniques. Transl Neurodegener. 2020;9(1):44. doi: 10.1186/s40035-020-00224-z.
8. DeFelipe J. Brain plasticity and mental processes: Cajal again. Nat Rev Neurosci. 2006;7(10):811–7. doi: 10.1038/nrn2005
9. Fox K, Stryker M. Integrating Hebbian and homeostatic plasticity: Introduction. Philos Trans R Soc Lond B Biol Sci. 2017;372(1715):20160413. doi: 10.1098/rstb.2016.0413
10. Lømo T. The discovery of long-term potentiation. Philos Trans R Soc Lond B Biol Sci. 2003;358(1432):617–20. doi: 10.1098/rstb.2002.1226
11. Dudek SM, Bear MF. Bidirectional long-term modification of synaptic effectiveness in the adult and immature hippocampus. J Neurosci. 1993;13(7):2910–8. doi: 10.1523/JNEUROSCI.13-07-02910.1993
12. Kennedy MJ, Ehlers MD. Organelles and trafficking machinery for postsynaptic plasticity. Annu Rev Neurosci. 2006;29:325–62. doi: 10.1146/annurev.neuro.29.051605.112808
13. Barroso-Flores J, Herrera-Valdez MA, Galarraga E, Bargas J. Models of short-term synaptic plasticity. Adv Exp Med Biol. 2017;1015:41–57. doi: 10.1007/978-3-319-62817-2_3
14. Yang Y, Calakos N. Presynaptic long-term plasticity. Front Synaptic Neurosci. 2013;5:8. doi: 10.3389/fnsyn.2013.00008
15. Baccus SA. Synaptic facilitation by reflected action potentials: Enhancement of transmission when nerve impulses reverse direction at axon branch points. Proc Natl Acad Sci USA. 1998;95(14):8345–50. doi: 10.1073/pnas.95.14.8345
16. O'Brien J. Design principles of electrical synaptic plasticity. Neurosci Lett. 2019;695:4–11. doi: 10.1016/j.neulet.2017.09.003

17. Kennedy MB. The protein biochemistry of the postsynaptic density in glutamatergic synapses mediates learning in neural networks. Biochemistry. 2018;57(27):4005–9. doi: 10.1021/acs.biochem.8b00496
18. Ko J, Zhang C, Arac D, Boucard AA, Brunger AT, Südhof TC. Neuroligin-1 performs neurexin-dependent and neurexin-independent functions in synapse validation. EMBO J. 2009;28(20):3244–55. doi: 10.1038/emboj.2009.249
19. Runge K, Cardoso C, de Chevigny A. Dendritic spine plasticity: Function and mechanisms. Front Synaptic Neurosci. 2020;12:36. doi: 10.3389/fnsyn.2020.00036
20. Rauti R, Cellot G, D'Andrea P, Colliva A, Scaini D, Tongiorgi E, Ballerini L. BDNF impact on synaptic dynamics: Extra or intracellular long-term release differently regulates cultured hippocampal synapses. Mol Brain. 2020;13(1):43. doi: 10.1186/s13041-020-00582-9
21. Bramham CR, Messaoudi E. BDNF function in adult synaptic plasticity: The synaptic consolidation hypothesis. Prog Neurobiol. 2005;76(2):99–125. doi: 10.1016/j.pneurobio.2005.06.003
22. Hunt DL, Castillo PE. Synaptic plasticity of NMDA receptors: Mechanisms and functional implications. Curr Opin Neurobiol. 2012;22(3):496–508. doi: 10.1016/j.conb.2012.01.007
23. Lutzu S, Alviña K, Puente N, Grandes P, Castillo PE. Target cell-specific plasticity rules of NMDA receptor-mediated synaptic transmission in the hippocampus. Front Cell Neurosci. 2023;17:1068472. doi: 10.3389/fncel.2023.1068472
24. Weinhold B. Epigenetics: The science of change. Environ Health Perspect. 2006;114(3):A160–7. doi: 10.1289/ehp.114-a160
25. Moore LD, Le T, Fan G. DNA methylation and its basic function. Neuropsychopharmacology. 2013;38(1): 23–38. doi: 10.1038/npp.2012.112
26. Ponnaluri VKC, Estève PO, Ruse CI, Pradhan S. S-adenosylhomocysteine hydrolase participates in DNA methylation inheritance. J Mol Biol. 2018;430(14):2051–65. doi: 10.1016/j.jmb.2018.05.014
27. Richa R, Sinha RP. Hydroxymethylation of DNA: An epigenetic marker. EXCLI J. 2014;13:592–610.
28. Wernig-Zorc S, Yadav MP, Kopparapu PK, Bemark M, Kristjansdottir HL, Andersson PO, Kanduri C, Kanduri M. Global distribution of DNA hydroxymethylation and DNA methylation in chronic lymphocytic leukemia. Epigenetics Chromatin. 2019;12(1):4. doi: 10.1186/s13072-018-0252-7
29. Mariño-Ramírez L, Kann MG, Shoemaker BA, Landsman D. Histone structure and nucleosome stability. Expert Rev Proteomics. 2005;2(5):719–29. doi: 10.1586/14789450.2.5.719
30. Zhang Y, Zhang Q, Zhang Y, Han J. The role of histone modification in DNA replication-coupled nucleosome assembly and cancer. Int J Mol Sci. 2023;24(5):4939. doi: 10.3390/ijms24054939.
31. Luger K, Dechassa ML, Tremethick DJ. New insights into nucleosome and chromatin structure: An ordered state or a disordered affair? Nat Rev Mol Cell Biol. 2012;13(7):436–47. doi: 10.1038/nrm3382
32. Ramazi S, Allahverdi A, Zahiri J. Evaluation of post-translational modifications in histone proteins: A review on histone modification defects in developmental and neurological disorders. J Biosci. 2020;45:135.
33. Slotkin RK, Martienssen R. Transposable elements and the epigenetic regulation of the genome. Nat Rev Genet. 2007;8(4):272–85. doi: 10.1038/nrg2072.
34. Wei JW, Huang K, Yang C, Kang CS. Non-coding RNAs as regulators in epigenetics (Review). Oncol Rep. 2017;37(1):3–9. doi: 10.3892/or.2016.5236
35. Li W, Yi J, Zheng X, Liu S, Fu W, Ren L, Li L, Hoon DSB, Wang J, Du G. miR-29c plays a suppressive role in breast cancer by targeting the TIMP3/STAT1/FOXO1 pathway. Clin Epigenetics. 2018;10:64. doi: 10.1186/s13148-018-0495-y. Erratum in: Clin Epigenetics. 2022 Jul 29;14(1):97.
36. Houston I, Peter CJ, Mitchell A, Straubhaar J, Rogaev E, Akbarian S. Epigenetics in the human brain. Neuropsychopharmacology. 2013;38(1):183–97. doi: 10.1038/npp.2012.78
37. Choi KY, Yoo M, Han JH. Toward understanding the role of the neuron-specific BAF chromatin remodeling complex in memory formation. Exp Mol Med. 2015;47:e155. doi: 10.1038/emm.2014.129.
38. Nie JH, Li TX, Zhang XQ, Liu J. Roles of non-coding RNAs in normal human brain development, brain tumor, and neuropsychiatric disorders. Noncoding RNA. 2019;5(2):36. doi: 10.3390/ncrna5020036
39. Kundakovic M, Champagne FA. Early-life experience, epigenetics, and the developing brain. Neuropsychopharmacology. 2015;40(1):141–53. doi: 10.1038/npp.2014.140
40. Taniguchi M, Carreira MB, Cooper YA, Bobadilla AC, Heinsbroek JA, Koike N, Larson EB, Balmuth EA, Hughes BW, Penrod RD, Kumar J, Smith LN, Guzman D, Takahashi JS, Kim TK, Kalivas PW, Self DW, Lin Y, Cowan CW. HDAC5 and its target gene, Npas4, function in the nucleus accumbens to regulate cocaine-conditioned behaviors. Neuron. 2017;96(1):130–44.e6. doi: 10.1016/j.neuron.2017.09.015
41. Nestler EJ. Epigenetic mechanisms of drug addiction. Neuropharmacology. 2014;76 Pt B:259–68. doi: 10.1016/j.neuropharm.2013.04.004
42. Lester BM, Conradt E, LaGasse LL, Tronick EZ, Padbury JF, Marsit CJ. Epigenetic programming by maternal behavior in the human infant. Pediatrics. 2018;142(4):e20171890. doi: 10.1542/peds.2017-1890

43. Jiang Z, Zhu Z, Zhao M, Wang W, Li H, Liu D, Pan F. H3K9me2 regulation of BDNF expression in the hippocampus and medial prefrontal cortex is involved in the depressive-like phenotype induced by maternal separation in male rats. Psychopharmacology. 2021;238(10):2801–13. doi: 10.1007/s00213-021-05896-7
44. Li M, Fu X, Xie W, Guo W, Li B, Cui R, Yang W. Effect of early life stress on the epigenetic profiles in depression. Front Cell Dev Biol. 2020;8:867. doi: 10.3389/fcell.2020.00867
45. Fachim HA, Corsi-Zuelli F, Loureiro CM, Iamjan SA, Shuhama R, Joca S, Menezes PR, Heald A, Louzada-Junior P, Dalton CF, Del-Ben CM, Reynolds GP. Early-life stress effects on BDNF DNA methylation in first-episode psychosis and in rats reared in isolation. Prog Neuropsychopharmacol Biol Psychiatry. 2021;108:110188. doi: 10.1016/j.pnpbp.2020.110188
46. Anacker C, O'Donnell KJ, Meaney MJ. Early life adversity and the epigenetic programming of hypothalamic-pituitary-adrenal function. Dialogues Clin Neurosci. 2014;16(3):321–33. doi: 10.31887/DCNS.2014.16.3/canacker
47. Gibney ER, Nolan CM. Epigenetics and gene expression. Heredity. 2010;105(1):4–13. doi: 10.1038/hdy.2010.54
48. Xu C, Zhang M, Zu L, Zhang P, Sun L, Liu X, Fang M. Repressor element-1 silencing transcription factor regulates glutamate receptors and immediate early genes to affect synaptic plasticity. Aging. 2021;13(11):15569–79. doi: 10.18632/aging.203118
49. Hsieh MC, Lai CY, Ho YC, Wang HH, Cheng JK, Chau YP, Peng HY. Tet1-dependent epigenetic modification of BDNF expression in dorsal horn neurons mediates neuropathic pain in rats. Sci Rep. 2016;6:37411. doi: 10.1038/srep37411
50. Kohli RM, Zhang Y. TET enzymes, TDG and the dynamics of DNA demethylation. Nature. 2013 Oct 24;502(7472):472–9. doi: 10.1038/nature12750.
51. Pon JR, Marra MA. MEF2 transcription factors: Developmental regulators and emerging cancer genes. Oncotarget. 2016;7(3):2297–312. doi: 10.18632/oncotarget.6223
52. Harrington AJ, Raissi A, Rajkovich K, Berto S, Kumar J, Molinaro G, Raduazzo J, Guo Y, Loerwald K, Konopka G, Huber KM, Cowan CW. MEF2C regulates cortical inhibitory and excitatory synapses and behaviors relevant to neurodevelopmental disorders. Elife. 2016;5:e20059. doi: 10.7554/eLife.20059
53. Sánchez-Lafuente CL, Kalynchuk LE, Caruncho HJ, Ausió J. The role of MeCP2 in regulating synaptic plasticity in the context of stress and depression. Cells. 2022;11(4):748. doi: 10.3390/cells11040748
54. Bathina S, Das UN. Brain-derived neurotrophic factor and its clinical implications. Arch Med Sci. 2015;11(6):1164–78. doi: 10.5114/aoms.2015.56342
55. Marek R, Coelho CM, Sullivan RK, Baker-Andresen D, Li X, Ratnu V, Dudley KJ, Meyers D, Mukherjee C, Cole PA, Sah P, Bredy TW. Paradoxical enhancement of fear extinction memory and synaptic plasticity by inhibition of the histone acetyltransferase p300. J Neurosci. 2011;31(20):7486–91. doi: 10.1523/JNEUROSCI.0133-11.2011
56. Geng H, Chen H, Wang H, Wang L. The histone modifications of neuronal plasticity. Neural Plast. 2021;2021:6690523. doi: 10.1155/2021/6690523
57. Gómez-Palacio-Schjetnan A, Escobar ML. Neurotrophins and synaptic plasticity. Curr Top Behav Neurosci. 2013;15:117–36. doi: 10.1007/7854_2012_231
58. Yoshii A, Constantine-Paton M. Postsynaptic BDNF-TrkB signaling in synapse maturation, plasticity, and disease. Dev Neurobiol. 2010;70(5):304–22. doi: 10.1002/dneu.20765
59. Misiak B, Ricceri L, Sąsiadek MM. Transposable elements and their epigenetic regulation in mental disorders: Current evidence in the field. Front Genet. 2019;10:580. doi: 10.3389/fgene.2019.00580
60. Evrony GD, Lee E, Mehta BK, Benjamini Y, Johnson RM, Cai X, Yang L, Haseley P, Lehmann HS, Park PJ, Walsh CA. Cell lineage analysis in human brain using endogenous retroelements. Neuron. 2015;85(1):49–59. doi: 10.1016/j.neuron.2014.12.028
61. Carmichael RE, Henley JM. Transcriptional and post-translational regulation of Arc in synaptic plasticity. Semin Cell Dev Biol. 2018;77:3–9. doi: 10.1016/j.semcdb.2017.09.007
62. Furlanis E, Scheiffele P. Regulation of neuronal differentiation, function, and plasticity by alternative splicing. Annu Rev Cell Dev Biol. 2018;34:451–469. doi: 10.1146/annurev-cellbio-100617-062826
63. Lipscombe D, Lopez Soto EJ. Alternative splicing of neuronal genes: New mechanisms and new therapies. Curr Opin Neurobiol. 2019;57:26–31. doi: 10.1016/j.conb.2018.12.013
64. Quesnel-Vallières M, Irimia M, Cordes SP, Blencowe BJ. Essential roles for the splicing regulator nSR100/SRRM4 during nervous system development. Genes Dev. 2015;29(7):746–59. doi: 10.1101/gad.256115.114
65. Wang DQ, Fu P, Yao C, Zhu LS, Hou TY, Chen JG, Lu Y, Liu D, Zhu LQ. Long non-coding RNAs, novel culprits, or bodyguards in neurodegenerative diseases. Mol Ther Nucleic Acids. 2018;10:269–76. doi: 10.1016/j.omtn.2017.12.011

66. Sayad A, Najafi S, Hussen BM, Abdullah ST, Movahedpour A, Taheri M, Hajiesmaeili M. The emerging roles of the β-secretase BACE1 and the long non-coding RNA BACE1-AS in human diseases: A focus on neurodegenerative diseases and cancer. Front Aging Neurosci. 2022;14:853180. doi: 10.3389/fnagi.2022.853180
67. Zayed AA, Seleem MM, Darwish HA, Shaheen AA. Role of long noncoding RNAs; BDNF-AS and 17A and their relation to GABAergic dysfunction in Egyptian epileptic patients. Metab Brain Dis. 2023;38(4):1193–204. doi: 10.1007/s11011-023-01182-x
68. Luo M, Wu L, Zhang K, Wang H, Zhang T, Gutierrez L, O'Connell D, Zhang P, Li Y, Gao T, Ren W, Yang Y. miR-137 regulates ferroptosis by targeting glutamine transporter SLC1A5 in melanoma. Cell Death Differ. 2018;25(8):1457–72. doi: 10.1038/s41418-017-0053-8
69. Ye Y, Xu H, Su X, He X. Role of MicroRNA in governing synaptic plasticity. Neural Plast. 2016; 2016:4959523. doi: 10.1155/2016/4959523
70. Ravi B, Kannan M. Epigenetics in the nervous system: An overview of its essential role. Indian J Hum Genet. 2013;19(4):384–91. doi: 10.4103/0971-6866.124357
71. Hwang JY, Aromolaran KA, Zukin RS. The emerging field of epigenetics in neurodegeneration and neuroprotection. Nat Rev Neurosci. 2017;18(6):347–61. doi: 10.1038/nrn.2017.46. Erratum in: Nat Rev Neurosci. 2018 Dec;19(12):771.
72. Das B, Yan R. Role of BACE1 in Alzheimer's synaptic function. Transl Neurodegener. 2017;6:23. doi: 10.1186/s40035-017-0093-5
73. Armstrong MJ, Jin Y, Vattathil SM, Huang Y, Schroeder JP, Bennet DA, Qin ZS, Wingo TS, Jin P. Role of TET1-mediated epigenetic modulation in Alzheimer's disease. Neurobiol Dis. 2023;185:106257. doi: 10.1016/j.nbd.2023.106257
74. Báez-Becerra C, Filipello F, Sandoval-Hernández A, Arboleda H, Arboleda G. Liver X receptor agonist GW3965 regulates synaptic function upon amyloid Beta exposure in hippocampal neurons. Neurotox Res. 2018;33(3):569–79. doi: 10.1007/s12640-017-9845-3
75. Xiao X, Liu X, Jiao B. Epigenetics: Recent advances and its role in the treatment of Alzheimer's disease. Front Neurol. 2020;11:538301. doi: 10.3389/fneur.2020.538301
76. Asadi MR, Hassani M, Kiani S, Sabaie H, Moslehian MS, Kazemi M, Ghafouri-Fard S, Taheri M, Rezazadeh M. The perspective of dysregulated LncRNAs in Alzheimer's disease: A systematic scoping review. Front Aging Neurosci. 2021;13:709568. doi: 10.3389/fnagi.2021.709568
77. Ng CW, Yildirim F, Yap YS, Dalin S, Matthews BJ, Velez PJ, Labadorf A, Housman DE, Fraenkel E. Extensive changes in DNA methylation are associated with expression of mutant huntingtin. Proc Natl Acad Sci USA. 2013;110(6):2354–9. doi: 10.1073/pnas.1221292110
78. Rahman A. The role of adenosine in Alzheimer's disease. Curr Neuropharmacol. 2009;7(3):207–16. doi: 10.2174/157015909789152119
79. Miranda-Morales E, Meier K, Sandoval-Carrillo A, Salas-Pacheco J, Vázquez-Cárdenas P, Arias-Carrión O. Implications of DNA methylation in Parkinson's disease. Front Mol Neurosci. 2017;10:225. doi: 10.3389/fnmol.2017.00225
80. Matsumoto L, Takuma H, Tamaoka A, Kurisaki H, Date H, Tsuji S, Iwata A. CpG demethylation enhances alpha-synuclein expression and affects the pathogenesis of Parkinson's disease. PLoS One. 2010;5(11):e15522. doi: 10.1371/journal.pone.0015522
81. Göttlicher M, Minucci S, Zhu P, Krämer OH, Schimpf A, Giavara S, Sleeman JP, Coco Lo F, Nervi C, Pelicci PG, Heinzel T. Valproic acid defines a novel class of HDAC inhibitors inducing differentiation of transformed cells. EMBO J. 2001;20(24):6969–78. doi: 10.1093/emboj/20.24.6969

6 Synaptic Plasticity and Neuroinflammation

A Therapeutic Approach to Neurological Illnesses

Vineeta Singh
Banaras Hindu University
Varanasi, India

Arshdeep Kaur Sethi
Sri Guru Ram Das University of Health Sciences
Sri Amritsar, Punjab, India

Shivani Rath
National Institute of Mental Health and Neurosciences
Bengaluru, India

*Vijaya Nath Mishra**
Banaras Hindu University
Varanasi, India

6.1 INTRODUCTION

The central nervous system (CNS) relies on activity-dependent forms of synaptic remodeling to translate and preserve memories. While it was previously assumed that the lack of usual immune responses within its boundaries rendered it immune-privileged, recent research has cast doubt on this notion. It is widely recognized that the neurological and immunological systems have the ability to communicate with each other bidirectionally. The central nervous system (CNS) actively participates in immune surveillance, leading to inflammatory responses inside the brain following various forms of injury [1]. This series of events includes the generation and release of pro-inflammatory chemicals as cytokines and nitric oxide (NO), as well as the penetration of immune cells into the CNS and stimulation of cells that are already present there [2].

Recent discoveries propose that glial cells have an active role in immunological responses inside the CNS, making them a significant component of the synapse along with pre- and postsynaptic neurons. These cells dynamically contribute to the transfer of signals between neurons, the ability of neurons to generate electrical impulses, and different types of alterations in the strength of connections among neurons, such as long-term potentiation (LTP). Significantly, there is an ongoing connection between the nervous and immune systems during the process of LTP induction. Recent research has found that chemicals commonly linked to immune function, such as cytokines or major histocompatibility complex (MHC) molecules, vigorously influence synaptic memory processes [2,3]. The hitherto underestimated impact of immunological diffusible mediators on synaptic remodeling

DOI: 10.1201/9781003464648-7

might involve the malfunctioning observed in neuroinflammatory brain disorders. According to this viewpoint, we can observe abnormalities in the initiation, continuation, and reversal of important changes in the brain's ability to adapt in different human diseases that are associated with cognitive deterioration and different levels of inflammation in the CNS, such as Parkinson's disease (PD) and Alzheimer's disease (AD). One may reasonably infer that a combination of atypical expression of inflammatory substances during inflammation in the CNS, together with defects particular to the illness, leads to a disruption in the capability of synapses to alter and adapt, which in turn destabilizes networks of neurons [4,5].

6.2 INFLAMMATION

When faced with dangerous stimuli like infections, damaged cells, or irritants, inflammation is an essential defense mechanism. Immune cells coordinate this process by generating a variety of soluble agents, as cytokines, chemokines, and free radicals. Nevertheless, an uncontrolled or disproportionate reaction might jeopardize the integrity of the tissue. Certain mediators have the capacity to move across the blood-brain barrier and disturb the functioning of the CNS by directly interacting with neurons and glia, such as microglia, that are the immune cells found within brain tissue. Existing investigation has centered on the harmful effects of immune system activation in the brain, demonstrating connections between neuroinflammation and several clinical situations, such as acute and chronic neurodegenerative disorders (NDs) [6].

In recent times, there has been a correlation shown between mediators that stimulate inflammation and the onset and advancement of severe depression and other psychiatric diseases in persons who are sensitive to them. Excessive inflammatory mediators are thought to have a neurotoxic effect on brain function and can negatively impair neural plasticity, which refers to the ability of the CNS to experience structural and functional alterations [7]. Increased concentrations of inflammatory chemicals have been found to reduce many indicators of brain plasticity, including LTP, levels of brain-derived neurotrophic factor (BDNF), and the process of neurogenesis. In contrast, a decrease in the natural formation of well-known inflammatory signals, like prostaglandins and cytokines, has a harmful consequence on synaptic plasticity and LTP. Endogenous baseline levels of prostaglandin E 2 (PGE 2) have a regulatory effect on membrane excitability, synaptic transmission, and plasticity in hippocampal cornu ammonis (CA1) pyramidal neurons. This acts as a vital role in the processes of memory acquisition, consolidation, and retention [8].

Additionally, cytokines as interleukin (IL)-1β, IL-1 receptor antagonist (IL-1RA), and tumor growth factor (TGF)-β all take part in the processes of synaptic remodeling and cognition restoration. According to this developing theory, inflammatory mediators and brain immune cells might critically take part in regulating neuronal remodeling, which includes processes such as the growth and removal of dendrites and axons, the development of synapses and related structures, cell death, and the generation of new neurons. These processes produce the remnants of nerve fibers, which are then eliminated by the immune system to aid in the rebuilding of brain tissue and preserve the balance of neurons. Nevertheless, this maintenance procedure is only efficient when microglial activation and immune mediator production are within a specific physiological range. Excessively high or low levels of these molecules might hinder neuronal functioning and disturb the normal role of microglia in brain function [9,10].

6.3 NEURONAL INTERPLAY: DYNAMICS OF SYNAPTIC COMMUNICATION

Using chemical neurotransmitters, synapses function as the main sites for quick communication among neurons and, on occasion, between neurons and astrocytes or microglia. Synaptic plasticity is known as the process in which the strength of synapses may be either increased or decreased. The length of these changes may be temporary (seconds to minutes) or longer lasting (hours to years), referred to as short-term or long-term plasticity, respectively. These factors describe several types of plasticity, including LTP, long-term depression (LTD), short-term potentiation, and short-term depression. Synaptic plasticity affects the function or presence of membrane proteins by mechanisms, including endocytosis, exocytosis, degradation, or synthesis. These activities are regulated by many intracellular signaling pathways [11–13].

In addition, neurons can primarily communicate in volume via diffusing peptide compounds like neuropeptides, cytokines, and growth factors. Neuronal communication occurs inside and between several cell types in the CNS, involving neurons, astrocytes, microglia, oligodendrocytes, endothelial cells, and immune cells. These peptides have a slower degradation or capture rate compared to smaller transmitters. As a result, they can spread further from their release point and activate intracellular signaling pathways when they reach their target. This has the potential to modify synaptic functioning [14–16].

6.3.1 Cell Signaling at Synapses

Synaptic events or the diffusion of substances can trigger the initiation of intracellular signaling pathways at synapses. This can lead to specific adjustments in synaptic strength and more widespread alterations in neuronal function, which involve modulation in gene expression. The repercussions of this signaling may be restricted to a localized effect inside or in close proximity to the activated synapse, or it may entail interactions between the soma (cell body) and synapses. Kinase cascades, such as calcium–calmodulin-dependent protein kinase II during long-term potentiation (LTP), exert targeted effects on altering the synaptic cytoskeleton, regulating the movement of α-amino-3-hydroxy-5-methyl-4-isoxazolepropionic acid receptors (AMPAR), and promoting localized protein synthesis. On the other hand, the protein kinase-A (PKA) cascade has the ability to influence the characteristics of glutamatergic receptors in a specific area and control the process of transcription and translation in the cell body [17,18]. There is communication between this series of events during both LTP and LTD. In addition to glutamatergic receptors, synapses also possess a variety of additional receptors, including neuropeptide, cytokine, and growth factor receptors. These receptors have the potential to enable a wide range of regulation of synaptic remodeling and other activities through the action of neuromodulators [19].

6.3.2 Immune System Signaling

The immune system defends the body by external infections and restores balance after tissue injury. The immune systems, both innate and adaptive, are differentiated by their ability to respond rapidly and non-specifically, or slowly and specifically, respectively. They identify patterns of molecules linked with damage or pathogens, known as damage-related molecular patterns and pathogen-associated molecular patterns [20]. Stimulation of toll-like receptors results in the formation of cytokines and inflammatory chemicals via pathways such as nuclear factor-kappa B (NF-kB), mitogen-activated protein kinases (MAPK), and caspase-1 [21]. The secreted substances attract immune cells and stimulate multiple receptors, triggering several signaling pathways that regulate genes controlling cell growth or programmed cell death. Cytokine receptors stimulate the Janus kinase/signal transducers and activators of transcription (JAK/STAT) pathway, chemokines induce G protein-coupled receptors (GPCRs) that regulate the PK-A/cAMP response element-binding protein (CREB) pathway, and growth factors activate phosphatidylinositol 3-kinase (PI3K)/protein kinase B (AKT) as well as the Ras/Raf/mitogen-activated protein kinase (MEK)/extracellular signal-regulated kinase (ERK) pathway. Significant communication occurs between these channels during the process of inflammation and repair [22].

6.3.3 The Central Nervous System and Its Immune System

Traditionally known as immune-privileged, the brain's degree of immunological privilege differs among its compartments. Within a healthy brain, there are circulating immune cells present in the subarachnoid space, cerebrospinal fluid (CSF), and vasculature. Furthermore, the brain parenchyma comprises tissue-resident macrophages called microglia [23]. Previously considered inactive, microglia are now acknowledged as actively monitoring the brain tissue, engaging with synapses, and reacting to neuronal activity. Astrocytes, the main glial cells in the CNS, undergo activation in response to brain damage, resulting in the formation of reactive astrogliosis. This procedure isolates the affected region but hinders the regrowth of nerve fibers. Astrocytes also regulate the passage of peripheral immune cells into the brain tissue. When neurons are injured, they produce substances that stimulate different cells in the CNS that are occupied in neuroinflammation. The connections among these cell types encompass extracellular communication and direct contacts, which influence synaptic activity and alter synaptic function [24].

6.4 SYNAPTO-IMMUNOLOGICAL MECHANISMS IN PHYSIOLOGICAL PROCESSES

Evidence shows that immunological signals play important roles in normal synaptic functioning, contributing to synaptic plasticity and stability. Originally they were thought to be engaged only in inflammatory processes.

6.4.1 Neuroplasticity at the Synapse

The majority of synaptic plasticity takes place at glutamatergic synapses, which rely on both ionotropic receptors (GluN) and metabotropic receptors (mGlu) for transmission. Transporters located on neurons and astrocytes play a crucial role in controlling the activation of glutamate receptors. Post-translational modifications have a substantial influence on these receptors, especially in terms of protein/protein interactions and phosphorylation [25]. Cytokines, including IL-1β, IL-2, IL-6, IL-8, IL-18, IFNα, IFNγ, and tumor necrosis factor (TNF)-α, are essential in regulating the initiation of synaptic plasticity.

On the other hand, the JAK2/STAT3 pathway has to be activated to produce GluN-dependent LTD in the hippocampus during normal physiological circumstances. Cytokine signaling exhibits metaplasticity and is an integral component of the physiological system [26]. The IL-1β receptor directly interacts with GluN receptors, controlling GluN function, whereas TNF-α affects GluA-dependent function in hippocampal neurons and might cause GluA internalization in striatal Gamma-aminobutyric acid (GABA)-energic neurons. MHC class I molecules and the complement system, namely complement component-C3, can directly influence glutamatergic synaptic transmission. The interactions described demonstrate the intricate manner in which cytokines and immune-related components modify synapse function [27].

6.4.2 Synaptic Scaling

Synaptic scaling refers to the procedure by which the stability of synaptic connections between neurons is adjusted to keep a stable level of neuronal activity. Synaptic scaling refers to the homeostatic processes that are activated in glutamatergic synapses in response to protracted periods of increased or decreased neuronal activity. This mechanism regulates the strength of all synapses in relation to bring the activity back to an ideal level [28]. Prolonged inhibition of neuronal action potential impulse or glutamatergic synaptic transduction for a period of 24 to 48 hours significantly enhances the size of miniature excitatory postsynaptic potentials (mEPSCs) and the presence of AMPARs on the cell surface. TNFα, which is released by glial cells, has a vital function in increasing the number of postsynaptic AMPARs (α-amino-3-hydroxy-5-methyl-4-isoxazolepropionic acid receptors). Additionally, β3 integrins have been recognized as essential for this particular activity. The study found that glial cells release TNFα to counter the decreased network activity. This release of TNFα results in a higher expression of β3 integrin on the surface of cells, which in turn leads to the accumulation of AMPARs at the synapse [29]. Kaneko and colleagues examined the visual organization to evaluate the physiological consequence of TNFα-induced synaptic scaling. In animals lacking TNFα, LTP in the striate cortex was unaffected, but there was a lack of increase in AMPAR-mediated miniature excitatory postsynaptic currents (mEPSCs). When studying live organisms, it was shown that mice lacking the TNFα gene showed a decrease in the ability to change the dominance of one eye over the other in response to covering one eye, particularly resulting in a lack of increased brain activity in the non-covered eye. This discovery indicates that synaptic-immunological variables play a critical role in both stages of ocular dominance remodeling. The complement cascade is responsible for the decrease in cortical responsiveness to the deprived eye, whereas TNFα is responsible for the higher cortical responsiveness to the non-deficient eye as a compensatory mechanism [30].

6.4.3 Structural Plasticity

Organisms possess an abundance of synapses throughout the CNS from birth. As the organism develops, synaptic pruning, which is influenced by experiences, eliminates unnecessary connections. Extensive research has been conducted on complement-mediated synaptic pruning in the visual system. The traditional complement cascade plays a crucial role in eliminating retinal ganglion cell inputs onto lateral geniculate nucleus neurons [31,32]. Neuronal activity regulates the process of complement-mediated synaptic pruning, indicating its role in synaptic plasticity. Synaptic pruning is also influenced by immunological pathways, such as purinergic signaling and MHC class I proteins. Research has shown that in the growing hippocampus, the process of microglia-mediated synaptic pruning occurs. This process comprises the elimination of synapses, which subsequently allows for the strengthening of the synapses that remain. Evidence supports the idea that this synapse removal is important for the future strengthening of the remaining synapses [33,34]. Microglia activation in the adult brain causes the displacement of inhibitory synapses and plays a role in restructuring neural networks by removing synapses. Synaptic pruning can also happen without glial involvement, being regulated by substances produced by astrocytes and microglia. Although microglia have traditionally been linked to synapse removal, current research suggests that they also play a role in spine creation, affecting changes in both directions of dendritic spine shape [35–37].

6.5 MECHANISMS OF SYNAPTO-IMMUNOLOGY IN ACUTE BRAIN DISORDERS

Multiple factors contribute to brain illnesses associated with neuroinflammation and disrupted synaptic function. Certain illnesses may manifest after a short-lived infection, whilst others have uncertain etiology or hereditary predisposition. The processes that modify synapse function are contingent upon the specific nature of the underlying cause.

6.5.1 Acute Infection

An immunological response to systemic or direct brain infections is coordinated by the immune system and the brain. Brain and spinal cord infections can occur as a result of several reasons, including traumatic damage, parasites, intoxication, and systemic infectious disorders caused by viruses, bacteria, fungi, and parasites [38,39]. All kinds of brain cells and processes can be affected by a widespread inflammatory response that is brought on by an overreaction of the immune system in sepsis. Pathogenic infections, such as influenza, HIV, herpes, and West Nile virus (WNV), have a direct or indirect effect on synaptic functioning, resulting in cognitive impairment. Research suggests that synapse loss after WNV infection is caused by the activation of the classical complement cascade in the hippocampus [40]. Viral proteins also impact the activity of voltage-dependent calcium channels and the functioning of GluA. Certain viral effects need the activation or modification of the N-methyl-D-aspartate receptor (NMDAR) via PKA and PK-C-dependent pathways. Furthermore, several defects in synapses caused by viruses are characterized by the production or disruption of proteins linked to synaptic impairment, such as amyloid beta (Aβ) or amyloid precursor protein (APP) [41].

6.5.2 Acute Brain Injuries without Pathogens

Sterile inflammation frequently occurs alongside acute brain injuries, which are caused by events such as mechanical trauma, ischemia, stress, or alcohol intake. Traumatic brain traumas (TBI), epilepsy, and stroke are all caused by sterile traumas and have similar pathways that lead to cell death [42]. These conditions are considered fundamental diseases. There is a connection between inflammation caused by injuries that are not infected and the worsening of diseases and the healing process [41]. The modification of synapse functions during sterile inflammation may vary depending on the individual disease, and the complex structure of synapses, which includes the postsynaptic density (PSD) and adhesion molecules, plays a crucial role in the spread of mechanical stress during TBI. Inflammation directly affects epilepsy by causing changes in protein expression or synaptic channel activity, which in turn leads to synapse deletion, sprouting, and alterations in synaptic strength [43,44]. Computational network modeling suggests that the production of TNFα by glial cells during inflammation can cause epileptogenesis by increasing the size of synapses. The impact of sterile inflammation on synaptic function may potentially be influenced by alterations in the neuronal environment during disease [45,46]. For instance, TNFα promotes LTP in the presence of reduced blood flow (ischemia), but inhibits LTP under normal physiological conditions. Excessive use of alcohol, especially in the form of binge drinking, is a major contributor to brain inflammation. Research has demonstrated that alcohol has an influence on glutamate receptors and synaptic plasticity, which has deleterious effects on brain growth, synaptic refinement, and cognitive functioning [47].

6.5.3 Dysregulation of Synapto-Immunological Processes in Neurodegenerative and Autoimmune Brain Disorders

There is a growing reconsideration of the conventional distinction between inflammation and neurodegeneration, since data indicates a complex relationship in neurodegenerative illnesses such as AD, PD and typical neuroinflammatory disorders like multiple sclerosis (MS) [48]. Empirical data indicates that synapses are the specific site where problems occur in these disorders. Disruptions in the initiation, continuation, or reversal of LTP and LTD are common characteristics in several models of brain diseases and human illnesses linked to inflammation [49]. Every illness has unique processes that impact the structure and function of synapses. The varied etiopathogenesis of various illnesses is likely caused by a combination of aberrant expression of immune mediators and disease-specific characteristics [50].

6.6 SYNAPTIC PLASTICITY MODULATION BY IMMUNOLOGICAL MECHANISMS

6.6.1 Cytokines and Synaptic Plasticity

In addition to being rely involve in both innate and adaptive immune responses and the regulation of inflammatory processes, cytokines are regulatory proteins that are released by different types of cells in the body. Although cytokines were formerly thought to only modulate the immune system, they are now recognized as active neuromodulators in the adult neurological system [51]. The vast array of identified cytokine genes and proteins, numbering in the hundreds, has facilitated the creation of therapeutic substances such as interferon (IFN)-b, which imitate the structure and function of cytokines [52]. The production of cytokines following a CNS injury is vital in determining the development of future neuronal diseases. Current research focuses on investigating the pro- and anti-inflammatory effects of cytokines. It defends the idea that pro-inflammatory cytokines have a disadvantageous influence on the function, survival, and adaptability of neurons [53].

Experimental data indicates that pro- and anti-inflammatory cytokines have important physiological functions in regulating communication between glial and neuronal cells and influencing synaptic plasticity. These cytokines are produced at reduced levels in the undamaged brain. Two thoughts have been suggested to resolve the apparent contradictory impacts of cytokines on the existence and adaptability of neurons [54]. The downstream properties of cytokines on remodeling and neuronal survival may be influenced by the synaptic concentration of these immune molecules [55]. These mediators are likely crucial for initiating and sustaining neuroplasticity at minimal physiological levels. On the other hand, in the case of neuroinflammation, the excessive expression of these genes might result in neurodegeneration and a decline in the ability of synapses to adapt and change [56]. Furthermore, the overall synaptic and neuronal effects of cytokines may be determined via equilibrium between pro-inflammatory and anti-inflammatory molecules [57]. Research shows that the decline in hippocampus LTP associated with aging is linked to reduced levels of the anti-inflammatory cytokine IL-(interleukin) 4 and elevated levels of the pro-inflammatory cytokine IL-1b. The next sections examine the data that supports the important role of cytokines in synaptic plasticity, with a particular emphasis on the processes that underlie the impact of IL-1 on the development of LTP [58].

6.6.1.1 Interleukin-1

IL-1, a pro-inflammatory cytokine, has been intensively researched in the CNS. IL-1, first known for its ability to cause fever, is formed by two ligands, IL-1a and IL-1b, that are associated with the type-I IL-1 receptor located on the cell membrane [59]. Although the IL-1 family members are naturally present within brain at diminished levels, their expression significantly increases once an injury occurs [60]. IL-1 expression is resulted by different cell types, such as monocytes, macrophages, astrocytes, glial cells, neurons, and vascular endothelial cells. Over the last twenty years, there has been evidence suggesting that this cytokine, which promotes inflammation, is involved in regulating long-term alterations in the connections between neurons. Initial research shown that IL-1b had a suppressive influence on the development of LTP in the hippocampus of mice [61]. The specific methods by which IL-1 affects synaptic plasticity are still a subject of debate, due to its many activities in neurons and glia, which frequently have conflicting effects.

The expression of the IL-1b gene significantly rises during LTP, and inhibiting IL-1 receptors hinders the maintenance of LTP, indicating a physiological role in synaptic plasticity. Research conducted on mice lacking IL-1 receptors demonstrated difficulties in memory tests that rely on the hippocampus, as well as the lack of LTP in the dentate gyrus. These outcomes showed more indication for the role of IL-1 in synaptic plasticity [62]. Induction of the IL-1 receptor triggers the conventional signaling pathway of IL-1, resulting in the initiation of MAPKs and NF-kB. As a consequence, this causes the secretion of additional pro-inflammatory substances as PGE2, IL-6, and NO [63].

IL-1 affects important molecular processes that are required for the initiation of synaptic plasticity, as the signaling of N-methyl-D-aspartate (NMDA) and α-amino-3-hydroxy-5-methyl-isoxazole-4-propionic acid (AMPA) glutamate receptors, the release of glutamate, the levels of calcium inside cells, the modulation of calcium channel currents, and cholinergic signaling [64]. The IL-1 receptor activation initiates the MAPK cascade, which is crucial for regulating synaptic plasticity and impacting transcriptional processes. Mediators such as NO and PGE2 play an active role in regulating brain synaptic plasticity.

The adverse consequence of IL-1b on LTP is linked to heightened action of c-Jun N-terminal kinase (JNK) and p38 MAPK, indicating their role in impaired LTP [65]. Importantly, the use of a JNK inhibitor effectively counteracts the IL-1b-induced decreased LTP [66]. IL-10, an anti-inflammatory cytokine, might counteract the impact of IL-1b by reversing the activation of JNK activity and the suppression of LTP. The MAPK family, particularly p38 and JNK, was identified as possible mediators of the IL-1b-induced suppression of LTP [67]. Under times of stress or as one ages, increased levels of IL-1b are linked with a decline in LTP in the hippocampus. This highlights the dual function of IL-1, where it is essential for LTP under normal circumstances but hinders it when present in excessive amounts under pathological conditions or aging [68]. Research on older individuals has shown that genetic variation in the IL1-b-converting enzyme (ICE) gene, which leads to lower IL-1b production, has a positive impact on cognitive function. This advises that having reduced levels of IL-1b can help protect against memory and learning problems in older people [69]. This underscores the potential therapeutic importance of suppressing ICE to preserve cognitive function in older individuals [70].

6.6.1.2 Tumor Necrosis Factor-Alpha (TNF-α)

TNF, a highly inflammatory cytokine, has a crucial function in facilitating communication between the immune and neurological systems, affecting the strength of synapses regulated by glial cells. More precisely, TNF-α raises the existence of AMPA glutamate receptors on the surface of cells, which in turn boosts synaptic effectiveness and has a role in specific types of homeostatic plasticity, such as synaptic scaling, when there is sustained activity blocking [71]. Nevertheless, there is ongoing dispute over the influence of TNF-α on the synaptic capacity to undergo rapid, activity-dependent phenomena such as LTP or LTD. Both LTP and LTD are exhibited in hippocampal synapses following TNF-α incubation. This phenomenon is also observed in hippocampal slices from mice that do not have TNF-α or its receptors. Both findings imply that TNF-α may not be a crucial mediator for both processes [72]. However, several researchers suggest that TNF-α may have a suppressive impact on the initiation of LTD and LTP. This impact is probably promoted by the group I metabotropic glutamate receptor and a process that relies on P38 MAPK [73]. TNF-α is thought to have a vital function in the physiological interaction between neurons and astrocytes.

Remarkably, the alteration of glia-neuron communication by TNF-α involves the participation of PGE2, a chemical that has already been demonstrated to play a vital role in synaptic plasticity. When stromal-derived factor-1 binds to its receptor C-X-C motif chemokine receptor 4 (CXCR4) in astrocytes, it induces a sequence of events inside the cell that ultimately results in the secretion of soluble TNF-α. Afterwards, TNF-α binds to its receptors on the surface of cells, triggering internal signaling processes that lead to the production of PGE2, which is reliant on the presence of calcium, and the release of glutamate from astrocytes, which is also reliant on PGE2 [74].

6.6.1.3 Additional Cytokines

IL-18, a cytokine that promotes inflammation and is part of the IL-1 family, seems to impede the initiation of LTP in the hippocampus. This is likely due to its negative modulation of NMDR-mediated transmission [75]. There is a hypothesis that suggests metabotropic glutamate receptors and downstream inflammatory mediators, such as cyclo-oxygenase 2 (COX-2) and inducible NO synthase, could have a role in mediating the inhibitory impact of IL-18 on the development of LTP [76].

Regarding LTP, there is evidence indicating that IL-6 may have a role in inhibiting the alteration of this kind of synaptic plasticity [77]. Suppressing the natural production of IL-6 results in a large increase in the duration of LTP and a major development in long-term memory [78]. IL-6 is believed to hinder the initiation of LTP via influencing the MAPK-ERK pathway [79]. It is worth mentioning that some substances found in the blood, specifically IL-6, are linked to a deterioration in cognitive function in older individuals who are otherwise healthy [80]. This supports the idea that abnormal levels of cytokines might have a negative impact on the brain's capability to adapt and change.

The impact of applying exogenous IFN on hippocampus function have been investigated. IFN reduces the strength of NMDA-induced currents, inhibits the initiation of LTP, and weakly reduces the enhancement of synaptic strength during the maintenance phase of LTP [81]. In addition, IL-2 affects LTP by inhibiting the formation of both post-tetanic potentiation and LTP in hippocampal slices [82].

6.6.2 The Main Histocompatibility Complexes

The MHC plays a crucial function in the immune system by facilitating the display of antigens to T cells, which in turn triggers and sustains the immunological response. Neurons also express class I MHC molecules, which are controlled by both spontaneous and evoked neuronal activity. In adult mice with a genetic deficiency in either cell surface class I MHC molecules or a class I MHC receptor component called cluster of differentiation (CD)-3 zeta, there is an increase in NMDAR dependent LTP in the hippocampus, and LTD is not observed [83].

6.6.3 Other Inflammatory Mediators: NO and COX Products

These findings suggest that this immune molecule has a function in controlling neuroplasticity. Specifically, there is a suggestion that MHC located in the post-synaptic region plays a crucial role in maintaining the balance of synaptic function and structure. It is believed that MHC acts in a retrograde fashion, transmitting signals across the synapse to bring about persistent changes in the structure of the synapse based on its activity [84]. As mentioned earlier, the inflammatory insult leads to the release of NO and COX products [85]. This article does not provide an in-depth explanation of the specific functions of these "transmitters" in synaptic remodeling. However, it is crucial to note that the involvement of NO and COX products, such as PGE2, in neurotransmission and plasticity has been proven [86].

6.7 APPROACHES IN HARNESSING NEUROPLASTICITY AGAINST NEUROLOGICAL DISORDERS

6.7.1 Molecular Compounds and Targets

6.7.1.1 For Induction of Synaptogenesis, Neurogenesis, and Repair

Neurodegeneration and other disorders like depression and dementia are associated with synaptic dysfunction. Compounds that promote the growth of new synapses following substantial loss of neurons have the potential to fill the gaps caused by neurodegenerative processes [87]. Quercetin and bilobalide, which are produced from Ginkgo biloba, have been shown to have the ability to induce the growth of synapses and neurons in the hippocampus. They also increase the synthesis of BDNF, a protein that supports the development and existence of neurons. These effects have been shown both in living organisms and in laboratory settings and are likely achieved via regulating the CREB pathway [88]. The administration of U0126 and PD169316, inhibitors of the MAPK/ERK pathway, resulted in the phosphorylation of CREB and led to enhanced cognitive and memory functions in rats injected with Aβ. This suggests a mutually beneficial interaction between the preservation of neurons and the formation of synapses. The study performed in a 5XFAD mice model provided more evidence to establish the association between neuroprotective properties and synapse regeneration. It demonstrated a decrease in synapse regeneration in the RNA-binding motif (RBM) 3-knockdown model, highlighting the importance of the RNA binding protein RBM3 [89]. Additional constituents of Ginkgo biloba, such as ginkgolide J, have been investigated for their potential to restore the impairment of LTP caused by Aβ. Nevertheless, they do not regulate the process of neurogenesis, indicating the existence of separate pathways involved in the restoration of LTP [90]. Manipulating proteins involved in the movement and direction of neurons, such as doublecortin (DCX), has repeatedly shown positive consequences in stimulating the formation of connections between neurons and the generation of new neurons in cases of TBI when these proteins are overproduced [90]. BDNF, a protein that has been widely researched, contributes to the control of neurogenesis in a dose-dependent manner through many pathways. These pathways include tropomyosin-related kinase B linked with phospholipase Cγ (PLCγ), PI3K, Akt, Ras, and Etv1 [91]. The antidepressant spadin specifically inhibits the activity of the potassium channel protein TREK-1, which is crucial for signal transmission. This inhibition leads to the beginning of neurogenesis through the CREB pathway [92]. Other antidepressants, such as sertraline, have been studied for their potential to promote the growth of new brain cells, not only for treating depression but also for raising the production of immature and mature neurons through the activation of glucocorticoid receptors. Fluoxetine, an SSRI, has been found to induce the formation of neurons and increases the ability of hippocampus granule cells to change and adapt their connections with other cells. This effect is particularly noticeable in mouse models that have been exposed to chronic corticosterone [93].

Docosahexaenoic acid (DHA), uridine, and choline, which are precursors of synaptic membranes, have been shown to enhance the development of synapses and the growth of new neurons in instances of neurodegenerative illnesses and mechanical injuries. These effects have been observed even when these substances are taken together with the health food Souvenaid [94,95]. DHA, more precisely, stimulates the synthesis of synapsin in the development of nerve fibers while augumenting the activity of p-Akt and lowering the activity of phosphatase and tensin homolog (PTEN), which are important factors in the ability of synapses to change and adapt [96]. Uridine has been found to counterbalance the lack of P2Y receptors exhibited in individuals with AD by enhancing the formation of neurites, in addition to facilitating the differentiation of neurons. Epigenetic modulation can influence molecular targets that show promise for synaptogenesis. For example, the reduction of histone deacetylase 2 (HDAC2), an enzyme that suppresses gene expression inside synapses by involving proteins like Sp3, is a potential approach. By inhibiting HDAC2 in mouse models, the activation of genes that promote synaptogenesis and restore long-term memory creation was shown [97]. A randomized, double-blind experiment was conducted using valproate (VPA), which is an HDAC inhibitor. The outcomes displayed that the patients in the trial had better control over perfect pitch compared to the control group. This highlights the important function of HDAC in synaptogenesis [98]. Additional epigenetic methods that involve manipulating microRNA, such as lin-4 and miR-181c, were for their potential use in treating schizophrenia patients [99–101] (Table 6.1).

6.7.1.2 For Reduction and Inhibition of Toxic Protein Aggregates

Demolition of neurons and synapses is firmly facilitated by the cumulation of toxic protein aggregates, which have been described to target neural cell adhesion molecule 2 in AD models [102]. An additional strategy for enhancing neuroplasticity involves depleting these toxic proteins that disrupt synaptic signal relays. While the accumulation of protein aggregates is not their natural etiology, it may be induced in cases of TBI or acquired over time in conditions like epilepsy or ataxia, albeit not at the same level as in NDs [103,104]. These findings suggest possibilities for ameliorating toxic protein aggregates beyond the scope of treating AD or PD. Comprehensive reviews have focused on the antioxidative capabilities of naturally occurring substances, which have received considerable attention. Squalamine, known for its ability to enhance neuroprotective effects in a Caenorhabditis elegans animal model of PD, was subjected to both in vitro and in vivo experiments. These studies demonstrated an important reduction in the incidence and harmfulness of tau protein collections in PD models [105]. This impact was ascribed to the competitive suppression of toxic aggregates on neural structures. Bilobalide, extracted from Ginkgo biloba, was discovered to regulate the PI3K pathway, resulting in a decrease in APP and Aβ [106]. Oral administration of epigallocatechin-3-gallate (EGCG) from Camelia sinensis and Ceratonia siliqua inhibited the ERK and NF-kB pathways. This led to a decrease in Aβ and tau levels in mouse models of AD and subsequently improved cognitive abilities. Curcumin, the primary component found in Curcuma longa, has shown significant effectiveness in improving Aβ aggregates at a concentration of one μM, surpassing the inhibitory effects of naproxen and ibuprofen. Additionally, it concurrently reduced the

TABLE 6.1
Compounds and Mechanisms for Harnessing Neuroplasticity

	Natural Products
Azaphilone	In-vitro reduction of tau polymer assembly via microtubule disruption.
Bilobalide	Amplified CREB phosphorylation, BDNF, neurogenesis, synaptogenesis, and prohibited Aβ-stimulated synaptic loss.
	Decrease of APP and Aβ burden by alteration of PI3K pathway.
Curcumin	Direct attachment to Aβ to stop establishment in vitro and in vivo while cumulative IL-4 and IL-2 aside from reducing tau.
	Noteworthy decrease of LTD, lipid peroxide, lipofuscin, and COX-2, while decreased BDNF, SOD, and GPx in vivo.
EGCG	Decreased β- and γ-secretases activity together with APP and Aβ improved memory activity in vivo.
	Diminished formation of soluble and insoluble toxic Aβ alongside sarkosyl-soluble phosphorylated tau in vivo.
	Marked decreased pro-inflammatory cytokines and dampen LPS-induced neurotoxicity by neuroprotection.
Ginkgolide J	Encourages LTP rescue in CA1 hippocampal area and defends culture from Aβ-induced neuronal loss/death.
Ginsenoside	Diminished harmful Aβ through the regulation of BACE1, β-CTF, PPARγ translocation, and α-synuclein inhibition.
	Raised astrocyte Aβ clearance through autophagy through mTOR phosphorylation suppression.
Resveratrol	It is being found that these anti-neuroinflammatory benefits are achieved by raising the levels of IL-4 and FGF-2, in addition to reducing the hazardous levels of Aβ in the CSF.
	There was a significant decrease in the harmful effects of Aβ, but no alteration in the speed at which Aβ forms oligomers.
Squalamine	In vitro, it effectively displaces α-synuclein inside carrier vesicles, and in vivo, it lessens paralysis.
	Drugs
Edaravone	The process of salvaging and activating SOD and Nrf2 in a living organism to decrease the accumulation of tau protein and improve cognitive impairment.
Spadin	Inhibition of the K+ channel protein TREK-1 and upregulation of the CREB pathway to promote neurogenesis and synaptogenesis.
Sertraline	The expression of CDKi genes, DCX+ neuroblasts, and neurogenesis is boosted in a glucocorticoid-dependent manner.
Fluoxetine	Chronic fluoxetine administration leads to an increase in the branching of dendrites in DCX+ neuroblasts and enhances LTP.
	Epigenetic Modulation
HDAC2	The restoration of GLUR1, BDNF, and other associated pro-synaptic genes and proteins was made possible by the gene's deficiency.
lin-4	Through suppressing LIN-14, it leads to a higher level of synaptogenesis compared to branch development.
miR-181c	The downregulation of this microRNA alters the expression of genes and signaling pathways that enhance the formation of neurites and synapses.
miR-214	The upregulation of this microRNA enhances the development of dendrites and the development of synapses by selectively regulating the Qki gene.
BDNF	Neurogenesis can be induced in a dose-dependent manner through the activation of PLCγ, PI3K, Akt, Ras, and Etv1 signaling pathways.
DHA	Elevated levels of synapsin in the neurite, enhanced activation of p-Akt, reduced expression of PTEN, and modulation of synaptic plasticity.
DCX	Neural progenitor cells that express this chemical are necessary for the process of injury-induced remodeling in the hippocampus.
IGF-II	Neuronal cultures are protected against long-term oxidative stress by the action of the IGFII/M6P receptor, which provides neuroprotection.

(Continued)

TABLE 6.1 *(Continued)*
Compounds and Mechanisms for Harnessing Neuroplasticity

	Natural Products
TGF-β	Administered an additional NMDAR ligand, D-serine, which stimulated the growth of synapses in cerebral cortical neurons.
AAV8	Overexpression caused by induction led to the recovery of miniature mEPSC and considerable clearance of amyloid-beta (Aβ).
Uridine	Lack of P2Y receptors in AD is often accompanied by enhanced development of neurites and the development of synapses.
VEGF	Neural stem cells that express VEGF receptor-3 are stimulated, leading to formation of new neurons inside the subventricular zone in a living organism.
Wnt	Neurogenesis in dentate gyrus of a neuroinflammatory in vivo model through different cytokines.
Lentiviral vector	Wnt3a-HA expression improved the healing process in an in vivo model of localized ischemia injury.
Immunization	The user's text is "AADvac1". In vivo, there was a significant 95% decrease in pathogenic tau levels, as opposed to healthy tau. Additionally, the phase I trial, which included 30 participants, demonstrated the safety of the treatment.
ACI-35	The long-term vaccine demonstrated a favorable safety profile and effectively reduced tauopathy indicators and bodyweight loss in vivo.
RG7345	Chronic vaccinations, which may be shown in lysosomes through lipid rafts, decreases the presence of tau/pS422 disease.
43D; 7739	Administering two doses of tau reduction treatment resulted in a decrease in total tau levels, whereas administering six doses of the treatment reduced hyperphosphorylated tau levels. Additionally, this treatment led to an improvement in spatial and short-term memory.
Physical Exercise	1. Exercise at a high intensity promoted hippocampus neurogenesis. 2. Exercise at a moderate to low intensity improved the neuronal survival, development, and spatial memory of newborns. 3. In the same region where neurogenesis is discernible in vivo, there are elevated electrophysiological alterations. 4. In vivo mouse models that were wheeled for three weeks showed improvement in previously identified cognitive impairment. 5. Increased blood supply to the brain, which may cause a boost in hippocampus neurogenesis. 6. When compared to control patients, practitioners of Tai Chi exhibit a significant decrease in TNF-α and IL-6.
Mental Training	1. Meditation has been found to enhance grey matter density in the human brainstem, which is related to learning and memory. 2. There is a notable augmentation in the size of white matter and ACC among individuals who practice meditation. 3. Long-term meditation practitioners with insomnia exhibit significant increases in NF-kB and TNF-α levels. 4. HDAC2 silencing was aided through meditation practices. 5. The effects of meditation may vary depending on the duration of time individuals spend practicing meditation.

burden of phosphorylated tau in in vivo models [107]. Several bioactive substances, such as ginsenosides, azaphilones, and resveratrol, have demonstrated promising pre-clinical efficacy in treating neurodegenerative disorders [108,109]. Investigations have been conducted on immunotherapeutic methods that aim to eliminate both Aβ and tau proteins. ACI-35, which employs a liposome-based mechanism for administering phosphorylated segments of tau as a vaccine, showed encouraging levels of safety and effectiveness in both pre-clinical and clinical environments [110]. The tau vaccine, AADvac1, demonstrated a significant 95% decrease in pathogenic tau proteins in in vivo trials, while also retaining a noteworthy level of safety. The phase I clinical trials of AADvacl demonstrated positive results in terms of safety and effectiveness. Only a small number of participants withdrew from the trials because of major side effects related to injection site responses [111]. The utilization of passive vaccination techniques, specifically the implementation of 77E9 and 43D, has shown promise in improving both tauopathies and amyloid load in 3xTg-AD mice models. Roche conducted a significant passive vaccination study utilizing RG7345, which successfully cleared tau pathology in TauPS2APP mouse models. This was achieved by specifically targeting phosphorylated tau (pS422) and potentially allowing internalization through the interaction of the antibody epitope with Fcγ receptors. Nevertheless, like traditional drug research, the progress of immunotherapy in promoting neuroplasticity encounters obstacles in clinical environments, particularly concerning safety and toxicity concerns. This is because the effects shown in vivo may not always accurately reflect human responses [112] (Table 6.1).

6.7.1.3 For Neuroprotection and Anti-Neuroinflammation

Prolonged inflammation and oxidative stress can exert an impact on the degradation of synapses and neural structures. Substances that might provide neuroprotection or reduce neuroinflammation may provide an alternate method for sustaining neuroplasticity [113]. Natural products with antioxidant properties, known for their ability to protect and reduce inflammation, have applications beyond just the nervous system and can be used to battle neurodegeneration, depressive disorders, and autism [114]. The aforementioned natural compounds, such as curcumin, ginsenosides, and EGCG, have shown interconnected effects in enhancing neuroplasticity, specifically in terms of neuroprotection [115–117].

Edaravone, an injectable pharmaceutical with potent antioxidant properties, has been authorized for the management of stroke and amyotrophic lateral sclerosis. Recent research has demonstrated that edaravone can effectively replenish superoxide dismutase (SOD) levels and significantly reduce tau excessive phosphorylation [118]. It functions in conjunction with the antioxidative Nrf2 signaling pathway, resulting in neuroprotective advantages in several animal models. IGF-II possesses long-lasting antioxidative properties that assist in repairing neuronal damage in oxidative damage models induced by glucocorticoids [119]. IGF-II overexpression, achieved by the use of adeno-associated virus 8 (AAV8) vectors mitigates impairments in dendritic spine, mEPSC, and memory in Tg2576 AD animal models [120–122]. Individuals with AD exhibit decreased levels of IGF-II, which is linked to an increase in the buildup of tau protein. Furthermore, researchers have also examined other growth factors such as Wnt, the growth factor for vascular endothelial cells (VEGF), and TGF-β [123]. The continued use of estrogen therapy in postmenopausal women induces alterations in the synaptic connections between neurons and offers neuroprotective effects by decreasing inflammation in older female animal models and humans. This impact is accomplished by modulating certain proteins known as B-cell lymphocyte 2 (Bcl-2) or Bcl-2-associated X (BAX). This information is summarized in Table 6.1 and is supported by reference [124].

6.7.2 Psychological Activity-Dependent Therapy and Stimuli

Inducing neurogenesis, which is significantly correlated with increased memory and attention and is primarily documented in rats, is a crucial therapy option that comes with the reputation of exercise, which offers several health advantages [125]. Scientific evidence shows that high-intensity exercise promotes the growth of new brain cells in the hippocampus, whereas medium-low-intensity exercise improves the survival and development of these new cells, as well as spatial memory [126]. The electrophysiological alterations seen in brain areas triggered by neurogenesis in running mice suggest a connection between exercise-induced neurogenesis and the processes of learning and memory. The inhibition of hippocampus neurogenesis caused by focal irradiation results in cognitive impairment, which can be improved by engaging in three weeks of wheel running [127,128].

The process by which physical activity stimulates the growth of new nerve cells in the subgranular zone (SGZ) of the hippocampus involves increasing the availability of energy and the release of regulatory substances, which in turn enhances the development of new nerve cells in the hippocampus [129,130]. Physical exercise is being identified as a possible treatment for NDs such as AD and PD, which are related to age and linked to a decline in the formation of new neurons in the hippocampus. Physical activity rejuvenates the area of the brain responsible for generating new neurons, promoting better blood circulation and improving the growth of nerve cells in the hippocampus. Deep breathing activities, such as Tai Chi, have been proven to significantly reduce stress and inflammation. This is evidenced by the drop in TNF-α and IL-6 levels observed in individuals with insomnia [131].

The dentate gyrus granule cells (DGCs) in rats undergo a crucial period of maturation that lasts around 4–6 weeks. In humans, this period is somewhat longer. During this time, newly formed neurons are more sensitive to physical activity and show stronger synaptic plasticity compared to fully mature DGCs [132,133]. Comprehensive research is essential to investigate the use of exercise as a treatment for slowing brain aging, either as a primary or supplementary therapy for neurodegenerative illnesses. In addition to its practicality, physical exercise is a cost-effective, non-intrusive, technologically simple, and non-pharmaceutical intervention. Engaging in extended periods of meditation and consistent mindfulness exercises enhances the density of grey matter in the brain stem, which is related to enhancements in learning and memory. The anterior cingulate cortex (ACC), which is linked to meditation, contributes to the enlargement of white matter in those who practice meditation. This finding supports the notion that mental training may effectively restructure neural connections [134]. Meditation demonstrates anti-inflammatory effects by effectively reducing the levels of TNF-α and has the ability to reduce pro-inflammatory NF-kB levels. The impact of meditation, which involves the suppression of HDAC2, may vary based on the length of each individual's meditation sessions. Exercises frequently incorporate elements of attention and focus that are also present in meditation. This implies that both activities work together to enhance activity-dependent neuroplasticity, as indicated by research studies [135,136] (Table 6.1).

6.8 PATHOPHYSIOLOGICAL INSIGHTS: POTENTIAL SUGGESTIONS OF IMMUNE-MEDIATED CONTROL OF NEUROPLASTICITY DURING CNS DISEASES

6.8.1 Parkinson's Disease

It is still mostly unclear what precise processes in PD cause symptoms to arise. Nevertheless, impaired synaptic plasticity at basal ganglia neuronal circuits plays a pivotal role in various ideas that seek to elucidate the anomalies observed in the basal ganglia network throughout this ND [137].

Multiple types of synaptic plasticity, such as LTD, LTP, and synaptic depotentiation, have been identified at cortico-striatal synapses. These procedures are all negatively affected by the pathogenic mechanisms linked to PD [138]. The main reason for the decline in striatal LTP and LTD in PD seems to be the compromised control of synaptic remodeling by dopamine. The development of PD has been linked to immune mechanisms, namely the presence of elevated amounts of cytokines (such as TNF-a, IL-1, IL-6, IL-2, and the anti-inflammatory cytokine IL-4) in the striatum and CSF fluid of individuals with PD [139]. It is worth mentioning that the excessive production of cytokines (namely TNF-a, IL-1, and IL-6) in the striatum of individuals with PD is known to impede the initiation of normal synaptic remodeling, which is essential for healthy brain activity [140].

The activation of inflammatory procedures in PD, specifically the increase in MHC molecules in the striatum, can potentially affect the capability of striatal neurons to undergo long-lasting synaptic alterations due to dopamine deprivation. As a result, the normal procedure of neural impulses within the basal ganglia neuronal circuit is disrupted [140]. However, despite these discoveries, there is presently a lack of empirical proof substantiating the exact function of cytokines in striatal synaptic plasticity during diseased circumstances. Consequently, additional experimental inquiries are necessary to authenticate this theory.

Similar to how oligomeric amyloid β (oAβ) in AD has a synaptotoxic effect, extracellular alpha-synuclein oligomers also have an impact on synaptic transmission and hinder LTP [141]. Nevertheless, the reliability of the findings obtained from a study that utilizes alpha-synuclein oligomers at concentrations higher than normal physiological levels is uncertain. The accumulation of α-synuclein stimulates microglia to secrete TNFα and IL-1β, which may result in harmful effects on dopaminergic cells. The existence of amplified levels of TNF, IL-1, IL-6, IL-2, and elevated MHC molecules in the striatum and CSF of patients with PD provides evidence that the immune response contributes to the development of PD [142]. Significantly, administration of the flavonoid baicalein has demonstrated encouraging outcomes in reducing the overexpression of TNFα and IL-1β, as well as restoring normal striatal glutamatergic transmission in a rodent model of PD [143]. The chronic and ongoing inflammation in individuals with PD may hinder the normal ability of synapses in the corticostriatal region to adapt and change, which in turn disrupts the communication routes in the basal ganglia network of neurons and contributes to the symptoms of PD. Additional research is essential to clarify the exact function of cytokines in the synaptic remodeling of the striatum, both in normal and abnormal situations [144].

6.8.2 Alzheimer's Disease

The amyloid hypothesis suggests that the incorrect processing and rapid accumulation of oligomeric forms of the amyloid-β peptide (Aβ) are key factors that drive pathogenic processes in AD [145]. The concept proposes that AD, particularly in its early stage, can be described as a "synaptic failure," as anomalies in the effectiveness of connections between neurons in the hippocampus occur before noticeable neuronal loss. Aβ oligomers have a detrimental effect on hippocampus LTP and the memory of learnt behaviors [146,147]. Moreover, its direct harmful impact on Aβ stimulates microglial cells and astrocytes, leading to neuronal dysfunction and degeneration through similar pathways. The activation of microglia and astrocytes by Aβ causes localized acute-phase response mediated by cytokines, as well as the production of inflammatory enzymes such as iNOS and COX-2 [148]. The inflammatory response can be observed in the brains, CSF, and peripheral blood of patients with AD. This response is characterized by glial activation and increased expression of molecules such as IL-1, IL-6, and TNF-α. These molecules are related with immune cell induction and have a negative impact on the induction of LTP [149]. Molecules associated with inflammation, like as TNF-α, NO, and superoxide, are suggested to be responsible for Aβ's ability to hinder LTP. Targeted suppression of COX-2 can effectively counteract the negative impact of Aβ on hippocampus LTP. Thus, the reported early decline in hippocampus LTP in AD is probably a consequence of the presence of both amyloid-beta (Aβ) oligomers and ongoing inflammatory processes. AD is a long-term degenerative disorder characterized by the gradual death of nerve cells and a reduction in cognitive function. OAβ is thought to play a part in the progression of AD by interfering with the synaptic plasticity, which is the ability of synapses to undergo changes and adapt through various mechanisms [150]. Patients with AD exhibit inflammatory features, such as the stimulation of glial cells and the existence of messengers including IL-1, IL-6, and TNFα, in the brain, CSF, and peripheral blood. TNFα and IL-1β are important cytokines that have been identified as key players in transmitting the detrimental effects of oAβ on LTP [151]. Evidence obtained from epidemiological investigations and clinical trials suggests that non-steroidal anti-inflammatory drugs offer neuroprotection in AD, therefore supporting the role of inflammation in the formation of the disease. Research has shown that selective COX-2 inhibitors are successful in preventing the negative impact on LTP generated by synthetic soluble Aβ [152]. The involvement of microglia in the progression of AD is becoming increasingly worrisome. Recently research suggests that the lack of proper signaling to keep microglia in a dormant state contributes to the negative effects of synaptic activity. The stimulation of microglia caused by Aβ is related with the impairment of LTP, and the inhibition of microglia activation can avoid the inhibition of LTP initiation by Aβ. In addition, the decrease in synapses in AD is linked to the activity of microglia, which engulf and remove cellular debris. This indicates a pathological reactivation of a process called synaptic pruning, which normally occurs throughout brain development, in the initial phases of AD [153].

6.8.3 Multiple Sclerosis

Individuals with MS often have demyelinated regions located in white matter areas, which show signs of inflammatory infiltrates [154]. Elevated concentrations of inflammatory cytokines, inclusive of IL-1β, TNF-α, IL-6, IL-2, and IFN-γ, are observed in the blood plasma or serum, cerebrospinal fluid, cultured peripheral blood cells, and brain tissue of individuals with MS as compared to control people [155]. In addition, chemicals and enzymes associated with inflammation, including as MHC gene products, endothelial adhesion molecules, COX-2, and iNOS, are present in MS plaques [156,157]. MS, especially its relapsing-remitting type, is an intricate autoimmune illness. The presence of neuroinflammation in MS has a substantial effect on the ability of neuronal systems to exhibit normal plasticity. This can potentially result in a decreased ability to maintain homeostatic balance, leading to negative effects on cognitive performance [158,159]. In patients with MS, and in the case of EAE, a widely recognized rat model of MS, the process of inflammation initiates synaptic damage at an early stage. Recent research has shown that transcranial magnetic stimulation, namely employing intermittent theta burst stimulation or continuous theta burst stimulation, might affect the brain's cortex's capacity to modify and adjust during the initial phases of inflammation in MS. In persons with MS, there is a tendency for LTP to be favored over LTD in response to recurrent synaptic activation in the brain. This impact is directly related to the levels of IL-1β in the CSF. Similar findings were observed in hippocampus slices obtained from mice with EAE. Regarding this matter, enhancement of CA1-LTP was caused by increased release of IL-1β from infiltrating CD3+ T lymphocytes or triggered microglia in hippocampus affected by EAE. Significantly, interventions aimed at pro-inflammatory cytokines and oxidative stress effectively reversed synaptic changes in the EAE model [160].

6.9 CONCLUSION

In conclusion, the traditional view of the CNS as an immune-privileged site has been challenged by a growing body of evidence revealing a profound interplay between the nervous and immune systems. This emerging understanding highlights the active role of glial cells, once considered mere support cells, in mediating immune responses within the CNS and participating in synaptic remodeling and plasticity processes crucial for learning and memory formation. Significantly, immune mediators such as cytokines and MHC have been found to exert a profound influence on synaptic memory processes, an impact that has been largely underestimated until recently. This neuroimmune crosstalk and the hitherto unrecognized impact of immunological factors on synaptic remodeling may contribute to the pathophysiology observed in various neuroinflammatory brain disorders associated with cognitive decline, such as PD and AD. The interplay between the abnormal expression of inflammatory substances during CNS inflammation and disease-specific defects may disrupt the ability of synapses to adapt and remodel, destabilizing the intricate neuronal networks that underlie cognitive functions. This emerging understanding challenges the long-standing notion of the CNS as an immune-privileged site and opens new avenues for understanding and potentially treating neuroinflammatory and neurodegenerative diseases characterized by cognitive impairment and synaptic dysfunction. As our knowledge of the bidirectional communication between the nervous and immune systems continues to evolve, it holds the promise of shedding light on the complex mechanisms underlying neurological disorders and paving the way for novel therapeutic strategies targeting the neuroimmune interface.

REFERENCES

1. Norris GT, Kipnis J. Immune cells and CNS physiology: Microglia and beyond. J Exp Med. 2019;216(1): 60–70. doi: 10.1084/jem.20180199.
2. Martella G. Molecular mechanisms of synaptic plasticity: Dynamic changes in neuron functions. Int J Mol Sci. 2023;24(16):12567. doi: 10.3390/ijms241612567.
3. Matejuk A, Vandenbark AA, Offner H. Cross-talk of the CNS with immune cells and functions in health and disease. Front Neurol. 2021;12:672455. doi: 10.3389/fneur.2021.672455.
4. DiSabato DJ, Quan N, Godbout JP. Neuroinflammation: The devil is in the details. J Neurochem. 2016;139(Suppl 2):136–53. doi: 10.1111/jnc.13607.
5. Schwab N, Wiendl H. Learning CNS immunopathology from therapeutic interventions. Sci Transl Med. 2023;15(721):eadg7863. doi: 10.1126/scitranslmed.adg7863.
6. Sochocka M, Diniz BS, Leszek J. Inflammatory response in the CNS: Friend or Foe? Mol Neurobiol. 2017;54(10):8071–89. doi: 10.1007/s12035-016-0297-1.
7. Abdulkhaleq LA, Assi MA, Abdullah R, Zamri-Saad M, Taufiq-Yap YH, Hezmee MNM. The crucial roles of inflammatory mediators in inflammation: A review. Vet World. 2018;11(5):627–35. doi: 10.14202/vetworld.2018.627-635.
8. Hein AM, O'Banion MK. Neuroinflammation and memory: The role of prostaglandins. Mol Neurobiol. 2009;40(1):15–32. doi: 10.1007/s12035-009-8066-z.
9. Miller AH, Haroon E, Raison CL, Felger JC. Cytokine targets in the brain: Impact on neurotransmitters and neurocircuits. Depress Anxiety. 2013;30(4):297–306. doi: 10.1002/da.22084.
10. Morimoto K, Nakajima K. Role of the immune system in the development of the Central nervous system. Front Neurosci. 2019;13:916. doi: 10.3389/fnins.2019.00916.
11. Collingridge GL, et al. Long-term depression in the CNS. Nat Rev Neurosci. 2010;11(7):459–73.
12. Nicoll RA, Roche KW. Long-term potentiation: Peeling the onion. Neuropharmacology. 2013;74:18–22.
13. Feldman DE. The spike-timing dependence of plasticity. Neuron. 2012;75(4):556–71.
14. Borbely E, Scheich B, Helyes Z. Neuropeptides in learning and memory. Neuropeptides. 2013;47(6): 439–50.
15. van den Pol AN. Neuropeptide transmission in brain circuits. Neuron. 2012;76(1):98–115.
16. Casillas-Espinosa PM, Powell KL, O'Brien TJ. Regulators of synaptic transmission: Roles in the pathogenesis and treatment of epilepsy. Epilepsia. 2012;53(Suppl 9):41–58.
17. Lisman J, Yasuda R, Raghavachari S. Mechanisms of CaMKII action in long-term potentiation. Nat Rev Neurosci. 2012;13(3):169–82. doi: 10.1038/nrn3192.
18. Zhou Q, Xiao M, Nicoll RA. Contribution of cytoskeleton to the internalization of AMPA receptors. Proc Natl Acad Sci USA. 2001;98(3):1261–6. doi: 10.1073/pnas.98.3.1261.
19. Impey S, Obrietan K, Wong ST, Poser S, Yano S, Wayman G, Deloulme JC, Chan G, Storm DR. Cross talk between ERK and PKA is required for Ca^{2+} stimulation of CREB-dependent transcription and ERK nuclear translocation. Neuron. 1998;21(4):869–83. doi: 10.1016/s0896-6273(00)80602-9.
20. Tang D, Kang R, Coyne CB, Zeh HJ, Lotze MT. PAMPs and DAMPs: Signal 0s that spur autophagy and immunity. Immunol Rev. 2012;249(1):158–75. doi: 10.1111/j.1600-065X.2012.01146.x.
21. Zhang G, Ghosh S. Toll-like receptor-mediated NF-kappaB activation: A phylogenetically conserved paradigm in innate immunity. J Clin Invest. 2001;107(1):13–9. doi: 10.1172/JCI11837.
22. Druszczyńska M, Godkowicz M, Kulesza J, Wawrocki S, Fol M. Cytokine receptors-regulators of antimycobacterial immune response. Int J Mol Sci. 2022;23(3):1112. doi: 10.3390/ijms23031112.
23. Ransohoff RM, Brown MA. Innate immunity in the central nervous system. J Clin Invest. 2012;122(4): 1164–71. doi: 10.1172/JCI58644.
24. Sun M, You H, Hu X, Luo Y, Zhang Z, Song Y, An J, Lu H. Microglia-astrocyte interaction in neural development and neural pathogenesis. Cells. 2023;12(15):1942. doi: 10.3390/cells12151942.
25. Vose LR, Stanton PK. Synaptic plasticity, metaplasticity and depression. Curr Neuropharmacol. 2017; 15(1):71–86. doi: 10.2174/1570159x14666160202121111.
26. Jankowsky JL, Derrick BE, Patterson PH. Cytokine responses to LTP induction in the rat hippocampus: A comparison of in vitro and in vivo techniques. Learn Mem. 2000;7(6):400–12. doi: 10.1101/lm.32600.
27. Patel RR, Khom S, Steinman MQ, Varodayan FP, Kiosses WB, Hedges DM, Vlkolinsky R, Nadav T, Polis I, Bajo M, Roberts AJ, Roberto M. IL-1β expression is increased and regulates GABA transmission following chronic ethanol in mouse central amygdala. Brain Behav Immun. 2019;75:208–19. doi: 10.1016/j.bbi.2018.10.009.
28. Collingridge GL, Abraham WC. Glutamate receptors and synaptic plasticity: The impact of Evans and Watkins Neuropharmacology. 2022;206:108922. doi: 10.1016/j.neuropharm.2021.108922.

29. Leonoudakis D, Braithwaite SP, Beattie MS, Beattie EC. TNFalpha-induced AMPA-receptor trafficking in CNS neurons; relevance to excitotoxicity? Neuron Glia Biol. 2004;1(3):263–73. doi: 10.1017/S1740925X05000608.
30. Liu Y, Zhou LJ, Wang J, Li D, Ren WJ, Peng J, Wei X, Xu T, Xin WJ, Pang RP, Li YY, Qin ZH, Murugan M, Mattson MP, Wu LJ, Liu XG. TNF-α differentially regulates synaptic plasticity in the hippocampus and spinal cord by microglia-dependent mechanisms after peripheral nerve injury. J Neurosci. 2017;37(4):871–81. doi: 10.1523/JNEUROSCI.2235-16.2016. Erratum in: J Neurosci. 2024 Mar 4.
31. Hooks BM, Chen C. Distinct roles for spontaneous and visual activity in remodeling of the retinogeniculate synapse. Neuron. 2006;52(2):281–91.
32. Stevens B, et al. The classical complement cascade mediates CNS synapse elimination. Cell. 2007;131(6): 1164–78.
33. Schafer DP, et al. Microglia sculpt postnatal neural circuits in an activity and complement-dependent manner. Neuron. 2012;74(4):691–705.
34. Sipe GO, et al. Microglial P2Y12 is necessary for synaptic plasticity in mouse visual cortex. Nat Commun. 2016;7:2–15.
35. Paolicelli RC, et al. Synaptic pruning by microglia is necessary for normal brain development. Science. 2011;333(6048):1456–8.
36. Ramiro-Cortés Y, Israely I. Long lasting protein synthesis-and activity-dependent spine shrinkage and elimination after synaptic depression. PLoS One. 2013;8(8):e71155.
37. Perry VH, O'Connor V. The role of microglia in synaptic stripping and synaptic degeneration: A revised perspective. ASN Neuro. 2010;2(5):e00047.
38. Sonneville R, et al. Understanding brain dysfunction in sepsis. Ann Intensive Care. 2013;3(1):15.
39. Wang GF, Li W, Li K. Acute encephalopathy and encephalitis caused by influenza virus infection. Curr Opin Neurol. 2010;23(3):305–11.
40. De Chiara G, et al. Infectious agents and neurodegeneration. Mol Neurobiol. 2012;46(3):614–38.
41. Piacentini R, et al. HSV-1 promotes Ca^{2+} –mediated APP phosphorylation and Aβ accumulation in rat cortical neurons. Neurobiol Aging. 2011;32(12):2323–26.
42. Liou AK, et al. To die or not to die for neurons in ischemia, traumatic brain injury and epilepsy: A review on the stress-activated signaling pathways and apoptotic pathways. Prog Neurobiol. 2003;69(2):103–42.
43. Hemphill MA, et al. Traumatic brain injury and the neuronal microenvironment: A potential role for neuropathological mechanotransduction. Neuron. 2015;85(6):1177–92.
44. Wall AM, et al. Tumor necrosis factor-alpha potentiates long-term potentiation in the rat dentate gyrus after acute hypoxia. J Neurosci Res. 2015;93(5):815–29.
45. Ward RJ, Lallemand F, de Witte P. Biochemical and neurotransmitter changes implicated in alcohol-induced brain damage in chronic or 'binge drinking' alcohol abuse. Alcohol. 2009;44(2):128–35.
46. LeMarquand D, Pihl RO, Benkelfat C. Serotonin and alcohol intake, abuse, and dependence: Findings of animal studies. Biol Psychiatry. 1994;36(6):395–421.
47. Bava S, et al. Longitudinal changes in white matter integrity among adolescent substance users. Alcohol Clin Exp Res. 2013;37(Suppl 1):E181–9.
48. Amor S, et al. Inflammation in neurodegenerative diseases–an update. Immunology. 2014;142(2):151–66.
49. Pignatelli M, et al. Synaptic plasticity as a therapeutic target in the treatment of autism-related single-gene disorders. Curr Pharm Des. 2013;19(36):6480–90.
50. Nisticò R, et al. Synaptic plasticity in multiple sclerosis and in experimental autoimmune encephalomyelitis. Philos Trans R Soc Lond Ser B Biol Sci. 2013;369(1633):20130162.
51. Vilcek J, Feldmann M. Historical review: Cytokines as therapeutics and targets of therapeutics. Trends Pharmacol Sci. 2004;25:201–9.
52. Allan SM, Rothwell NJ. Cytokines and acute neurodegeneration. Nat Rev Neurosci. 2001;2:734–44.
53. Bezzi P, et al. CXCR4-activated astrocyte glutamate release via TNFalpha: Amplification by microglia triggers neurotoxicity. Nat Neurosci. 2001;4:702–10.
54. Bezzi P, et al. Prostaglandins stimulate calcium-dependent glutamate release in astrocytes. Nature. 1998;391:281–5.
55. Schneider H, et al. A neuromodulatory role of interleukin-1beta in the hippocampus. Proc Natl Acad Sci USA. 1998;95:7778–83.
56. Avital A, et al. Impaired interleukin-1 signaling is associated with deficits in hippocampal memory processes and neural plasticity. Hippocampus. 2003;13:826–34.
57. Maher FO, et al. Downregulation of IL-4-induced signalling in hippocampus contributes to deficits in LTP in the aged rat. Neurobiol Aging. 2005;26:717–28.

58. Nolan Y, et al. Role of interleukin-4 in regulation of age-related inflammatory changes in the hippocampus. J Biol Chem. 2005;280:9354–62.
59. Allan SM, et al. Interleukin-1 and neuronal injury. Nat Rev Immunol. 2005;5:629–40.
60. Rothwell NJ, Luheshi GN. Interleukin 1 in the brain: Biology, pathology and therapeutic target. Trends Neurosci. 2000;23:618–25.
61. Katsuki H, et al. Interleukin-1 beta inhibits long-term potentiation in the CA3 region of mouse hippocampal slices. Eur J Pharmacol. 1990;181:323–26.
62. Vereker E, et al. The inhibitory effect of interleukin-1beta on long-term potentiation is coupled with increased activity of stress-activated protein kinases. J Neurosci. 2000;20:6811–19.
63. Calabrese V, et al. Nitric oxide in the central nervous system: Neuroprotection versus neurotoxicity. Nat Rev Neurosci. 2007;8:766–75.
64. Plata-Salamán CR, ffrench-Mullen JM. Interleukin-1β inhibits Ca^{2+} channel currents in hippocampal neurons through protein kinase C. Eur J Pharmacol. 1994;266:1–10.
65. Gahring LC, et al. Pro-inflammatory cytokines modify neuronal nicotinic acetylcholine receptor assembly. J Neuroimmunol. 2005;166:88–101.
66. Curran BP, et al. A role for c-Jun N-terminal kinase in the inhibition of long-term potentiation by interleukin-1beta and long-term depression in the rat dentate gyrus in vitro. Neuroscience. 2003;118:347–57.
67. Kelly A, et al. The anti-inflammatory cytokine, interleukin (IL)-10, blocks the inhibitory effect of IL-1 beta on long term potentiation. A Role for JNK. J Biol Chem. 2001;276:45564–72.
68. Murray CA, Lynch MA. Evidence that increased hippocampal expression of the cytokine interleukin-1 beta is acommon trigger for age- and stress-induced impairments in long-term potentiation. J Neurosci. 1998;18:2974–81.
69. Goshen I, et al. Dual role for interleukin-1 in hippocampal-dependent memory processes. Psychoneuroendocrinology. 2007;32:1106–15.
70. Trompet S, et al. Genetic variation in the interleukin-1 beta-converting enzyme associates with cognitive function. The PROSPER Study. Brain. 2008;131:1069–77.
71. Beattie EC, et al. Control of synaptic strength by glial TNFalpha. Science. 2002;295:2282–5.
72. Stellwagen D, Malenka RC. Synaptic scaling mediated by glial TNF-alpha. Nature. 2006;440:1054–9.
73. Albensi BC, Mattson MP. Evidence for the involvement of TNF and NF-kappaB in hippocampal synaptic plasticity. Synapse. 2000;35:151–9.
74. Butler MP, et al. Dissection of tumor-necrosis factor-alpha inhibition of long-term potentiation (LTP) reveals a p38 mitogen-activated protein kinase-dependent mechanism which maps to early-but not late-phase LTP. Neuroscience. 2004;124:319–26.
75. Curran B, O'Connor JJ. The pro-inflammatory cytokine interleukin-18 impairs long-term potentiation and NMDA receptor-mediated transmission in the rat hippocampus in vitro. Neuroscience. 2001; 108:83–90.
76. Cumiskey D, et al. Interleukin-18 mediated inhibition of LTP in the rat dentate gyrus is attenuated in the presence of mGluR antagonists. Neurosci Lett. 2007;412:206–10.
77. Cumiskey D, et al. A role for inflammatory mediators in the IL-18 mediated attenuation of LTP in the rat dentate gyrus. Neuropharmacology. 2007;52:1616–23.
78. Balschun D, et al. Interleukin-6: A cytokine to forget. FASEB J. 2004;18:1788–90.
79. Tancredi V, et al. The inhibitory effects of interleukin-6 on synaptic plasticity in the rat hippocampus are associated with an inhibition of mitogen-activated protein kinase ERK. J Neurochem. 2000;75:634–43.
80. John CC, et al. Cerebrospinal fluid cytokine levels and cognitive impairment in cerebral malaria. Am J Trop Med Hyg. 2008;78:198–205.
81. D'Arcangelo G, et al. Interferon inhibits synaptic potentiation in rat hippocampus. Brain Res. 1991;564: 245–8.
82. Tancredi V, et al. Interleukin-2 suppresses established long-term potentiation and inhibits its induction in the rat hippocampus. Brain Res. 1990;525:149–51.
83. Huh GS, et al. Functional requirement for class I MHC in CNS development and plasticity. Science. 2000;290:2155–9.
84. Goddard CA, et al. Regulation of CNS synapses by neuronal MHC class I. Proc Natl Acad Sci USA. 2007;104:6828–33.
85. Nathan C, Xie QW. Nitric oxide synthases: Roles, tolls, and controls. Cell. 1994;78:915–8.
86. O'Dell TJ, et al. Endothelial NOS and the blockade of LTP by NOS inhibitors in mice lacking neuronal NOS. Science. 1994;265, 542–6.
87. Tracy TE, et al. Acetylated tau obstructs KIBRA-mediated signaling in synaptic plasticity and promotes tauopathy-related memory loss. Neuron. 2016;90(2):245–60.

88. Ashabi G, et al. ERK and p38 inhibitors attenuate memory deficits and increase CREB phosphorylation and PGC-1α levels in Aβ-injected rats. Behav Brain Res. 2012;232(1):165–73.
89. Peretti D, et al. RBM3 mediates structural plasticity and protective effects of cooling in neurodegeneration. Nature. 2015;518(7538):236–9.
90. Vitolo O, et al. Protection against β- amyloid induced abnormal synaptic function and cell death by Ginkgolide J Neurobiol Aging. 2009;30(2):257–65.
91. Leal G, Comprido D, Duarte CB. BDNF-induced local protein synthesis and synaptic plasticity. Neuropharmacology. 2014;76:639–56.
92. Mazella J, et al. Spadin, a sortilin-derived peptide, targeting rodent TREK-1 channels: A new concept in the antidepressant drug design. PLoS Biol. 2010;8(4):e1000355. https://doi.org/10.1371/journal.pbio.1000355
93. Wang J-W, David DJ, Monckton JE, Battaglia F, Hen R. Chronic fluoxetine stimulates maturation and synaptic plasticity of adult-born hippocampal granule cells. J Neurosci. 2008;28(6):1374–84.
94. Liu Z-H, et al. A single bolus of docosahexaenoic acid promotes neuroplastic changes in the innervation of spinal cord interneurons and motor neurons and improves functional recovery after spinal cord injury. J Neurosci. 2015;35(37):12733–52.
95. Rijpma A, et al. Effects of Souvenaid on plasma micronutrient levels and fatty acid profiles in mild and mild-to moderate Alzheimer's disease. Alzheimers Res Ther. 2015;7(1):51.
96. Lai MKP, et al. Selective loss of P2Y2 nucleotide receptor immunoreactivity is associated with Alzheimer's disease neuropathology. J Neural Transm. 2008;115(8):1165–72.
97. Guan J-S, et al. HDAC2 negatively regulates memory formation and synaptic plasticity. Nature. 2009; 459(7243):55–60.
98. Gervain J, et al. Valproate reopens critical-period learning of absolute pitch. Front Syst Neurosci. 2013;7:102.
99. Miller BH, et al. MicroRNA-132 dysregulation in schizophrenia has implications for both neurodevelopment and adult brain function. Proc Natl Acad Sci. 2012;109(8):3125–30.
100. Jadhav AD, Li W, Xu Z, Shi P. Compartmentalized Synapse Microarray for High-Throughput Screening. In: Biffi E, editor. Microfluidic and Compartmentalized Platforms for Neurobiological Research. New York, NY: Springer New York; 2015. p. 231–45.
101. Takizawa H, Hiroi N, Funahashi A. Mathematical modeling of sustainable synaptogenesis by repetitive stimuli suggests signaling mechanisms in vivo. PLoS One. 2012;7(12):e51000.
102. Leshchyns'ka I, et al. Aβ-dependent reduction of NCAM2-mediated synaptic adhesion contributes to synapse loss in Alzheimer's disease. Nat Commun. 2015;6:8836.
103. Gerson J, et al. Tau oligomers derived from traumatic brain injury cause cognitive impairment and accelerate onset of pathology in htau mice. J Neurotrauma. 2016;33(22):2034–43.
104. Kang J-Q, Shen W, Zhou C, Xu D, Macdonald RL. The human epilepsy mutation GABRG2(Q390X) causes chronic subunit accumulation and neurodegeneration. Nat Neurosci. 2015;18(7):988–96.
105. Perni M, et al. A natural product inhibits the initiation of α-synuclein aggregation and suppresses its toxicity. Proc Natl Acad Sci. 2017;114(6):E1009–17.
106. Shi C, Wu F, Xu J, Zou J. Bilobalide regulates soluble amyloid precursor protein release via phosphatidyl inositol 3 kinase-dependent pathway. Neurochem Int. 2011;59(1):59–64.
107. Rezai-Zadeh K, et al. Green tea epigallocatechin-3-gallate (EGCG) reduces β-amyloid mediated cognitive impairment and modulates tau pathology in Alzheimer transgenic mice. Brain Res. 2008; 1214:177–87.
108. Yang F, et al. Curcumin inhibits formation of amyloid β oligomers and fibrils, binds plaques, and reduces amyloid in vivo. J Biol Chem. 2005;280(7):5892–901.
109. Chen L, et al. Ginsenoside Rgl attenuates β- amyloid generation via suppressing PPARγ-regulated BACE1 activity in N2a-APP695 cells. Eur J Pharmacol. 2012;675(1–3):15–21.
110. Theunis C, et al. Efficacy and safety of a liposome-based vaccine against protein Tau, assessed in tau. P301L mice that model tauopathy. PLoS One. 2013;8(8):e72301.
111. Novak P, et al. Safety and immunogenicity of the tau vaccine AADvac1 in patients with Alzheimer's disease: A randomised, double-blind, placebo-controlled, phase 1 trial. Lancet Neurol. 2017;16(2):123–34.
112. Collin L, Bohrmann B, Göpfert U, Oroszlan-Szovik K, Ozmen L, Grüninger F. Neuronal uptake of tau/pS422 antibody and reduced progression of tau pathology in a mouse model of Alzheimer's disease. Brain. 2014;137(10):2834–46.
113. Kamat PK, et al. Mechanism of oxidative stress and synapse dysfunction in the pathogenesis of Alzheimer's disease: Understanding the therapeutics strategies. Mol Neurobiol. 2016;53(1):648–61.
114. Liu T, et al. A meta-analysis of oxidative stress markers in depression. PLoS One. 2015;10(10):e0138904.

115. Choi G-Y, et al. Curcumin alters neural plasticity and viability of intact hippocampal circuits and attenuates behavioral despair and COX-2 expression in chronically stressed rats. Mediators Inflamm. 2017;2017:1–9.
116. Zhang X, et al. Ginsenoside Rd and ginsenoside Re offer neuroprotection in a novel model of Parkinson's disease. Am J Neurodegener Dis. 2016;5(1):52–61.
117. Liu J-B, et al. Neuroprotective activity of (–) -epigallocatechin gallate against lipopolysaccharide-mediated cytotoxicity. J Immunol Res. 2016;2016:1–10.
118. Petrov D, Mansfield C, Moussy A, Hermine O. ALS clinical trials review: 20 years of failure. Are We Any Closer to Registering a New Treatment? Front Aging Neurosci. 2017;9:68.
119. Yoshida H, et al. Edaravone and Nrf2-inducers as neuroprotective agents in human astrocytes exposed to hypoxia/reoxygenation. Hirosaki Med J. 2010;61:S147–56.
120. Martín-Montañez E, et al. IGF-II promotes neuroprotection and neuroplasticity recovery in a long-lasting model of oxidative damage induced by glucocorticoids. Redox Biol. 2017;13:69–81.
121. Pascual-Lucas M, et al. Insulin-like growth factor 2 reverses memory and synaptic deficits in APP transgenic mice. EMBO Mol Med. 2014;6(10):1246–62.
122. Hertze J, Nägga K, Minthon L, Hansson O. Changes in cerebrospinal fluid and blood plasma levels of IGF-II and its binding proteins in Alzheimer's disease: An observational study. BMC Neurol. 2014; 14(1):64.
123. Heupel K, et al. Loss of transforming growth factor-beta 2 leads to impairment of central synapse function. Neural Dev. 2008;3(1):25.
124. Sharma K, Mehra RD. Long-term administration of estrogen or tamoxifen to ovariectomized rats affords neuroprotection to hippocampal neurons by modulating the expression of Bcl-2 and Bax. Brain Res. 2008;1204:1–15.
125. van Praag H. Exercise enhances learning and hippocampal neurogenesis in aged mice. J Neurosci. 2005;25(38):8680–5.
126. Lezi E, Burns JM, Swerdlow RH. Effect of high-intensity exercise on aged mouse brain mitochondria, neurogenesis, and inflammation. Neurobiol Aging. 2014;35(11):2574–83.
127. van Praag H, Kempermann G, Gage FH. Neural consequences of environmental enrichment. Nat Rev Neurosci. 2000;1(3):191–8.
128. Ji J, et al. Forced running exercise attenuates hippocampal neurogenesis impairment and the neurocognitive deficits induced by whole brain irradiation via the BDNF-mediated pathway. Biochem Biophys Res Commun. 2014;443(2):646–51.
129. Shihabuddin LS, Horner PJ, Ray J, Gage FH. adult spinal cord stem cells generate neurons after transplantation in the adult dentate gyrus. J Neurosci. 2000;20(23):8727–35.
130. Hüttenrauch M, et al. Physical activity delays hippocampal neurodegeneration and rescues memory deficits in an Alzheimer disease mouse model. Transl Psychiatry. 2016;6(5):e800.
131. Irwin MR, et al. Tai Chi, cellular inflammation, and transcriptome dynamics in breast cancer survivors with insomnia: A randomized controlled trial. JNCI Monogr. 2014;2014(50):295–301.
132. Bergami M, et al. A critical period for experience-dependent remodeling of adult-born neuron connectivity. Neuron. 2015;85(4):710–7.
133. Tang Y-Y, Lu Q, Fan M, Yang Y, Posner MI. Mechanisms of white matter changes induced by meditation. Proc Natl Acad Sci. 2012;109(26):10570–4.
134. Epel ES, et al. Meditation and vacation effects have an impact on disease-associated molecular phenotypes. Transl Psychiatry. 2016;6(8):e880.
135. Kaliman P, Álvarez-López MJ, Cosín-Tomás M, Rosenkranz MA, Lutz A, Davidson RJ. Rapid changes in histone deacetylases and inflammatory gene expression in expert meditators. Psychoneuroendocrinology. 2014;40:96–107.
136. Brefczynski-Lewis JA, Lutz A, Schaefer HS, Levinson DB, Davidson RJ. Neural correlates of attentional expertise in long-term meditation practitioners. Proc Natl Acad Sci. 2007;104(27):11483–8.
137. Calabresi P, et al. Levodopa-induced dyskinesia: A pathological form of striatal synaptic plasticity? Ann Neurol. 2000;47:S60–68.
138. Calabresi P, et al. Dopamine-mediated regulation of corticostriatal synaptic plasticity. Trends Neurosci. 2007;30:211–9.
139. Lang AE, Lozano AM. Parkinson's disease. First of two parts. N Engl J Med. 1998;339:1044–53.
140. Hunot S, Hirsch EC. Neuroinflammatory processes in Parkinson's disease. Ann Neurol. 2003;53(Suppl 3), S49–58.
141. Diogenes MJ, et al. Extracellular alpha-synuclein oligomers modulate synaptic transmission and impair LTP via NMDA-receptor activation. J Neurosci. 2012;32(34):11750–62.

142. Hirsch EC, Hunot S. Neuroinflammation in Parkinson's disease: A target for neuroprotection? Lancet Neurol. 2009;8(4):382–97.
143. Hunot S, Hirsch EC. Neuroinflammatory processes in Parkinson's disease. Ann Neurol. 2003;53(Suppl 3): S49–58. discussion S58–60
144. Xue X, et al. Baicalein ameliorated the upregulation of striatal glutamatergic transmission in the mice model of Parkinson's disease. Brain Res Bull. 2014;103:54–9.
145. Walsh DM, Selkoe DJ. Deciphering the molecular basis of memory failure in Alzheimer's disease. Neuron. 2004;44:181–93.
146. Selkoe, D.J. Alzheimer's disease is a synaptic failure. Science. 2002;298:789–91.
147. Klyubin I, et al. Amyloid beta protein immunotherapy neutralizes Abeta oligomers that disrupt synaptic plasticity in vivo. Nat Med. 2005;11:556–61.
148. Heneka MT, O'Banion MK. Inflammatory processes in Alzheimer's disease. J Neuroimmunol. 2007; 184:69–91.
149. Wang Q, et al. Beta-amyloid inhibition of long-term potentiation is mediated via tumor necrosis factor. Eur J Neurosci. 2005;22:2827–32.
150. Nisticò R, et al. Targeting synaptic dysfunction in Alzheimer's disease therapy. Mol Neurobiol. 2012; 46(3):572–87.
151. Schmid AW, Lynch MA, Herron CE. The effects of IL-1 receptor antagonist on beta amyloid mediated depression of LTP in the rat CA1 in vivo. Hippocampus. 2009;19(7):670–6.
152. Kotilinek LA, et al. Cyclooxygenase-2 inhibition improves amyloid-beta-mediated suppression of memory and synaptic plasticity. Brain. 2008;131(Pt 3):651–64.
153. Hong S, et al. Complement and microglia mediate early synapse loss in Alzheimer mouse models. Science. 2016;352(6286):712–6.
154. Martino G, et al. Inflammation in multiple sclerosis: The good, the bad, and the complex. Lancet Neurol. 2002;1:499–509.
155. Merrill JE, Benveniste EN. Cytokines in inflammatory brain lesions: Helpful and harmful. Trends Neurosci. 1996;19:331–8.
156. Ransohoff RM, Estes ML. Astrocyte expression of major histocompatibility complex gene products in multiple brain tissue obtained by stereotactic biopsy. Arch Neurol. 1991;48:1244–6.
157. Rose JW, et al. Inflammatory cell expression of cyclooxygenase-2 in the multiple sclerosis lesion. J Neuroimmunol. 2004;149:40–9.
158. Bagasra O, et al. Activation of the inducible form of nitric oxide synthase in the brains of patients with multiple sclerosis. Proc Natl Acad Sci USA. 1995;92:12041–5.
159. Hoffmann S, et al. Cognitive impairment in multiple sclerosis. Curr Opin Neurol. 2007;20:275–80.
160. Mori F, et al. Interleukin-1beta promotes long-term potentiation in patients with multiple sclerosis. NeuroMolecular Med. 2014;16(1):38–51.

7 Recent Therapeutic Approaches for Neurodegenerative Disorders

Anindita Ghosh
Indian Institute of Science Education and Research Bhopal (IISERB)
Bhopal, Madhya Pradesh, India

Anita Pal
University of Allahabad
Prayagraj, Uttar Pradesh, India

Ram Sagar
The Johns Hopkins University
Baltimore, MD

Utkarsh Gaharwar
TechInvention Lifecare Ltd
Mumbai, Maharashtra, India

Himanshu Singh
Memorial Sloan Kettering Cancer Center
New York, NY

*Deepak Kumar**
University of Allahabad
Prayagraj, Uttar Pradesh, India

7.1 INTRODUCTION

Neurodegenerative disorders, which include Alzheimer's disease (AD), Parkinson's disease (PD), amyotrophic lateral sclerosis (ALS), and Huntington's disease (HD), are progressive degenerations of neurons in the brain or spinal cord that damage both their structure and function. The clinical symptoms of a particular neurodegenerative disorder are a representation of a specific brain region being affected [1]. For instance, PD is distinguished by the destruction of dopaminergic neurons in the substantia nigra, while AD is characterized by substantial deterioration of the cortical and hippocampal regions. Similarly, ALS impacts upper and lower motor neurons, whereas HD causes specific neuronal decline in the basal ganglia, and spinocerebellar ataxias (SCAs) often involve degeneration of Purkinje cells [2–5] (Figure 7.1). There is also evidence that some neurodegenerative disorders can be accompanied by multiple disorders. Researchers have reported cases linking ALS to dementia and Parkinsonism. There have been reports of cases where ALS is linked to dementia and parkinsonism [6]. Similarly, ALS and PD can also occur alongside spinocerebellar ataxia type 2 (SCA2). It is widely believed that neurodegenerative diseases are primarily caused by factors associated with

* Anindita Ghosh and Anita Pal contributed equally to this work, and both are the joint first authors)

DOI: 10.1201/9781003464648-8

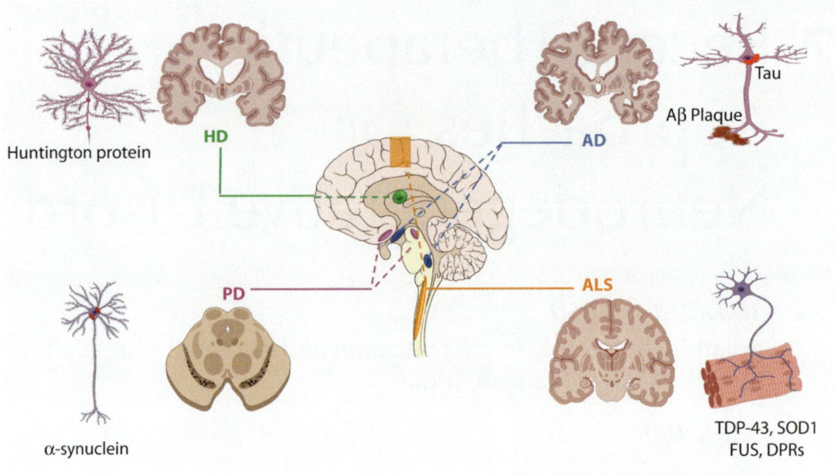

FIGURE 7.1 Schematic representation of major brain areas and specific neuronal type affected in AD, PD, ALS, and HD.

ageing, genetics, and the environment. Neurodegenerative diseases exhibit considerable clinical and pathological heterogeneity at the cellular level, and many have overlapping pathological events.

The core shared neuronal pathways include protein aggregation, ubiquitin-proteasome impairments, mitochondrial dysfunctions, oxidative stress, impaired autophagy-lysosome function, glutamate excitotoxicity, protein seeding, inflammation, and immune responses [7]. Intriguingly, some of these overlapping characteristics could provide useful perspective on potential therapeutic targets. Neurological illnesses not only result in economic losses for individuals and families but also impose significant costs on society as a whole. Moreover, as human life expectancy continues to rise, there is growing concern about the escalating burden of neurodegenerative diseases in the coming years, making a therapeutic perspective crucial [8]. In recent decades, we have gained a better understanding of the causes of diseases, but complex mechanisms of neuronal loss, combined with physiological factors that are incompatible, pose challenges in describing the pathogenic processes and devising effective treatments. Currently, there are no known treatments for most of these diseases, rendering them incurable. At present, there are no medications available that can protect the nervous system. Although a few drugs that can alleviate specific symptoms are used clinically, the disease progresses unabatedly. Therefore, there is a pressing need for therapeutic interventions that have the potential to effectively treat neurodegeneration. This chapter provides a concise overview of commonly occurring neurodegenerative disorders and the current state of treatments. Additionally, it provides an update on the progress made in addressing neurodegeneration through immunotherapy, gene therapy, and stem cell therapy.

7.1.1 Alzheimer's Disease

AD primarily affects the elderly. Symptoms include progressive memory deterioration, cognitive decline, and decreased capacity to perform daily tasks. With the progression of the disease, memory deteriorates, cognitive functions decline, and daily living activities become more difficult [4]. AD is rare in individuals under the age of 50, but a small subgroup of patients develop the disease early. Most cases are diagnosed after 60 years of age, and people over 65 are most likely to develop dementia, with AD being the main cause [9]. Globally, AD affects 50 million people and is expected to grow significantly by 2050 [10]. In the coming years, we expect to see an increase in AD as well as AD-related dementia globally as longevity increases. Several factors are thought to contribute to AD development, including genetic and non-genetic factors. Mutations in the APP and Presenilin 1 and 2 genes are most commonly associated with early-AD development. There are a plethora of putative risk genes that are associated with the late-onset AD phenotype (Table 7.1). For instance,

TABLE 7.1
Genetic Risk Loci and Molecular Pathways Implicated in AD, PD, ALS, and HD

Diseases	Genes Involved/ Affected	Pathway(s) Altered	Downstream Effects	Reference(s)
AD	APP	Aggregation	Increased generation as well as aggregation of β-amyloid peptide.	[55]
	PSEN1, PSEN2	Protein processing/ signaling	Mutations interfere with the processing of gamma secretase, leading to aggregation of β-amyloid.	[55,56]
	ADAM10, ABCA7, BIN1, CD2AP, CLU, FERMT2, CASS4, SLCA10A2, PLD3, SLC24A4/RIN3, DSG2, BCHE, CTSD, MEF2C, APH1B, MAPT, HS3ST1	Increased production/ aggregation of tau or β-amyloid	Aggregations induce swellings in neuronal projections, impeding neural circuits	[57,58]
	PICALM, SORL1, CR1, CD33, CST3, ABI3, PLCG2, SHARPIN, ACE, IGF1, LKB1	Autophagy/ clearance of αβ and tau	Impairment of autophagy-lysosome leads to accumulation of aggregates of Aβ & tau, which further exacerbate autophagy-lysosome dysfunction.	[57,59]
	FERMT2, CASS4, CD2AP, COBL	Axonal/ dendritic dysfunction	Increases levels of mature APP on the cell surface, resulting in increased production of Aβ-peptide. A defect in synaptic transmission, calcium signaling, and microtubule stabilization, as well as a reduced number of neuronal branch points in the dendritic tree.	[57,60–62]
	APOE	Cholesterol & fatty acid metabolism	Impaired clearance of chylomicrons & VLDL remnants.	[56]
	APOE, TREM2, APOJ, CLU, ECHDC3, PICALM, ABCA1, ABCA7, SLC10A2, SLC24A4	Lipid metabolism	Multi-regional changes in free fatty acid levels and glycerophospholipid contents. Changes in glycerolipids, sphingomyelin, and ceramide	[57,63]
	TOMM40	Mitochondrial dynamics	Mitochondrial membrane dysfunction and oxidative stress.	[57,64]
	TOMM40, MINK1, ACE, MEF2C	Oxidative stress	The outcome is the occurrence of breaks in the DNA strands, an elevation in reactive oxygen species (ROS) within the neuronal and glial populations, and the formation of harmful compounds, lipid peroxidation and the buildup of advanced glycation products.	[57,65]
	BIN1, CLU, TREM2, MTHFR, MEF2C, ABI3, PLCG2, SHARPIN, MILK1, ACE	Neuroinflammation	A long lasting neuroinflammation is suggested to cause a decline in homeostatic functions of neuroglia leading to neuronal loss & neurodegeneration.	[57,66]
PD	SNCA, PARK1	Aggregation (alpha synuclein)	Modified α-synuclein binds to mitochondrial protein TOM20 with high affinity preventing its binding with co-receptor TOM22, thereby impairing mitochondrial protein transport. Mutant α synuclein also impairs CMA & macroautophagy.	[1,67–70]

(Continued)

TABLE 7.1 *(Continued)*
Genetic Risk Loci and Molecular Pathways Implicated in AD, PD, ALS, and HD

Diseases	Genes Involved/ Affected	Pathway(s) Altered	Downstream Effects	Reference(s)
	SNCA, PRKN, PINK1, DJ1, LRRK2	Oxidative stress	The protein products of these genes are linked to disrupted oxidorective homeostasis in PD. Enhanced ROS leads to accumulation of oxidized proteins, lipids & nucleic acids in brain of PD patients.	[71]
	PRKN, PINK1, PARK2, PARK5, PARK6, PARK7, FBX07, POLG	Mitochondrial dynamics & mitophagy	Disruption of mitochondrial transport & autophagic turnover of mitochondria.	[72–77]
	ATP13A2, SNCA (A53T, A30P), LRRK2, GBA-1	Autophagy	Mutant α synuclein binds to LAMP-2, hindering not only their own processing & degradation but also that of other CMA substrates. Mutant LRRK2 disrupts cytoskeletal dynamics and impairs protein translation & vesicular trafficking.	[67,70,74,78–80]
	VPS35	Cargo sorting & vesicular transport	Impairs endolysosome maturation & autophagy; recycling of membrane receptors is disturbed.	[81–83]
	ASCL4, DJ1, FTH1, GPX4, IL13, IL13RA1, PLA2G6, miR-355, NRF2, TP53, SQSTM1	Ferroptosis	Iron builds up in the substantia nigra and binds to α synuclein, causing it to clump together by promoting structural alterations.	[84–86]
ALS	TDP43, SOD1, FUS, DPRs	Aggregation	Stimulates inflammation, impairs autophagosome & autolysosome formation and mitophagy. Defects in RNA processing, DNA damage repair, stress granule function, and nucleocytoplasmic transport.	[1,87–89]
	SOD1, SQSTM1, GLTD81	Oxidative stress/ cytotoxicity	Alterations in redox balance and Ca^{2+} homeostasis, mitochondrial dysfunction, protein aggregation, axonal transport defects, glial dysfunction. Mutated GLT8D1 disrupts neurotrophin signaling within membrane lipid rafts.	[87,88,90–93]
	OPTN, TBK1, C9orf72, UBQLN2, CYLD	Inflammatory	Activated microglia during the gradual increase in inflammatory response further promote degeneration of neurons.	[88,94–99]
	TBK1	Signaling/kinase	TBK1 impairs p62 mediated autophagy, mitochondrial dysfunction & inflammation.	[87,95,100,101]
	C9orf72, OPTN, SPG11, VABP, CAV1, NEK1	Dysregulated vesicular transport/ trafficking	Mutant transcripts trap proteins that bind RNA, forming foci. Additionally, the repeat-associated non-AUG (RAN) translated RNA produces dipeptides that tend to form aggregates. Defects in intracellular & neurotrophic signaling, trafficking, microtubule stability, reorganization of actin & microtubules, vesicle transport, cell cycle control, proteosome impairment.	[41,87,88]

(Continued)

TABLE 7.1 *(Continued)*
Genetic Risk Loci and Molecular Pathways Implicated in AD, PD, ALS, and HD

Diseases	Genes Involved/ Affected	Pathway(s) Altered	Downstream Effects	Reference(s)
	TDP43, SOD1, FUS, C9orf72, C21orf2, UBQLN2, ATXN2, SEXT, ANG, MATR3, HnRNPA1	DNA/RNA metabolism	Mutations affect DNA damage repair, splicing, transcription, nuclear & mRNA transport & miRNA biogenesis. Mislocalization of TDP-43 disrupts its own autoregulation, interrupts several RNA processing pathways, thereby affecting RNA stability as well as gene expression.	[88,102,103]
	CHCHD10, VCP, TBK1, SOD1	Mitochondrial dysfunction	Stress granule defects, mitochondrial and endoplasmic reticulum dysfunctions, synaptic dysfunctions, and proteosome defect.	[87,88]
	TIA1, DNAJC7, COorf72, SQSTM1, CCNF, TARDBP, CHMP2B, VCP, UBQLN2, OPTN, SOD1, SIGMAR1, TBK1, FUS, VABP	Impaired protein homeostais & ER stress	DNA repair, RNA processing, transport, and stress granule function defects. Delays in stress granule disassembly, accumulation of stress granules, transport defects, aggregation, and proteasome dysfunction.	[87,88]
	TBK1, C9orf72, VCP, SPG11, OPTN, CYLD, SOD1, UBQLN2	Autophagy	Protein aggregation, inflammation, axonal transport & trafficking defects, reorganization of actin & microtubules.	[87–89,104,105]
	ALS2, ANXA11, SIGMAR1, FIG4, VAPB, TUBA4A, SQSTM1, PRN1, CCNF, SPG11, KIF5A, KANK1, C21orf2, OPTN, PFN1, DCTN1	Axonopathy	Hyperactivity, aggregation in neurite terminals, defects in mitochondrial function, oxidative stress, microtubule assembly, vesicular transport, intracellular trafficking, proteosome impairment, axonal & cytoskeletal defect.	[87,88,106–109]
HD	HTT	Aggregation	*mhtt* aggregates usually sequester essential proteins including transcription factors, chaperones, proteasome subunits, and also the wild type huntington protein.	[1,110]
	HTT	Protein homeostasis	*mhtt* aggregates sequester proteins, transcription factors, chaperones, proteosome subunits as well as wild type htt.	[111,112]
	HTT	Transcriptional dysregulation	*mhtt* (a) Inhibits CREB and TFIID transcription complexes, decreases activation of PGC-1α gene, (b) represses transcription of Sp-1 dependent promoters involved in DNA damage response & apoptosis, (c) fails to interact with REST/NRSF, which suppresses the expression of neuronal genes including BDN, (d) reduces the expression of SREBP affecting cholesterol biosynthesis, and (e) affects HSF1 transcription factor. Reduction of BDNF affects the TrkB/p75NTR signaling, that impairs synaptogenesis, neuron survival, and induces apoptosis, neuroinflammation & degeneration.	[113–117]

(Continued)

TABLE 7.1 *(Continued)*

Genetic Risk Loci and Molecular Pathways Implicated in AD, PD, ALS, and HD

Diseases	Genes Involved/ Affected	Pathway(s) Altered	Downstream Effects	Reference(s)
	HTT	Epigenetic alterations & ncRNAs	PolyQ repeats bind to the histone acetyltransferase domain (HAC) of a transcriptional coactivator, CBP, which disrupts their histone acetyltransferase activity. Huntingtin has also been reported to be found in RNA structures like P bodies, stress granules, and dendritic RNA granules that govern protein expression at a posttranslational level.	[110,118–120]
	HTT	Mitochondrial dysfunction	*mhtt* suppresses the transcription factor PGC-1 alpha that is involved in mitochondria biogenesis. *mhtt* also induces abnormalities in respiratory chain, hinders in translocation of mitochondrial DNA & proteins. *mhtt* inhibits mitophagy by sequestering mTOR.	[114]
	HTT	Synaptic plasticity & defects in axonal transport	Huntintin aggregates sequester motor proteins and slow down the transport of organelles along axons by blocking axons, which in turn affects vesicle trafficking. *mhtt* also interrupts the interaction between synaptic receptors and their corresponding adaptors & motor proteins affecting synaptic transmission.	[121–125]
	HTT	Neuroglia dysfunction	Glial cells affected show altered transcription profile, altered electrical properties, increased expression of inflammatory cytokines, reduced chemoattractant response & process motility, thinner myelin sheaths, and impaired glutamate uptake by astrocytes.	[110,126]
	HTT	Cell-to-cell transfer of aggregates	*mhtt* aggregates have been shown to spread from cell to cell through tunneling nanotubes	[127]

the APOE4 gene makes people more susceptible to disease development [11]. In addition to genes, chronic exposure to toxins, prolonged exposure to several heavy metals, preexisting medical conditions, and lifestyle habits can contribute to disease. Although the exact pathophysiology of AD is largely unknown, multiple theories have been proposed for understanding its mechanisms. The hallmark of AD is the accumulation and deposition of abnormal proteins such as amyloid plaques and tau aggregates. Amyloid plaques, a type of protein called amyloid beta (Aβ), form when Aβ proteins aggregate. The precise mechanism behind amyloid plaques and their role in AD is still unknown, but they disrupt neuronal function [12]. APP is a transmembrane glycoprotein that can be cleaved by proteases or sequentially fragmented by enzymes. Most of the monomers are Aβ40/42. When APP is not processed evenly, toxic monomers build up, resulting in Aβ oligomers and plaques [13]. It leads to ion channel obstructions, calcium level disturbances, and oxidative stress in mitochondria, compromised energy metabolism, and glucose regulation dysregulation, eventually causing synaptic dysfunction and neuron death. Another characteristic feature of AD pathogenesis is the neurofibrillary tangles [4]. Tau proteins found in neurons stabilize microtubules, ensuring their proper function in healthy neurons. However, abnormal tau protein aggregates form neurofibrillary tangles. GSK-3α/β, PKC4, PKA, and ESK2 are some of the kinases responsible for the typical phosphorylation of tau in these tangles [14]. Hyperphosphorylation of tau causes it to dissociate from microtubules, leading to the destabilization of microtubules and the formation of oligomers [15]. The accumulation of these entanglements disrupts cellular processes and triggers neuronal death. Additionally, glial cells exhibit both defensive and harmful effects on amyloid monomers and oligomers (Figures 7.2 and 7.3).

7.1.2 Parkinson's Disease

PD is characterized by the death of dopaminergic neurons in the substantia nigra, but some neurons in other brain regions are also affected by the disease. Symptoms of PD include tremors, stiffness, slowness of movement, postural instability, and other psychiatric and autonomic dysfunctions. In the course of the disease, these symptoms worsen and affect the quality of life [5,16]. On a global scale, PD has a significant economic and societal impact, ranking second behind Alzheimer's. It is estimated that 7–10 million people suffer from this disease worldwide, and that number could reach over 10 million by 2040 [17]. The etiology of PD remains uncertain, but it is hypothesized that a multifaceted interplay between genetic and environmental factors contributes to the pathogenesis of the disease. Retrospective epidemiological studies have established a correlation between pesticide exposure, smoking, and dietary factors and the risk of disease development [18]. Ageing is also a leading risk factor for nigrostriatal dopamine dysfunction, with a notable increase in PD symptoms after 60 [19]. From a genetic standpoint, the disease can be classified as either familial or sporadic. Genes such as LRRK2, SNCA, and PARKIN have been linked to PD [20]. Many genome-wide association studies also report other genetic risk factors and candidate genes [21,22] (Table 7.1). Alpha-synuclein interacts with presynaptic axon terminals to release neurotransmitters, and its mutations are linked to familial PD cases. Several familial forms of PD, notably the autosomal dominant, are partly explained by parkin and ubiquitin C-terminal hydrolase L1. These mutations are responsible for a minor proportion of hereditary cases, while the majority of PD cases are still sporadic. These mutations account for only a small percentage of PD hereditary cases, and most cases remain sporadic. It is interesting to note that PD is 1.5 times more common in males than in females [23]. The precise pathophysiology of PD remains unknown; however, the accumulation of Lewy bodies (LBs) in the substantia nigra is considered a characteristic feature of the disease. The protein α-synuclein is involved in presynaptic signaling and membrane trafficking and regulates neurotransmitter release from presynaptic terminals. Soluble monomers of synuclein combine into protofibrils, and then a large, insoluble fibril of synuclein (Lewy bodies) acquires neurotoxic properties during pathogenesis.

LBs are bound to ubiquitin, which effectively limits proteasomal action. Additional mechanisms involve seed aggregation, impaired functioning of the lysosomal autophagy system, mitochondrial dysfunction, oxidative stress, and iron load [5,24] (Figures 7.2 and 7.3). Neuronal loss in the substantia nigra can also be accompanied by astrocyte death and an elevation in microglial cells. There is evidence that synuclein aggregation influences both innate and adaptive immunity [25].

7.1.3 Amyotrophic Lateral Sclerosis

ALS is a progressive, fatal degenerative disease that affects the upper and lower motor neurons and the corticospinal tract [3,26]. The disease-affected individuals present with muscle weakness, slowness, speech slurring, hyperreflexia, spasticity, and respiratory failure. ALS prevalence is estimated at 4–8 per 100,000 individuals globally, with Europe having 2–3 cases per 100,000 [27]. However, incidence varies based on ancestral origin, with European populations showing a higher incidence, while east and south Asia have lower rates [28]. The cause of ALS is often unknown, but there are instances where it runs in families, possibly due to genetic components. Several genes are known causes of familial forms of ALS (fALS), while over 100 genes are associated with sporadic forms (sALS)

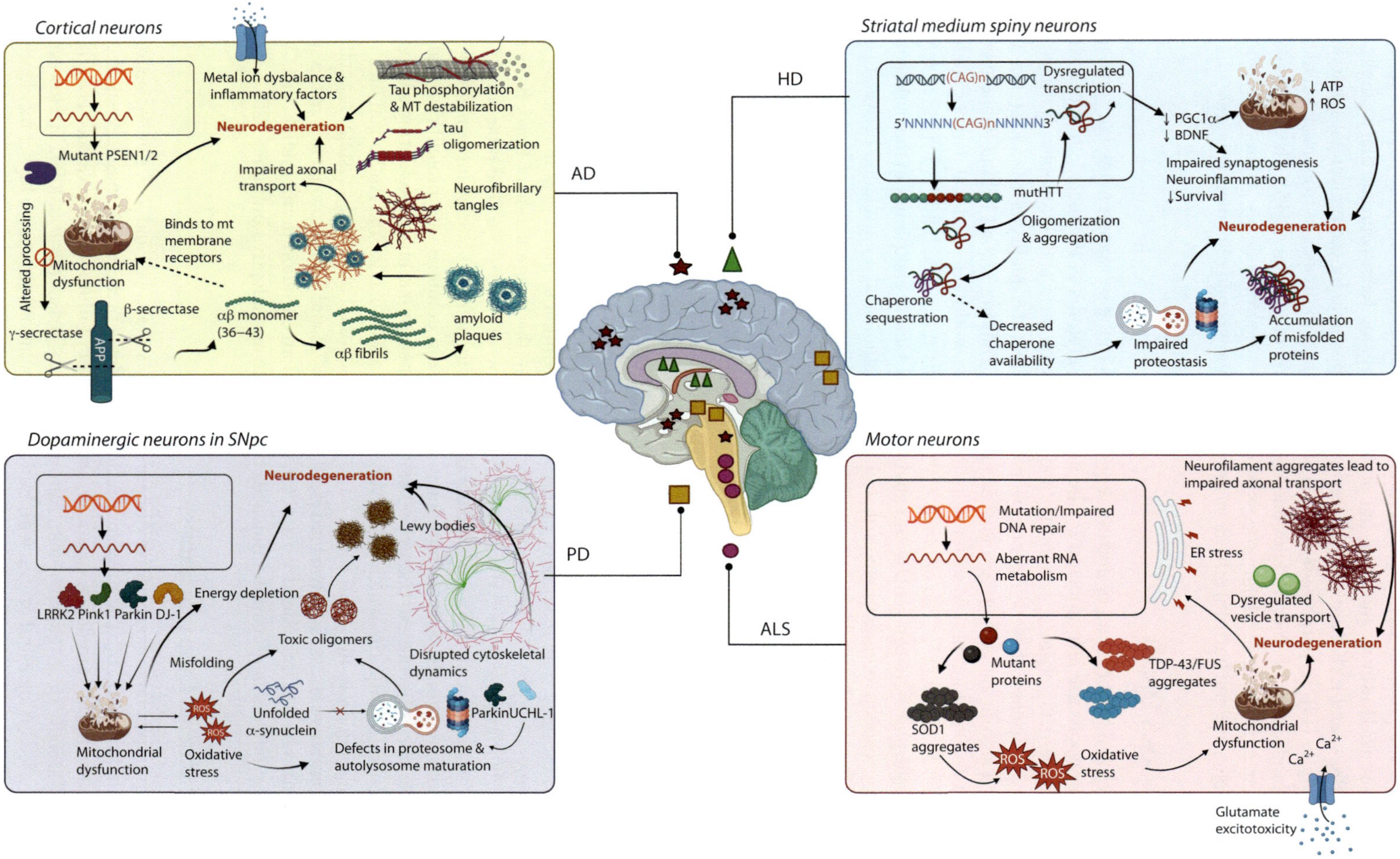

FIGURE 7.2 Schematic representation of major pathogenetic events in AD, PD, ALS, and HD.

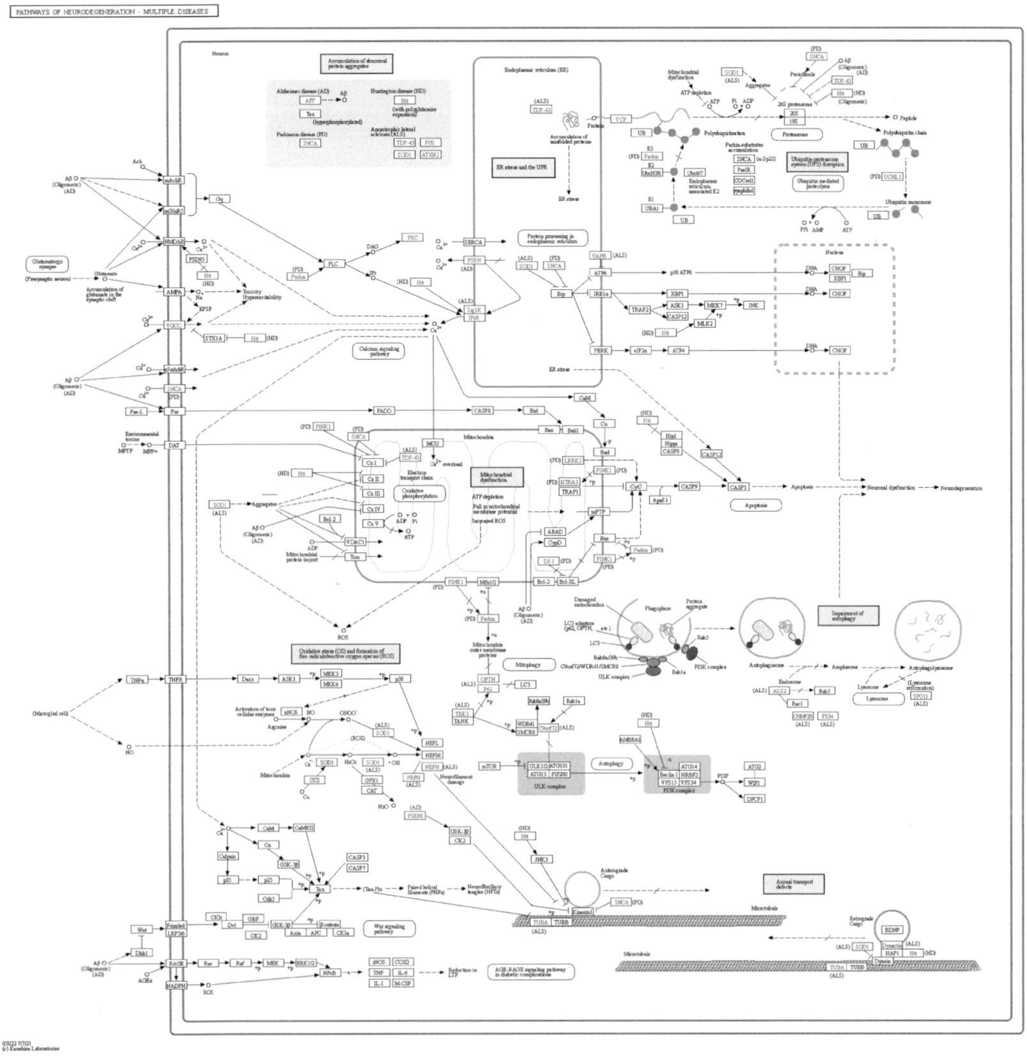

FIGURE 7.3 Schematic KEGG map representation of major pathways involved in AD, PD, ALS, and HD., (Image credit: Kanehisa Laboratories, KEGG: hsa05022. Permission to use the modified image was graciously granted by Kanehisa Laboratories, [128].)

[29–31]. Mutations in SOD1, FUS, TDP-43, and C9orf72 are associated with both familial and sporadic cases and account for up to 70% of familial ALS cases (Table 7.1). Most ALS patients exhibit motor impairments, but a significant number also exhibit subclinical signs of frontotemporal syndromes (FTD) associated with behavioral, cognitive, or memory dysfunctions [32]. Multiple genetic and environmental factors are implicated in conferring the disease state and two forms of ALS [33]. It is unclear what causes ALS, but there are multiple mechanisms depending on its origin, and mutations can exert their toxic effects in different ways. A variety of genes that influence cellular proteostasis, cytoskeletal dynamics, and RNA metabolism and processing play roles in ALS development. Several genes are associated with ALS that encode RNA-binding proteins. TDP-43 is one of these proteins, which aggregates in the cytoplasm of motor neurons in most cases. Mutations in these genes affect TARDBP and FUS aggregation in the cytoplasm [34]. This may also reduce the amount of RNA-binding protein in the nucleus, suggesting abnormal processing of the targeted RNA transcripts. In the majority of subtypes of the disease, motor neurons accumulate proteinaceous inclusions, including

TDP43 [3]. Additional types of protein aggregates may also occur, including neurofilaments, hyaline conglomerates, and misfolded SOD1. Prion-like proteins might be the source of the disease's spread [35]. Superoxide dismutase (SOD1) is a major antioxidant enzyme present in mitochondria, and more than 100 SOD1 mutations are detected in different forms of ALS [36]. The change in mitochondrial dynamics in terms of fusion fission events/biogenesis, mitochondrial membrane potential, and impairments has been reported in ALS [37]. A growing body of evidence suggests that mitochondria dysfunction and a number of oxidative and immune markers such as elevated cytokines, reactive oxygen species, and abrupt caspase activation have been reported in the blood and neuronal tissues of disease-affected patients [38,39]. But how they result in a combination of signatures and clinical and genetic heterogeneity in ALS and ALS-FTD remains poorly understood. Another common cause of ALS is the expansion of the hexanucleotide repeat in the C9orf72 gene [40]. The C9orf72 mutation is responsible for RNA foci, dipeptide repeat proteins, or loss of function mechanisms [41]. The predominant mechanisms observed in ALS include aberrant proteostasis, impaired RNA metabolism, vesicle trafficking, excitotoxicity, mitochondrial dysfunction, and neuroinflammation [3]. Non-neuronal cells like astrocytes and microglia, which contribute to neurodegenerative progression, secrete neurotoxic mediators and modulate glutamate receptor expression [42,43] (Figures 7.2 and 7.3).

7.1.4 Huntington's Disease

HD is a progressive neurodegenerative disorder caused by the expansion of trinucleotide repeats in the HTT gene, which encodes the Huntington protein [44]. The abnormal expansion of polyglutamine in the Huntington protein forms protein aggregates that are toxic to striatal neurons. Furthermore, the disease also affects other regions of the brain, not just the medium-spiny neurons in the striatum. HD patients present a confluence of cognitive deterioration, psychiatric manifestations, dementia, motor dysfunction, and chorea [2]. In most cases, the symptoms begin between the ages of 30 and 40 and gradually worsen, leading to death within 15 to 20 years following the onset of symptoms; however, juvenile cases have also been reported [45]. The global prevalence of HD is 5–10 per 100,000 inhabitants; however, white Europeans have higher incidences compared to Asians [46]. The disease initiation and severity may be affected by CAG repeat length and genetic modifiers, which affect the expression and processing of mutant proteins, but the underlying disease mechanism is largely undetermined [45,47] (Table 7.1). The huntingtin protein is essential for the development of the nervous system and regulates both neuronal and glial function [48]. A key part of the disease involves the breakdown of huntingtin into fragments of elongated glutamine sequences at its amino-terminus, which move into the nucleus and undergo aberrant post-translational modifications, such as phosphorylation and ubiquitination. [49]. The N-terminal phosphorylation of htt fragments is a key step in their aggregation and nuclear localization [50]. Mutant HTT interacts with multiple transcription factors, disrupting transcription, RNA processing, signaling cascades, and vesicle trafficking. Also, the brain-derived neurotropic factor (BDNF) or transport of axonal materials may be insufficient to support striatal neurons [51,52]. Additionally, Huntington's disease severely compromises the heat shock response, which protects cells from stressors [53,54]. In general, the mutant huntingtin causes cellular dysfunction that leads to a network imbalance, which can contribute to neurological and non-neurological symptoms (Figures 7.2 and 7.3).

7.2 TREATMENTS

The majority of neurological disorders are incurable, resulting in poor quality of life and high healthcare costs. To date, there are no definitive neuroprotective medications, and only symptomatic treatments are available. As a result of the inability of large- and small-molecule drugs to cross the blood-brain barrier and the limited capacity for nerve cell regeneration, it is difficult to target widespread neuronal cell death. Treatments currently focus on symptom relief, but the disease progresses. To combat neurodegeneration, advances in immunotherapy, gene therapy, and stem cell therapy, are being explored. An overview is presented in Figure 7.4.

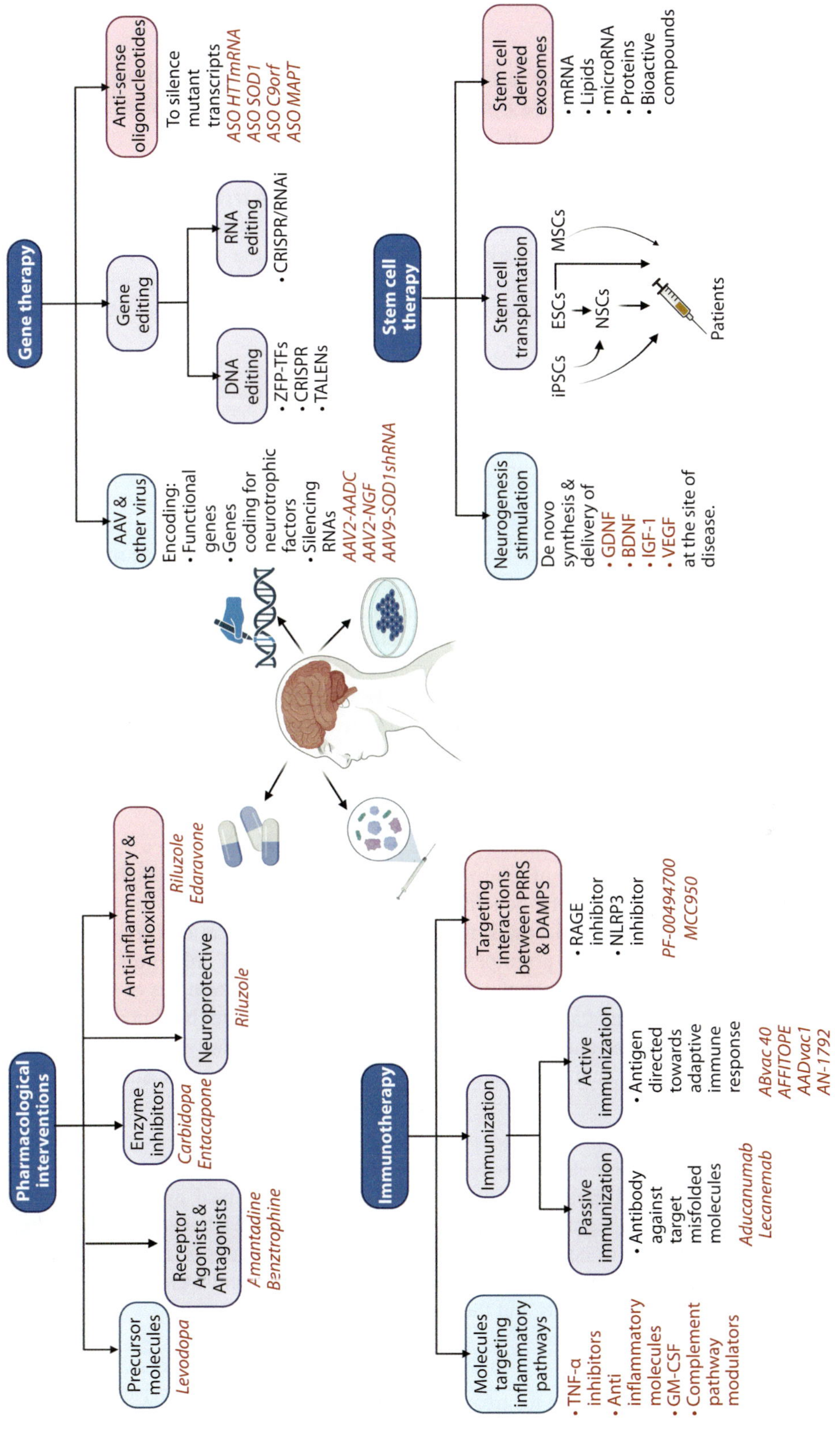

FIGURE 7.4 Schematic overview of advanced therapeutic approaches in AD, PD, ALS, and HD.

7.2.1 Pharmacological Treatments

7.2.1.1 Symptomatic Therapies

Neurodegenerative disorders can cause psychiatric, motor, gastrointestinal, and sleep disruptions. Selective serotonin reuptake inhibitors (SSRIs) and tricyclic antidepressants like nortriptyline and amitriptyline are frequently prescribed for depression. In the treatment of dementia, antipsychotics and stabilizers can be beneficial. It is recommended that SSRIs and anticonvulsants, such as carbamazepine, be administered to patients suffering from irritable and aggressive HD. Dantrolene, tizanidine, and benzodiazepines can relieve spasticity in ALS patients, but baclofen is usually preferred. Urinary incontinence has a profound effect on the daily activities and overall quality of life of patients. Tolterodine and oxybutynin can be used to relieve urinary urgency. For sleeping complaints (insomnia), sedating antidepressants like Trazadone, Mirtazapine, or tricyclics are helpful.

7.2.1.2 Disease Specific Therapies

Deficits in neurotransmitters are linked to neurodegenerative diseases like AD and PD. Replacement of these neurotransmitters temporarily alleviated symptoms and thus gained attention (Table 7.2).

7.2.1.2.1 Alzheimer's Disease

The cholinergic hypothesis suggests that acetylcholine substitution or an increase in acetylcholine could stabilize cholinergic transmission in AD. Many cholinesterase inhibitors against AD have been developed, such as galantamine, rivastigmine, and donepezil. Taking cholinesterase inhibitors as a first-line treatment depends on several factors, including each patient's unique circumstances; however, there is insufficient evidence that these agents are effective long-term. For the treatment of the advanced condition of AD, the FDA has approved the neuroprotective drug Memantine, which claims to improve brain functions such as learning and memory; however, a more comprehensive clinical trial is required to confirm its effectiveness.

7.2.1.2.2 Parkinson's Disease

Symptoms can be treated with levodopa, dopamine agonists, and monoamine oxidase inhibitors. The drug levodopa is very effective in treating bradykinesia and tremor. A peripheral decarboxylase, carbidopa, enhances therapeutic benefits. Dopamine agonists activate dopamine receptors, resulting in improved Parkinson's disease symptoms. Among the medications prescribed early to prevent levodopa complications are pramipexole and ropinirole. DOPT inhibitors are drugs that prevent dopamine breakdown and enhance the effect of dopamine. Examples of COMT inhibitors include tolcapone and entacapone. Amantadine is among the anticholinergic medications that promote dopamine release and block the NMDA receptor, potentially alleviating levodopa-associated dyskinesias. In the management of tremors, additional techniques, such as DBS, may be helpful.

7.2.1.2.3 Amyotrophic Lateral Sclerosis

In the current ALS treatment market, several drugs have been approved by the FDA: riluzole/Rilutek, edaravone/Radicava, Tiglutik, Exservan and Nuedexta, RELYVRIO and Qalsody but the choice of treatment is riluzole or edaravone. Originally intended for epilepsy, riluzole also protects ALS neurons from excitotoxicity-induced neurodegeneration. A clinical trial of riluzole in ALS showed modest survival improvements, but the efficacy remains unknown.

7.2.1.2.4 Huntington's Disease

The FDA has approved tetrabenazine for treating HD. The drug belongs to a class of drugs called synaptic vesicular amine transporter inhibitors. Pallidum stimulation is also in the research phase.

TABLE 7.2
Pharmacological Approaches for the Management of AD, PD, ALS, and HD

Disease	Drug	Adjunct Drug(s)	Nature	Mode of Action	Reference(s)
AD	**Donepezil**		Cholinesterase inhibitor (Piperidine derivative)	It is a specific inhibitor of acetylcholinesterase that hinders the activity of acetylcholinesterase. Preventing ACh hydrolysis therefore boosting ACh availability at synapses and improving cholinergic transmission.	[55,129]
	Rivastigmine		Cholinesterase inhibitor	The inhibitor is "pseudo-irreversible" for acetyl and butyrylcholinesterases, resulting in a net increase in ACh levels. Additionally, it selectively suppresses the G1 enzymatic variant of acetylcholinesterase, which is more prevalent in AD brain	[130,131]
	Galantamine		Cholinesterase inhibitor	Functions as an allosteric modulator that enhances the activity of nicotinic acetylcholine receptors, eliciting a greater response by increasing the sensitivity of the remaining nicotinic acetylcholine receptors. Additionally enhances the presynaptic release of acetylcholine and extends the duration of postsynaptic neurotransmission.	[129,132]
	Memantine		Glutamate regulator (anti-excitotoxic)	NMDA receptor antagonist. In the CNS, it blocks the NMDA glutamate receptor and inhibits an overactive glutaminergic system that plays a role in AD neurotoxicity.	[129,133]
PD	**Levodopa**		Dopamine precursor	Levodopa is the precursor to dopamine. Levodopa once converted to dopamine compensates for the decline in endogenous dopamine by activating postsynaptic dopaminergic receptors.	[129,134]
	Carbidopa	Used in combination with levodopa	Dopa decarboxylase inhibitor	Carbidopa inhibits the AADC enzyme by irreversibly binding to PLP. The AACD transforms 5-HTP into serotonin and L-dopa into dopamine. A decrease in peripheral levodopa metabolism allows for greater CNS uptake of L-dopa when combined with carbidopa.	[129,135]

(Continued)

TABLE 7.2 *(Continued)*
Pharmacological Approaches for the Management of AD, PD, ALS, and HD

Disease	Drug	Adjunct Drug(s)	Nature	Mode of Action	Reference(s)
	Entacapone *Levodopa + Carbidopa*	Used in combination with levodopa & carbidopa	Cathechol-O-methyltransferase (COMT) inhibitor	Carbidopa inhibits the decarboxylation of LD, causing COMT to become the primary enzyme responsible for metabolizing LD by catalyzing its conversion to 3-OMD. Entacapone inhibits the enzyme COMT, which is responsible for breaking down dopamine. This prevents the degradation of dopamine, which is converted from levodopa, both inside and outside the brain.	[129,136]
	Tolcapone	Used in combination with levodopa & carbidopa	Cathechol-O-methyltransferase (COMT) inhibitor	Tolacapone inhibits both peripheral & central COMT, which further decreases the degradation of LD in the periphery, allowing more LD to reach the brain, facilitating a lower dosage of LD with prolonged duration.	[129,137]
	Amantadine		Dopamine agonist/ anticholinergic	Possesses anticholinergic properties that counteract the imbalance between striatal dopamine and Ach levels brought on by degeneration of dopaminergic neurons. Furthermore, it stimulates dopamine release and inhibits the NMDA receptor, potentially alleviating dyskinesias caused by levodopa.	[129,138]
	Benztrophine		Anticholinergic	Antagonizes muscarinic ACh receptors in CNS and smooth muscles, consequently reducing central cholinergic effects and relieving muscle stiffness. Also blocks the reuptake of dopamine.	[129,139]
	Trihexyphenidyl		Anticholinergic	The precise mechanism is not fully comprehended, It inhibits efferent impulses directly by acting on the parasympathetic nervous system.	[140]
	Apomorphine		Dopamine agonist	Apomorphine exhibits a wide range of effects due to its catechol structure, which acts as a dopamine receptor agonist. Its mechanism of action is similar to that of dopamine or its precursor levodopa.	[141–143]

(Continued)

TABLE 7.2 ***(Continued)***
Pharmacological Approaches for the Management of AD, PD, ALS, and HD

Disease	Drug	Adjunct Drug(s)	Nature	Mode of Action	Reference(s)
	Bromocriptine		Ergot dopamine agonist	Stimulates dopamine-2 like (D_2) receptors and partially antagonizes D1 receptor. Activating dopamine receptors bypasses the need for presynaptic synthesis of dopamine. Upon binding to striatal D2 receptors, it stimulates locomotion and attenuates the bradykinetic symptoms.	[129,144–146]
	Cabergoline	Also used in combination with levodopa.	Ergot dopamine agonist	It preferably binds to a D_2 like receptor and is a long-acting agonist. It is used both as a monotherapy and as an adjunct therapy to levodopa.	[146,147]
	Lisuride	Also used in combination with levodopa.	Ergot dopamine agonist	Stimulates D_2 like receptors. Provides effective monotherapy, additionally permits levodopa dose when used in conjunction with levodopa.	[146,147]
	Pergolide	Also used in combination with levodopa.	Ergot dopamine agonist	It acts on both D_1 and D_2 like receptors. Shown to improve condition when used as monotherapy or in conjunction.	[146,147]
	Pramipexole		Non-ergot Dopamine agonist	Stimulates D2 like receptors with highest affinity for D3.	[146]
	Ropinirole	Also used in combination with levodopa.	Non-ergot dopamine agonist	It is a selective dopamine D_2 type receptor agonist. Effective as monotherapy and as an adjunct therapy.	[146,147]
	Rotigotine	Also used in combination with levodopa.	Dopamine agonist	Works as a partial or full agonist of an array of dopamine receptors thereby reducing the sensitivity developed over time from levodopa usage.	[148]
	Selegiline		Monoamine oxidase B (MOB) inhibitor	Upon binding to MAO-B, it prevents the enzyme from degrading dopamine. Inhibiting these enzymes prevents reuptake of dopamine in the CNS; thus, the synaptic cleft is saturated with biologically active monoamines.	[129,149]
	Rasagiline		Monoamine oxidase B (MOB) inhibitor	This is a highly selective, irreversible inhibitor of MAO-B that slows metabolism of endogenous and exogenous dopamine, hence providing symptomatic benefit.	[150–153]

(Continued)

TABLE 7.2 *(Continued)*
Pharmacological Approaches for the Management of AD, PD, ALS, and HD

Disease	Drug	Adjunct Drug(s)	Nature	Mode of Action	Reference(s)
ALS	**Riluzole**		Anti-excitotoxic	Prevents excitotoxicity-induced degeneration of motor neurons by interrupting glutamatergic transmission and decreasing glutamate concentration.	[154]
	AMX0035 (Relyvrio)			It prevents motor neuron cell death by mitigating mitochondrial dysfunction and endoplasmic reticulum stress.	[155,156]
	Edaravone (Radicava)		Antioxidant	In brains and spinal cords, edaravon protects neurons by scavenging free radicals.	[155,157]
HD	**Tetrabenzene**			Functions in reducing Huntington-associated chorea by irreversibly inhibiting the human vesicular monoamine transporter-2 (VMAT2), resulting in selective depletion of central monoamines from the nerve terminals.	[158,159]
	Deutetrabenazine			It is an isotopic isomer of tetrabenzene having the same chemical function. However, the 6 hydrogen atoms replaced by deuterium extends its half-life, requiring less frequent administration.	[158,159]

7.2.2 Advanced Therapies for Neurodegenerative Diseases

At present, the majority of neurodegenerative diseases are managed by focusing on alleviating symptoms rather than targeting the root causes. Emerging evidence indicates that immunotherapy, gene therapy, and regenerative medicine may be effective in disease management.

7.2.2.1 Immunotherapies

Immune therapy refers to the manipulation of an individual's immune system to combat diseases. Multiple reports indicate that immunological cells and their constituents have the potential to combat progressive neurodegenerative diseases, particularly those resulting from misfolded proteins and protein aggregates. In several initiatives, monoclonal antibodies and antigens have been used to fight plaques, tau, and syn aggregates (Table 7.3).

7.2.2.1.1 Alzheimer's Disease

AD is a prime example, with ongoing clinical studies exploring ways to address Aβ aggregates [160]. Active immunization employs a variety of Aβ antigens, whereas passive immunization utilizes purified anti-Aβ antibodies. The overall goal is to expedite the removal of Aβ. AN1792, the first anti-Aβ vaccine, was clinically tested. The study showed reduced Aβ deposition, improved CSF tau levels, and neuropsychological test scores, but also caused T cell-mediated meningoencephalitis in a few participants [161]. The vaccine amilomotide (CAD106) stimulates anti-Aβ

TABLE 7.3
Immunotherapeutic Drugs Listed for AD, PD, ALS, and HD

Disease	Molecule Name	Target	Class	Trial Phase	Clinical Trial Identifier
AD	Masitinib	Inhibitor (Tyrosin kinase)	Small molecule	III	NCT03127267
	Ibudilast (MN-166)	Inhibitor (Phosphodiesterase)	Small molecule	II / III	NCT04057898
	Fasudil	Inhibitor (Rho kinase)	Small molecule	II	NCT03792490
	Ravulizumab	Inhibitor (Complement component 5)	Monoclonal antibody	III	NCT04248465
	Zilucoplan	Inhibitor (Complement component 5)	Small molecule	II / III	NCT04297683
	Anakinra	Antagonist (IL-1 receptor)	Monoclonal antibody	II	NCT01277315
	Tocilizumab	Antagonist (IL-1 receptor)	Monoclonal antibody	II	NCT02469896
	Aducanumab	Epitope (N terminus of Aβ)	Ig1 monoclonal antibody	Ib	NCT01677572
	donanemab	Epitope (N terminus of Aβ)	Ig1 monoclonal antibody	II, III	NCT03367403 NCT05026866
	Lecanemab	Epitope (N terminus of Aβ)	Ig1 monoclonal antibody	II, III	NCT01767311 NCT03887455
	Crenezumab	Epitope (N terminus of Aβ)	Ig1 monoclonal antibody	II, III	NCT01998841
	Gantenerumab	Epitope (N terminus of Aβ)	Ig1 monoclonal antibody	II, III	NCT04339413 NCT04374253
	Tilavonemab	Epitope (N terminus of Tau)	Ig1 monoclonal antibody	II, III	NCT02880956, NCT03712787
PD	sargramostim	Protein (GF)	GM-CSF	Ib	NCT01882010
	AFFITOPE - PD01A	Mimics native C-terminal αSyn epitope	Monoclonal antibody	IB	NCT02267434
	AFFITOPE - PD03A	Mimics native C-terminal αSyn epitope	Monoclonal antibody	I	NCT02618941
	Prasinezumab	Targeting native C-terminal of αSyn	Monoclonal antibody	II	NCT03100149
	BIIB054	Targeting native N-terminal of αSyn	Monoclonal antibody	II	NCT03318523
	Lu AF82422	Targeting αSyn	Monoclonal antibody	I	NCT03611569
	MEDI1341	Targeting αSyn	Monoclonal antibody	I	NCT03272165
	ABBV-0805	Targeting αSyn	Monoclonal antibody	I	NCT04127695
ALS	Masitinib	Inhibitor (Tyrosin kinase)	Small molecule	III	NCT03127267
	Ibudilast (MN-166)	Inhibitor (Phosphodiesterase)	Small molecule	II / III	NCT04057898
	Fasudil	Inhibitor (Rho kinase)	Small molecule	II	NCT03792490
	Ravulizumab	Inhibitor (Complement component 5)	Monoclonal antibody	III	NCT04248465
	Zilucoplan	Inhibitor (Complement component 5)	Small molecule	II / III	NCT04297683
	Anakinra	Antagonist (IL-1 receptor)	Monoclonal antibody	II	NCT01277315
	Tocilizumab	Antagonist (IL-1 receptor)	Monoclonal antibody	II	NCT02469896
HD	Pepinemab/ VX15/2503	Anti SME4D	Monoclonal antibody	1/II	NCT02481674
	ANX005	Inhibitor (complement component 1q)	Monoclonal antibody	II	NCT04514367

antibody production without causing a T-cell response, but unexpected changes in body weight, brain volume, and cognitive performance forced an early termination of the research trial [162]. UB-311, a combination of artificial Aβ1-14-targeting peptides and a helper T-cell peptide epitope, has shown promise for cognitive improvement and dementia in AD patients [163]. Vaccines like AADvac1 target conformational epitopes in tau's microtubule-binding region to boost immune responses against tau pathology aggregation and promote tau clearance [164]. In an ongoing multicenter clinical trial (NCT04445831), AC Immune will evaluate the immunogenicity, safety, and tolerability of ACI-35, an active immunotherapy vaccine targeting hyperphosphorylated tau. It is expected that the study will conclude by 2023. Passive immunotherapy, utilizing humanized antibodies or polyclonal immunoglobulins, boosts Aβ clearance but is often linked to side effects like vasogenic edema and cerebral amyloid angiopathy. Aducanumab, a human IgG1 monoclonal antibody, reduces amyloid load by binding to the N terminus of Aβ. Based on its claim to improve cognitive function and the clinical development of AD, the FDA approved it in June 2021. The efficacy and safety of Lecanemab, Crenezumab, Gantenerumab, and Tilavonemab are also being studied.

7.2.2.1.2 Parkinson's Disease

In PD, acute or chronic damage to neurons can contribute to extracellular misfolded proteins and innate immune activation. As a result, glial cells become activated, producing pro-inflammatory cytokines that damage neuronal circuits. T cells or antibodies respond to deviations in immunological tolerance as well as the formation of effector memory T cells that could aggravate microglial responses and neuronal susceptibility [165]. There is evidence that macrophage-colony stimulating factor can modulate inflammatory responses and has shown therapeutic promise in Parkinson's disease [166,167].

Immunotherapies targeting aSyn aim to reduce extracellular aSyn load and prevent its spread in the brain. Anti-aSyn immunotherapy offers both active and passive immunization [168]. Multiple clinical trials involving a-syn are currently underway, and several are expected to begin shortly. In Phase I clinical trials, Affris has used its AFFITOPE PD03A vaccine to target a-syn-mimicking peptides for Parkinson's patients. In the study, most adverse events were mild, with no serious reactions observed. Biogen antibody BIIB-054, a drug directed against a-syn, is currently undergoing clinical trials (Alzforum, 2017). Similarly, phase 1a and 1b trials of humanized monoclonal antibodies (PRX002, Prothena) have been conducted, and multiple clinical immunotherapy initiatives focused on targeting a-syn in PD are expected to grow in the coming years.

7.2.2.1.3 Amyotrophic Lateral Sclerosis

SOD1, an enzyme responsible for reducing oxidative stress, has been linked to familial ALS. Studies have demonstrated that vaccines targeting extracellular SOD1 can alleviate symptoms and reduce ALS pathogenicity [169]. It has also been demonstrated that antibody therapy against mutant SOD1 oligomers reduces neurotoxic aggregates, inhibits gliosis, and reduces inflammation in motor neurons [170]. The EU-funded project is targeting the C90rf72 mutation and developing a poly-GA peptide vaccine for ALS treatment. The preclinical findings are promising, and clinical trials are expected to commence soon. More in-depth information about ALS immunotherapies can be found in recent reviews by Richard J. Mead [88].

7.2.2.1.4 Huntington's Disease

Several approaches have been used to reduce mutant huntingtin protein levels (mHTT), but immunomodulation is a potential alternative for disease modification [171]. As part of its HD immunotherapy, semaphorin 4D (SEM4D) is a potential therapeutic target since it is involved in cellular trafficking and is expressed in invading immune cells and neuroinflammatory processes in the CNS. SIGNAL research is currently evaluating a humanized monoclonal antibody, VX15/2503, which targets SEM4D in HD (ClinicalTrials.gov, NCT02481674). In rodent models of HD, active vaccination increases immunogenic responses to peptides from HTT exon 1, whereas DNA plasmid vaccination improves diabetic phenotypes [172,173]. There have been studies investigating recombinant anti-Htt-directed antibodies as a possible therapeutic intervention against mHtt. When HD R6/1 transgenic mice are treated with recombinant AAV expression of an anti-N-terminal Huntington fragment, mutant Huntington accumulates more slowly [174].

7.2.2.2 Gene-based Therapies for Neurodegenerative Diseases

Gene therapy is a promising solution for treating diseases by precisely targeting their causes and providing a permanent cure. It can modulate gene and RNA expression levels or the physiological properties of endogenous proteins, thus treating neurodegenerative diseases. The advancement of technologies like AAVs, ASOs, CRISPR/Cas9, TALENs, ZFNs, and small molecules has broad implications for therapeutics in AD, PD, ALS, and HD (Table 7.4).

7.2.2.2.1 Alzheimer's Disease

Presently, preclinical trials in the field of AD research are primarily focused on the amyloid pathway, tau protein, APOE variants, inflammatory cytokines, and neurotrophic factors. Three genes—APP, MAPT, and APOE—that are strongly linked to Alzheimer's disease are potential targets for genetic therapies. Obstacles to successful application include the complexity of administering monogenic genetic therapies, risks associated with intracranial injection, instability of genetic therapies, limited applicability of pre-clinical AD models, and substantial costs associated with producing genetic therapies.

TABLE 7.4
Gene Therapy Listed for AD, PD, ALS, and HD

Diseases	Target Proteins	Roles	Vectors	References
AD	NGF	Neurotrophic, synaptic plasticity	MLV (Ex vivo), rAAV	[210]
	BDNF	Neurotrophic, synaptic plasticity	Lentivirus	[177]
	Neprilysin	Aβ degradation, neuroprotection	rAAV, Lentivirus, HSV	[211]
	Neprilysin (secreted form)	Aβ degradation, neuroprotection	Lentivirus	[212,213]
	ECE	Aβ degradation	rAAV-5	[214]
	APOE2	Lipoprotein metabolism, Aβ burden	Lentivirus	[215]
	BACE1	Aβ generation	Lentivirus(siRNA)	[216]
	APP	Aβ generation	HSV (siRNA)	[217]
PD	GDNF	Induces neurite sprouting	AAV	[218]
	BDNF	Protective effect on dopaminergic neuron	HSV1	[219]
	CDNF	Neurotrophic effect on nigral dopaminergic neurons	AAV	[220]
	VEGF	Neuroprotective effect on nigral neurons	AAV	[221]
	Neurturin (NTRN)	Protective effect on dopaminergic neuron	AAV	[222]
	Aromatic acid decarboxylase (AADC)	Dopamine synthesis	AAV	[222,223]
	Tyrosine hydroxylase (TH)	Protective effect on dopaminergic neuron	AAV	[222,223]
ALS	ASO against SOD1 mRNA	Safety, pharmacokinetics, CSF-SOD1 concentration	I, II	[224]
	ASO against SOD1 mRNA	CSF-SOD1 concentration, NfL	III	[225]
	ASO against SOD1 mRNA	Safety and pharmacokinetics	I	[224]
	AAV-miR-SOD1	Safety and pharmacokinetics	Open-label	[226]
	ASO against C9orf72 mRNA	Safety	Literature	[227]
	ASO against FUS mRNA	FUS level	Pre-clinical	[228]
HD	ASO against HTT	Prevent neuronal death by mhtt	AAV	[228]
	AAV5-miHTT	Prevent neuronal death by mhtt	AAV5	[229]

7.2.2.2.1.1 Expressing Growth Factors In AD, cholinergic neuron loss destabilizes cholinergic transmission. Neurotrophic factors are crucial in maintaining neuronal health. It has been shown that nerve growth factor (NGF) enhances cholinergic function, memory, and transmission in animal models. Using the AAV2 vector, an exogenous NGF gene delivery trial was carried out to modify NGF to express in the forebrain of AD patients [175]. While preclinical studies have suggested potential benefits from exogenous NGF treatment, phase 1/2 clinical trials have not shown improvements in cognition [176]. In other clinical trials, NGF levels were increased in patients; however, their benefits varied. BDNF is on the horizon in future studies [177].

7.2.2.2.1.2 Targeting the APOE Gene The APOE gene confers the greatest genetic susceptibility to AD [11]. Distinct ApoE isoforms exert distinct effects on AD pathophysiology. While APOE4 is the most significant risk factor for AD development, APOE2 confers protection. Increasing APOE2 levels in the brain, converting APOE4 to APOE2/3, regulating APOE4 lipidation, targeting ApoE structural properties, and targeting ApoE receptors are all potential treatment strategies [178,179].

7.2.2.2.1.3 Targeting the APP Gene AD pathogenesis is heavily reliant on APP. There is evidence that targeting APP expression can improve physiological and behavioral abnormalities. To reduce Aβ production, Chang et al. utilized SSOs to precisely target and regulate APP expression [180]. The CRISPR/Cas9-mediated targeting of APP in vivo and ex vivo reduces pathogenic Aβ load [181]. An important component of synaptic complexity and maintenance is sAPPα, which is derived from the complete form of APP. Evidence suggests that increasing sAPP levels may have a beneficial impact on AD [182]. siRNA technology holds great potential as a therapeutic modality due to its ability to selectively inhibit target genes. RNAi targeting APP is a promising option for both familial and sporadic AD [183].

7.2.2.2.1.4 Targeting the MAPT Gene Tau protein plays an integral role in disease development. A growing body of evidence suggests that reducing tau levels may reduce the severity of AD. Several studies have successfully used ASOs to lower MAPT gene levels in animal models [184]. As part of a human AD clinical trial currently underway, tau-targeting antisense oligonucleotides are being tested to inhibit MAPT expression [185]. Further, the use of siRNA can also decrease tau levels [183].

7.2.2.2.1.5 Targeting the BACE1 Gene Inhibition of BACE1 reduces amyloid plaque production and improves behavioral and neuropathological signs in animal models [186]. A critical component of the secretase complex is PS1. PS1 mutations are linked to inherited AD. Using anti-PS1 siRNA, researchers were able to reduce A42 levels within IMR-32 cells [187] and it may have therapeutic potential [188].

7.2.2.2.2 Parkinson's Disease

Parkinson's disease can be treated with both non-disease-modifying and disease-modifying genetic therapies. A non-disease-modifying therapy expresses enzymes to correct abnormal basal ganglia firing, while a disease-modifying therapy restores neurons and prevents PD-mediated cells from dying. It has been shown that genetic approaches can promote GABA activation and dopaminergic signaling in the brain to alleviate PD symptoms. In human clinical trials, these therapies have proved to be well-tolerated and have helped alleviate the symptoms of Parkinson's disease [189]. One of the most extensively researched methods involves amplifying growth factors that have neuroprotective properties in the nigral cortex [190]. Table 7.4 lists promising candidates that express neurotrophic and regenerative factors in PD research. There are several genes associated with PD including SNCA, GBA, and LRRK2. Findings indicate that knocking down synuclein in mice effectively halts the process of neurodegeneration [191]. In a patient-based iPSC neuronal model, downregulation of synuclein improves the viability of dopaminergic neurons [192]. Reducing LRRK2 can also ameliorate PD phenotypes. LRRK2 ASOs reduced mRNA and protein levels without causing any adverse side effects, and its human clinical trials are underway [193]. PD can also be addressed using siRNA technology, which targets the α-syn gene. There is evidence that siRNA can reduce the expression of SNCA both in vitro and in vivo [194]. The utilization of anti-SNCA siRNA in the primate (monkey) model resulted in the successful reduction of α-synuclein mRNA and protein without adverse effects [195]. Further, anti-SNCA siRNA was reported to reduce the level of syn mRNA and protein in a primate model. To increase siRNA effectiveness, virus vectors containing α-syn siRNA are in current practice. An RNA interference technique, ExCont-RNAi, improved motor function in PD models by regulating SNCA overexpression [196].

7.2.2.2.3 Huntington's Disease

There are two basic strategies for preventing neural death in HD: 1) growth factors that provide nutrients and support for survival, as discussed previously in the PD gene-targeted therapy, and 2) reducing or eradicating mutant huntingtin proteins that cause cellular damage. A wide range of neurotrophic factors have been studied in animal models and clinical trials for Huntington's disease (HD) and other neurodegenerative disorders [197]. The use of ASOs can reduce overall levels of HTT protein and delay Huntington's disease progression. The administration of AAV5, along with a microRNA targeting HTT, in the brains of transgenic HD minipigs resulted in a significant reduction of mutant mHTT [197,198]. Recently, two therapies that are currently in the most advanced stage of clinical investigation have used ASOs. Intrathecally, ASOs are administered directly into the CSF. The purpose of administering the ASOs via CSF is to disperse them throughout the CNS, where they are absorbed by recipient cells. Once the ASOs reach the nucleus, they attach themselves to their target, in this case, HTT mRNA. This leads to the degradation of huntingtin protein bound to endogenous RNAse H1, which can be allele-specific (NCT04000594) or mutant allele-specific (NCT03225833, NCT03225846) [197,199]. It is currently being studied whether AAV5 injected directly into the striatum can treat HD (NCT04120493). As part of the AAV5 vector, a transgene produces microRNA. MicroRNA is a modified short RNA that blocks the translation of huntingtin mRNA into huntingtin protein [200].

7.2.2.2.4 Amylotrophic Lateral Sclerosis

Several neurotrophic factors, immunomodulators, and therapies targeting ALS-linked genetic mutations are being studied both in preclinical and clinical settings for the treatment of ALS. Clinical trials are currently testing gene therapies targeting SOD1 and C9orf72 mutations [201–203].

7.2.2.2.4.1 Expressing growth factors Neurotrophins are molecules that transmit signals to regulate the activity of neurons and influence their survival. Studies have shown that scAAV9-encoding IGF-1, 2, and VEGF injections can reduce motor neuron deterioration in the spinal cord, delay disease progression, increase lifespan, and prolong survival and motor strength in mice [204,205].

7.2.2.2.4.2 Targeting the SOD1 Gene Administering ASO-SOD1 into the cerebrospinal fluid (CSF) of animal models carrying the mutatations in the human SOD1 gene resulted in a reduction of SOD1 levels and an improvement in survival rates [206]. Initially, ASOs specific for ALS were tested in rats and rhesus monkeys and exhibited slow disease progression. In the human I phase trial, results were safe and well tolerated (NCT01041222). Phase I/II trials of BIIB067 showed a notable decrease in SOD1 level, particularly in patients experiencing rapid disease progression (NCT02623699). An ongoing phase 3 clinical trial of BIIB067 is in the planning stages (NCT03070119).

7.2.2.2.4.3 Targeting the C9orf72 Gene An expanded hexanucleotide repeat mutation in C9orf72 may lead to RNA foci, dipeptide repeat proteins, or loss of function mechanisms in ALS. There is evidence that ASO-C9orf72 injection reduces mRNA foci and improves cognitive function in mice [207]. In a clinical study, intrathecal administration of ASO5-2 led to the suppression of C9orf72 transcript levels in a single ALS patient. The study, however, showed no clinical benefits (NCT03626012). RNAi and CRISPR gene editing were both used for inactivating mutant SOD1. In an animal model, intravenous administration of AAV9-SOD1 shRNA decreased SOD1 levels, slowed disease progression, and extended life expectancy [208]. CRISPR/Cas-mediated disruption of SOD1 expression in mice leads to enhanced motor function and increased survival [209]. Similarly, siRNA injected into spinal motor neurons slows the progression of a disease. Foust et al. reported that intrathecal and peripheral administration of AAV-encoded shRNA against human SOD1 mRNA can significantly reduce disease progression and onset [208].

7.2.2.3 Stem Cell Therapy in Neurodegenerative Diseases

There are two types of stem cells: embryonic (ESC) and adult. ESCs have the potential to differentiate into any kind of cell, but their clinical application is limited due to ethical concerns. Adult stem cells, found in the brain (NSC), bone marrow (BMSC), and adipose tissue (AMSC), are extensively utilized in clinical research. The induced pluripotent stem cell is a type of stem cell derived from any somatic cell through reprogramming and can be used for both regenerative medicine and drug discovery. Stem cell therapy is a regenerative therapy model that uses stem cells and released extracellular vesicles or other derivatives to improve the repair response of degenerative cells. It aims to replace cellular tissues and enrich the environment for existing neurons by inducing neurotrophic factors, including GDNF, BDNF, IGF-1, and VEGF, and building auxiliary neural networks for treating neurodegenerative disorders [230,231]. Stem cell therapy has significantly transformed clinical practice by offering valuable and promising medical applications in the treatment of several neurodegenerative diseases [232]. Although stem cell research is still in its infancy, it is quickly emerging as a promising, exciting, and safe treatment option for neurodegenerative diseases. To develop and translate stem cells successfully from laboratory to patient, it is crucial to select a suitable cell type and deeply investigate mechanisms of support and neuronal pathology (Table 7.5).

7.2.2.3.1 Alzheimer's Disease

In recent years, preclinical and clinical studies have shown stem cell therapy can treat disease, improve neurogenesis, and alleviate cognitive impairment, suggesting it may become a viable treatment for AD, which is currently considered incurable. NSC transplants directly replace lost neurons in the recipient tissue. Transplanted NSCs can also stimulate neurogenesis indirectly via paracrine cytokines. NSCs from humans have been proven to improve synaptic plasticity, prevent pathology burden, and restore spatial learning and memory in AD mouse models. In a mouse model of AD, NSCs expressed growth factors and choline acetyltransferase, boosted neurogenesis, and enhanced cognitive performance [233,234]. The ability of transplanted NSCs to develop into cholinergic neurons may be crucial to the therapeutic benefit [235]. NSC transplantation in the hippocampus region improves memory function and cognitive function through the release of BDNF [236]. While NSCs play a key role in promoting neurogenesis and maintaining cognitive function in AD, there is still much to learn. For instance, NSC transplantation still poses challenges in generating non-neuronal cell types [235,237]. Transplanting human ESCs into mice led to improved cognitive function [238]. Recently, transgenic AD mice implanted with iPSC-derived cholinergic neuronal precursors showed improvement in their spatial memory. Eventually, these precursors become mature cholinergic neurons [239]. It is important to note, however, that any use of iPSCs for tissue regeneration may carry some risk of tumor development [240]. To overcome these problems, research is being conducted using trans-differentiation approaches for converting non-neuronal cells into neurons. The use of mesenchymal stem cells to treat Alzheimer's disease in animals has been extensively researched due to their easy access, versatility, and ability to change into different types of cells. There is evidence that the use of MSCs can enhance neurogenesis and synaptogenesis, reduce amyloid accumulation, and improve cognitive recovery [241] and the safety and efficacy of MSCs are currently underway in several AD clinical trials [242–245].

7.2.2.3.2 Parkinson's Disease

Dopaminergic neuronal loss is a contributing factor to PD; therefore, researchers have looked at alternative ways to supplement dopamine by using stem cell therapy to replace dopaminergic neuronal loss. ESCs, iPSCs, and NSCs are being used to differentiate into mature dopaminergic cells [246].

TABLE 7.5
Stem Cell Therapy Listed for AD, PD, ALS, and HD

Disorders	Stem Cells	Functional Outcome	Clinical Phase	Identifier
AD	hNSCs	Neurotrophic factors release microglial activation	I	ChiCTR2000028744
	hNSCs	Neuroprotection by activation of microglia	I	ChiCTR2000039011
	Autologous hNSCs	Neurotrophic factors release microglial activation	I	ChiCTR-ONC-17014171
	hUC-MSCs	Improved spatial learning and memory	I	NCT01297218
	hUC-MSCs	Improved spatial learning and memory	II	NCT01696591
	hUC-MSCs	Improved spatial learning and memory	I, IIa	NCT02054208
	hUC-MSCs	Improved spatial learning and memory	I, IIa	NCT03172117
	hUC-MSCs	Improved spatial learning and memory	I, II	NCT01547689
	hUC-MSCs	Improved spatial learning and memory	I	NCT04040348
	Human amniotic and UC-MSCs	Improved spatial learning and memory	I	NCT03899298
	Human amniotic and UC-MSCs	Improved neurodegeneration and immune protection	I, II	NCT04684602
	Human placenta MSCs (CB-AC-02)	Improved neurodegeneration and immune protection	I, II	NCT02899091
	Longeveron allogeneic hMSCs	Decrease in amyloid beta and tau amount in CSF, improve cognition	I	NCT02600130
	Allogeneic hMSCs	Improve cognition defects and amyloid beta accumulation	IIa	NCT02833792
	Autologous MSC-NTF cells	Decrease in amyloid beta and tau amount in CSF	IIa	EUCTR2020-002872-11-NL
	Filgrastim (G-CSF)	Improved spatial learning and memory	I, II	NCT01617577
PD	hNSCs	Relief from static tremor, bradykinesia, rigidity	II, III	NCT03128450
	hNSCs	Regulates the dopamin secretion	I	NCT02452723
	hNSCs	Improve movement disorder	I, II	NCT03309514
	UMSCs	Reduction of motor symptoms, replacing of dopamine	I	NCT03550183
	MSCs	Improve motor systems	I	NCT03684122
	MSCs	Improve psychomotor development, neurological regression	II	NCT04506073
	Human stem cells	Improvement of motor responses	I	NCT02780895
	Human amniotic epithelial stem cells	Improve the secretion of dopamine and various neurotrophic factors.	I	NCT04414813
	iPSCs	Relief from static tremor, bradykinesia, rigidity	I	NCT03815071
ALS	MSCs	Changes in muscle strength grading	I	NCT01777646
	MSCs	Delayed the disease onset and progression	I	NCT01609283
	MSCs	Changes in muscle strength grading	I	NCT01142856
	MSCs	Alterations in the assessment of muscle power	I	NCT01759784
	MSCs	Improve muscle spasticity	I	NCT01759797
	MSCs	Improve muscle atrophy and fasciculations	I	NCT01771640
	Human spinal cord derived neural stem cell	Attenuation of motor function loss, maintenance of respiratory capacity;	II	NCT01730716

(Continued)

TABLE 7.5 *(Continued)*
Stem Cell Therapy Listed for AD, PD, ALS, and HD

Disorders	Stem Cells	Functional Outcome	Clinical Phase	Identifier
	Neural stem cell	Reduction of spasticity/rigidity	I	NCT01640067
	Human spinal cord derived neural stem cell	Improve muscle atrophy	I	NCT01348451
HD	Dental pulp MSCs	Neuroprotection and neurogenesis	II	NCT03252535.
	Immature dental pulp MSCs	Long-term safety and efficacy evaluation	II/III	NCT04219241
	Genetically engineered MSCs	Improvements in motor and cognitive function, reduced anxiety-like behaviors	I	NCT01937923
	Fetal neurons	Clinically insignificant	II	NCT00190450

Human fetal brain cell transplantation is a promising treatment and is currently being rigorously tested in clinical trials to evaluate its effectiveness while minimizing adverse effects. Several studies explored the dopaminergic neurons derived from human fetuses and implanted them into the depleted strata of PD patients. Takagi and colleagues have shown that the monkey's ESC-derived dopaminergic neurons transplanted into a Parkinson's primate model improved their functional ability [247]. Meanwhile, Wang et al. tested the safety and efficacy of dopaminergic neurons produced from human ESCs for the treatment of PD by precise dopaminergic neuron transplantation in monkeys, which showed excellent outcomes [248]. The success of partial motor recovery has led to the goal of reversing motor degeneration. At the same time, initial studies showed promising results, but ethical concerns and the risk of the development of tumorigenesis led to their ban. However, with more research and advancements in technology, we can hope to overcome these challenges and find effective solutions for reversing motor degeneration in PD.

There has been some hope for PD symptoms to be alleviated with NSCs, since they have a dopaminergic interneuron trait that can release dopamine. In an animal model, NSCs migrate and differentiate site-specifically after transplantation [249]. Adult NSCs possess the ability to differentiate into DA neurons within the subventricular zone(SVZ). Human NSCs transplanted into rats improved symptoms [250]. SVZ-derived nuclear-1 receptor-engineered NSC differentiation into DA neurons successfully reversed the behavioral deficits in rats with PD. Stem cell therapy with MSCs has shown promising results in reducing dopamine depletion and restoring damaged nerve terminals in PD animal models and human patients [251]. Recent studies suggest that MSCs can improve bladder dysfunction and dyskinesias without adverse effects [251]. For patients with PD, implantation of intra-arterial autologous stem cells is highly safe and effective, which also eliminates the risk of tumorigenesis and immunological compromises [252,253]. iPSCs are very useful for treating Parkinson's disease by stimulating dopamine replacement in patients. In preclinical studies, iPSC-derived dopaminergic progenitors have been proven to be an effective and excellent source of patient-specific cells for regenerative medicine. Studies irrefutably demonstrated that the transplantation of DA neurons differentiated from iPSCs into animal models can not only survive but also seamlessly fit in host networks and lead to functional improvements [254]. In addition, in vivo studies have shown that transplantation of iPSC-derived DA neurons can be beneficial in improving functional outcomes. The compelling evidence strongly supports the use of iPSCs as a promising and effective cell therapy in PD [255,256]. Although replacing cells is a promising approach to treating PD, another method is to provide environmental enrichment by delivering growth factors. This can help existing dopamine neurons slow down or prevent further degeneration. Innovative research demonstrates that encapsulating the necessary cells in a semipermeable polymer, along with neurotrophic factors and neurotransmitters, can be a promising strategy for cell replacement and environmental enhancement in PD research [257,258].

7.2.2.3.3 *Amyotrophic Lateral Sclerosis*

Genetics and the environment interact to make ALS difficult to treat. Studies evaluating stem cell transplantation in ALS have successfully shown the safety and feasibility of stem cell therapy for ALS [259,260]. Transplantation of hESC-derived MNPs into the spinal cord resulted in a mixture of mature and immature motor neurons [261] Fetal neural progenitors are multipotent stem cells derived from the fetal brain or spinal cord that have the potential to differentiate into a neuroglial population. Several studies have investigated the therapeutic potential of NSC in ALS [262,263]. Exogenous transplantation of neural stem cells (NSCs) has demonstrated significant progress in treating ALS. NSC injections of fetal tissue have been shown to significantly slow down disease progression in clinical trials [264]. Transplanting human iPSC-derived NSC into a rat model results in the long-term survival of grafted cells and their differentiation into motor neuron-like cells [265]. Trials on MSC transplantation in the dorsal-spinal cord have shown no short-term or long-term complications, making it a safe and viable treatment option for ALS patients [266]. Clinical trials with MSC transplantation in ALS patients showed increased motor neurons and reduced inflammation and ubiquitin deposits [267,268]. Currently, an extensive assessment of growth factors' potential neuroprotective effects on motor neurons is underway. Researchers are also investigating stem cells' ability to produce growth factors that can improve the environment. As part of its ALS treatment program, Brainstorm Therapeutics administers intrathecal injections of MSCs with a novel approach and analyzes their components.

7.2.2.3.4 *Huntington's Disease*

Stem cell therapy for HD has shown great promise in recent years. A variety of tissues can be used as grafts in cell transplantation studies. Experimental studies on animals have shown promising results [269,270]. BDNF-overexpressing MSCs slow striatal atrophy and promote neurogenesis [271]. NSCs, derived from various sources, are used for HD treatment. Overexpressing GDNF in NSCs and NPCs reduces neuronal death in HD mice and improves motor functions [272]. Several studies have shown that MSCs promote cell survival, reduce apoptosis, and stimulate compensatory neurogenesis. Furthermore, transplanting MSCs overproducing BDNF or NGF decreased behavioral deficits in HD models [273]. Also, NPC progenitors derived from ESCs that overexpress BDNF alleviate neurodegenerative symptoms [274]. Further, transplantation of MSCs into the hippocampus of mice with HD can enhance neural stem cell growth and differentiation, leading to neuroprotection and neurorestorative effects [275]. Utilizing stem cells derived from dental pulp is also a viable option for treating HD [276]. The use of NSCs derived from iPSC for HD cell replacement therapy has been studied very little till now. As a first attempt at replacing HD cells using hiPSC-derived NSCs, Jeon et al. have demonstrated remarkable success. NSCs transplanted into HD mouse models showed promising results, improving functional outcomes and preventing human mutant huntingtin aggregates [277]. The transplanted cells not only survived their transplantation but also differentiated successfully into motor neurons, proving the effectiveness of this method.

7.3 CONCLUSIONS

Neurological disorders pose a significant financial burden for individuals, families, and society. Understanding the causes of these disorders is crucial for developing effective treatments. Recent research has led to the development of numerous potential medicinal substances for treating neurodegenerative disorders. Clinical trials are ongoing to test these medications for their ability to cure specific symptoms. However, none of these treatments have been successful in halting the progression of the disease. Advancements in neurodegenerative medicine offer a positive outlook for the future. Regenerative medicine, immune modulation, and genetic substitution are being developed as advanced treatments, each with its own advantages and disadvantages. Gene therapy aims to restore brain function by administering functional genes to replace defective or absent genes. However, it can result in adverse effects, including off-target effects and potential mutations. The field of stem cells is highly promising, but the use of embryonic stem cells (ESCs) raises ethical concerns and iPSCs have been linked with tumorigenesis. Despite these challenges, the potential of these treatments is significant, and their development is crucial for the future of neurodegenerative diseases.

ACKNOWLEDGMENT

Our sincere thanks go out to the University of Allahabad, Prayagraj, Uttar Pradesh for all the support we needed to complete this work.

CONFLICT OF INTEREST

The authors declare no conflict of interest.

ABBREVIATIONS

Aβ	Amyloid beta
AAV	Adeno-associated virus
AAVs	Adeno-associated viruses
AD	Alzheimer's disease
ALS	Amyotrophic lateral sclerosis
APOE4	Apolipoprotein E4
APP	Amyloid precursor protein
ASOs	Antisense oligonucleotides
BACE1	Beta-secretase 1
BDNF	Brain-derived neurotrophic factor
C9orf72	intronic GGGGCC hexanucleotide repeat expansion in the chromosome 9 open reading frame 72
CAG	Cytosine–Adenine–Guanine
CNS	Central Nervous System
COMT	Catechol-O-methyl transferase
CRISPR-Cas9	CRISPR-associated nucleases
CSF	Cerebrospinal fluid
DA	Dopamine
DBS	Deep brain stimulation
ESK	Extracellular signal-regulated kinases
FDA	Food and Drug Administration
FTD	Frontotemporal syndromes
Fus	Fused in sarcoma
GBA	Glucocerebrosidase
GSK-3	Glycogen synthase kinase-3
HD	Huntington's disease
hESC	Human embryonic stem cells
IGF-1	Insulin-like Growth Factor 1
iPSCs	Induced pluripotent stem cells
KO	Knock out
LRRK	Leucine-rich repeat kinase
LRRK2	Leucine-rich repeat kinase
MAPT	Microtubule-associated protein tau

mHTT	Mutant Huntington protein
MN	Motor neurons
MNPs	Motor neuron progenitors
MPTP	1-methyl-4-phenyl-1, 2, 3, 6-tetrahydropyridine
MS	Multiple sclerosis
MSC	Mesenchymal stem cell
MSN	Medium spiny neurons
NGF	Nerve growth factor
NMDA	N-methyl-D-aspartate
NRF	Nuclear respiratory factor
NSCs	Neural stem cells
PARK	Parkin
PD	Parkinson's disease
PGC1 α	Peroxisome proliferator-activated receptor gamma coactivator 1-alpha
PINK	PTEN-induced kinase
PKA	Protein kinase A
PKC	Protein kinase c
PNS	Peripheral nervous system
PS1	Presenilin-1
ROS	Reactive oxygen species
SCA 2	Spinocerebellar ataxia 2
shRNA	Short hairpin RNA
siRNA	Small interfering RNA
SN	Substantia nigra
SNCA	Synuclein alpha
SOD	Superoxide dismutase
SSRIs	Serotonin reuptake inhibitors
SVZ	Subventricular zone
TALENs	Transcription activator-like effector nucleases
TARDBP	TAR DNA binding protein
TDP-43	TAR DNA-binding protein 43
VEGF	Vascular endothelial growth factor
ZFNs	Zinc-finger nucleases

REFERENCES

1. Fu, H., Hardy, J., and Duff, K. E., 2018, "Selective Vulnerability in Neurodegenerative Diseases," Nat. Neurosci., **21**(10), pp. 1350–1358.
2. Bates, G. P., Dorsey, R., Gusella, J. F., Hayden, M. R., Kay, C., Leavitt, B. R., Nance, M., Ross, C. A., Scahill, R. I., Wetzel, R., Wild, E. J., and Tabrizi, S. J., 2015, "Huntington Disease," Nat. Rev. Dis. Primer, **1**, p. 15005.
3. Hardiman, O., Al-Chalabi, A., Chio, A., Corr, E. M., Logroscino, G., Robberecht, W., Shaw, P. J., Simmons, Z., and van den Berg, L. H., 2017, "Amyotrophic Lateral Sclerosis," Nat. Rev. Dis. Primer, **3**, p. 17071.
4. Knopman, D. S., Amieva, H., Petersen, R. C., Chételat, G., Holtzman, D. M., Hyman, B. T., Nixon, R. A., and Jones, D. T., 2021, "Alzheimer Disease," Nat. Rev. Dis. Primer, **7**(1), p. 33.
5. Poewe, W., Seppi, K., Tanner, C. M., Halliday, G. M., Brundin, P., Volkmann, J., Schrag, A.-E., and Lang, A. E., 2017, "Parkinson Disease," Nat. Rev. Dis. Primer, **3**, p. 17013.
6. Narain, P., Gomes, J., Bhatia, R., Singh, I., and Vivekanandan, P., 2017, "C9orf72 Hexanucleotide Repeat Expansions and Ataxin 2 Intermediate Length Repeat Expansions in Indian Patients with Amyotrophic Lateral Sclerosis," Neurobiol. Aging, **56**, p. 211.e9–211.e14.
7. Wilson, D. M., Cookson, M. R., Van Den Bosch, L., Zetterberg, H., Holtzman, D. M., and Dewachter, I., 2023, "Hallmarks of Neurodegenerative Diseases," Cell, **186**(4), pp. 693–714.

8. Deuschl, G., Beghi, E., Fazekas, F., Varga, T., Christoforidi, K. A., Sipido, E., Bassetti, C. L., Vos, T., and Feigin, V. L., 2020, "The Burden of Neurological Diseases in Europe: An Analysis for the Global Burden of Disease Study 2017," Lancet Public Health, **5**(10), pp. e551–e567.
9. Trevisan, K., Cristina-Pereira, R., Silva-Amaral, D., and Aversi-Ferreira, T. A., 2019, "Theories of Aging and the Prevalence of Alzheimer's Disease," BioMed Res. Int., **2019**, p. e9171424.
10. Nichols, E., Steinmetz, J. D., Vollset, S. E., Fukutaki, K., Chalek, J., Abd-Allah, F., Abdoli, A., Abualhasan, A., Abu-Gharbieh, E., Akram, T. T., Hamad, H. A., Alahdab, F., Alanezi, F. M., Alipour, V., Almustanyir, S., Amu, H., Ansari, I., Arabloo, J., Ashraf, T., Astell-Burt, T., Ayano, G., Ayuso-Mateos, J. L., Baig, A. A., Barnett, A., Barrow, A., Baune, B. T., Béjot, Y., Bezabhe, W. M. M., Bezabih, Y. M., Bhagavathula, A. S., Bhaskar, S., Bhattacharyya, K., Bijani, A., Biswas, A., Bolla, S. R., Boloor, A., Brayne, C., Brenner, H., Burkart, K., Burns, R. A., Cámera, L. A., Cao, C., Carvalho, F., Castro-de-Araujo, L. F. S., Catalá-López, F., Cerin, E., Chavan, P. P., Cherbuin, N., Chu, D.-T., Costa, V. M., Couto, R. A. S., Dadras, O., Dai, X., Dandona, L., Dandona, R., De la Cruz-Góngora, V., Dhamnetiya, D., da Silva, D. D., Diaz, D., Douiri, A., Edvardsson, D., Ekholuenetale, M., Sayed, I. E., El-Jaafary, S. I., Eskandari, K., Eskandarieh, S., Esmaeilnejad, S., Fares, J., Faro, A., Farooque, U., Feigin, V. L., Feng, X., Fereshtehnejad, S.-M., Fernandes, E., Ferrara, P., Filip, I., Fillit, H., Fischer, F., Gaidhane, S., Galluzzo, L., Ghashghaee, A., Ghith, N., Gialluisi, A., Gilani, S. A., Glavan, I.-R., Gnedovskaya, E. V., Golechha, M., Gupta, R., Gupta, V. B., Gupta, V. K., Haider, M. R., Hall, B. J., Hamidi, S., Hanif, A., Hankey, G. J., Haque, S., Hartono, R. K., Hasaballah, A. I., Hasan, M. T., Hassan, A., Hay, S. I., Hayat, K., Hegazy, M. I., Heidari, G., Heidari-Soureshjani, R., Herteliu, C., Househ, M., Hussain, R., Hwang, B.-F., Iacoviello, L., Iavicoli, I., Ilesanmi, O. S., Ilic, I. M., Ilic, M. D., Irvani, S. S. N., Iso, H., Iwagami, M., Jabbarinejad, R., Jacob, L., Jain, V., Jayapal, S. K., Jayawardena, R., Jha, R. P., Jonas, J. B., Joseph, N., Kalani, R., Kandel, A., Kandel, H., Karch, A., Kasa, A. S., Kassie, G. M., Keshavarz, P., Khan, M. A., Khatib, M. N., Khoja, T. A. M., Khubchandani, J., Kim, M. S., Kim, Y. J., Kisa, A., Kisa, S., Kivimäki, M., Koroshetz, W. J., Koyanagi, A., Kumar, G. A., Kumar, M., Lak, H. M., Leonardi, M., Li, B., Lim, S. S., Liu, X., Liu, Y., Logroscino, G., Lorkowski, S., Lucchetti, G., Saute, R. L., Magnani, F. G., Malik, A. A., Massano, J., Mehndiratta, M. M., Menezes, R. G., Meretoja, A., Mohajer, B., Ibrahim, N. M., Mohammad, Y., Mohammed, A., Mokdad, A. H., Mondello, S., Moni, M. A. A., Moniruzzaman, M., Mossie, T. B., Nagel, G., Naveed, M., Nayak, V. C., Kandel, S. N., Nguyen, T. H., Oancea, B., Otstavnov, N., Otstavnov, S. S., Owolabi, M. O., Panda-Jonas, S., Kan, F. P., Pasovic, M., Patel, U. K., Pathak, M., Peres, M. F. P., Perianayagam, A., Peterson, C. B., Phillips, M. R., Pinheiro, M., Piradov, M. A., Pond, C. D., Potashman, M. H., Pottoo, F. H., Prada, S. I., Radfar, A., Raggi, A., Rahim, F., Rahman, M., Ram, P., Ranasinghe, P., Rawaf, D. L., Rawaf, S., Rezaei, N., Rezapour, A., Robinson, S. R., Romoli, M., Roshandel, G., Sahathevan, R., Sahebkar, A., Sahraian, M. A., Sathian, B., Sattin, D., Sawhney, M., Saylan, M., Schiavolin, S., Seylani, A., Sha, F., Shaikh, M. A., Shaji, K. S., Shannawaz, M., Shetty, J. K., Shigematsu, M., Shin, J. I., Shiri, R., Silva, D. A. S., Silva, J. P., Silva, R., Singh, J. A., Skryabin, V. Y., Skryabina, A. A., Smith, A. E., Soshnikov, S., Spurlock, E. E., Stein, D. J., Sun, J., Tabarés-Seisdedos, R., Thakur, B., Timalsina, B., Tovani-Palone, M. R., Tran, B. X., Tsegaye, G. W., Tahbaz, S. V., Valdez, P. R., Venketasubramanian, N., Vlassov, V., Vu, G. T., Vu, L. G., Wang, Y.-P., Wimo, A., Winkler, A. S., Yadav, L., Jabbari, S. H. Y., Yamagishi, K., Yang, L., Yano, Y., Yonemoto, N., Yu, C., Yunusa, I., Zadey, S., Zastrozhin, M. S., Zastrozhina, A., Zhang, Z.-J., Murray, C. J. L., and Vos, T., 2022, "Estimation of the Global Prevalence of Dementia in 2019 and Forecasted Prevalence in 2050: An Analysis for the Global Burden of Disease Study 2019," Lancet Public Health, **7**(2), pp. e105–e125.
11. Huang, Y.-W. A., Zhou, B., Wernig, M., and Südhof, T. C., 2017, "ApoE2, ApoE3, and ApoE4 Differentially Stimulate APP Transcription and Aβ Secretion," Cell, **168**(3), pp. 427–441.e21.
12. Sun, J., and Roy, S., 2018, "The Physical Approximation of APP and BACE-1: A Key Event in Alzheimer's Disease Pathogenesis," Dev. Neurobiol., **78**(3), pp. 340–347.
13. O'Brien, R. J., and Wong, P. C., 2011, "Amyloid Precursor Protein Processing and Alzheimer's Disease," Annu. Rev. Neurosci., **34**, pp. 185–204.
14. Rawat, P., Sehar, U., Bisht, J., Selman, A., Culberson, J., and Reddy, P. H., 2022, "Phosphorylated Tau in Alzheimer's Disease and Other Tauopathies," Int. J. Mol. Sci., **23**(21), p. 12841.
15. Gong, C.-X., and Iqbal, K., 2008, "Hyperphosphorylation of Microtubule-Associated Protein Tau: A Promising Therapeutic Target for Alzheimer Disease," Curr. Med. Chem., **15**(23), pp. 2321–2328.
16. Buddhala, C., Loftin, S. K., Kuley, B. M., Cairns, N. J., Campbell, M. C., Perlmutter, J. S., and Kotzbauer, P. T., 2015, "Dopaminergic, Serotonergic, and Noradrenergic Deficits in Parkinson Disease," Ann. Clin. Transl. Neurol., **2**(10), pp. 949–959.
17. Dorsey, E. R., Sherer, T., Okun, M. S., and Bloem, B. R., 2018, "The Emerging Evidence of the Parkinson Pandemic," J. Park. Dis., **8**(Suppl 1), pp. S3–S8.

18. Ball, N., Teo, W.-P., Chandra, S., and Chapman, J., 2019, "Parkinson's Disease and the Environment," Front. Neurol., **10**, p. 218.
19. Pang, S. Y.-Y., Ho, P. W.-L., Liu, H.-F., Leung, C.-T., Li, L., Chang, E. E. S., Ramsden, D. B., and Ho, S.-L., 2019, "The Interplay of Aging, Genetics and Environmental Factors in the Pathogenesis of Parkinson's Disease," Transl. Neurodegener., **8**(1), p. 23.
20. Klein, C., and Westenberger, A., 2012, "Genetics of Parkinson's Disease," Cold Spring Harb. Perspect. Med., **2**(1), p. a008888.
21. Billingsley, K. J., Bandres-Ciga, S., Saez-Atienzar, S., and Singleton, A. B., 2018, "Genetic Risk Factors in Parkinson's Disease," Cell Tissue Res., **373**(1), pp. 9–20.
22. Funayama, M., Nishioka, K., Li, Y., and Hattori, N., 2023, "Molecular Genetics of Parkinson's Disease: Contributions and Global Trends," J. Hum. Genet., **68**(3), pp. 125–130.
23. Cerri, S., Mus, L., and Blandini, F., 2019, "Parkinson's Disease in Women and Men: What's the Difference?" J. Park. Dis., **9**(3), pp. 501–515.
24. Riederer, P., Nagatsu, T., Youdim, M. B. H., Wulf, M., Dijkstra, J. M., and Sian-Huelsmann, J., 2023, "Lewy Bodies, Iron, Inflammation and Neuromelanin: Pathological Aspects Underlying Parkinson's Disease," J. Neural Transm., **130**(5), pp. 627–646.
25. MacMahon Copas, A. N., McComish, S. F., Fletcher, J. M., and Caldwell, M. A., 2021, "The Pathogenesis of Parkinson's Disease: A Complex Interplay Between Astrocytes, Microglia, and T Lymphocytes?" Front. Neurol., **12**, p. 666737.
26. Chiò, A., Logroscino, G., Hardiman, O., Swingler, R., Mitchell, D., Beghi, E., Traynor, B. G., and Eurals Consortium, 2009, "Prognostic Factors in ALS: A Critical Review," Amyotroph. Lateral Scler. Off. Publ. World Fed. Neurol. Res. Group Mot. Neuron Dis., **10**(5–6), pp. 310–323.
27. Longinetti, E., and Fang, F., 2019, "Epidemiology of Amyotrophic Lateral Sclerosis: An Update of Recent Literature," Curr. Opin. Neurol., **32**(5), pp. 771–776.
28. Marin, B., Hamidou, B., Couratier, P., Nicol, M., Delzor, A., Raymondeau, M., Druet-Cabanac, M., Lautrette, G., Boumediene, F., Preux, P. M., and French register of ALS in Limousin, 2014, "Population-Based Epidemiology of Amyotrophic Lateral Sclerosis (ALS) in an Ageing Europe–the French Register of ALS in Limousin (FRALim Register)," Eur. J. Neurol., **21**(10), pp. 1292–1300, e78–79.
29. Megat, S., Mora, N., Sanogo, J., Roman, O., Catanese, A., Alami, N. O., Freischmidt, A., Mingaj, X., De Calbiac, H., Muratet, F., Dirrig-Grosch, S., Dieterle, S., Van Bakel, N., Müller, K., Sieverding, K., Weishaupt, J., Andersen, P. M., Weber, M., Neuwirth, C., Margelisch, M., Sommacal, A., Van Eijk, K. R., Veldink, J. H., Project Mine Als Sequencing Consortium, Lautrette, G., Couratier, P., Camuzat, A., Le Ber, I., Grassano, M., Chio, A., Boeckers, T., Ludolph, A. C., Roselli, F., Yilmazer-Hanke, D., Millecamps, S., Kabashi, E., Storkebaum, E., Sellier, C., and Dupuis, L., 2023, "Integrative Genetic Analysis Illuminates ALS Heritability and Identifies Risk Genes," Nat. Commun., **14**(1), p. 342.
30. Mejzini, R., Flynn, L. L., Pitout, I. L., Fletcher, S., Wilton, S. D., and Akkari, P. A., 2019, "ALS Genetics, Mechanisms, and Therapeutics: Where Are We Now?" Front. Neurosci., **13**, p. 1310.
31. Sabatelli, M., Conte, A., and Zollino, M., 2013, "Clinical and Genetic Heterogeneity of Amyotrophic Lateral Sclerosis," Clin. Genet., **83**(5), pp. 408–416.
32. Ferrari, R., Kapogiannis, D., Huey, E. D., and Momeni, P., 2011, "FTD and ALS: A Tale of Two Diseases," Curr. Alzheimer Res., **8**(3), pp. 273–294.
33. Bozzoni, V., Pansarasa, O., Diamanti, L., Nosari, G., Cereda, C., and Ceroni, M., 2016, "Amyotrophic Lateral Sclerosis and Environmental Factors," Funct. Neurol., **31**(1), pp. 7–19.
34. Birsa, N., Bentham, M. P., and Fratta, P., 2020, "Cytoplasmic Functions of TDP-43 and FUS and Their Role in ALS," Semin. Cell Dev. Biol., **99**, pp. 193–201.
35. Polymenidou, M., and Cleveland, D. W., 2011, "The Seeds of Neurodegeneration: Prion-like Spreading in ALS," Cell, **147**(3), pp. 498–508.
36. Ruffo, P., Perrone, B., and Conforti, F. L., 2022, "SOD-1 Variants in Amyotrophic Lateral Sclerosis: Systematic Re-Evaluation According to ACMG-AMP Guidelines," Genes, **13**(3), p. 537.
37. Smith, E. F., Shaw, P. J., and De Vos, K. J., 2019, "The Role of Mitochondria in Amyotrophic Lateral Sclerosis," Neurosci. Lett., **710**, p. 132933.
38. Robelin, L., and Aguilar, J., 2014, "Blood Biomarkers for Amyotrophic Lateral Sclerosis: Myth or Reality?" BioMed Res. Int., **2014**, p. 525097.
39. Wilkins, H. M., Dimachkie, M. M., and Agbas, A., 2021, "Blood-Based Biomarkers for Amyotrophic Lateral Sclerosis," *Amyotrophic Lateral Sclerosis*, T. Araki, ed., Exon Publications, Brisbane (AU).
40. DeJesus-Hernandez, M., Mackenzie, I. R., Boeve, B. F., Boxer, A. L., Baker, M., Rutherford, N. J., Nicholson, A. M., Finch, N. A., Flynn, H., Adamson, J., Kouri, N., Wojtas, A., Sengdy, P., Hsiung, G.-Y. R., Karydas, A., Seeley, W. W., Josephs, K. A., Coppola, G., Geschwind, D. H., Wszolek, Z. K.,

Feldman, H., Knopman, D. S., Petersen, R. C., Miller, B. L., Dickson, D. W., Boylan, K. B., Graff-Radford, N. R., and Rademakers, R., 2011, "Expanded GGGGCC Hexanucleotide Repeat in Noncoding Region of C9ORF72 Causes Chromosome 9p-Linked FTD and ALS," Neuron, **72**(2), pp. 245–256.

41. Gendron, T. F., and Petrucelli, L., 2018, "Disease Mechanisms of C9ORF72 Repeat Expansions," Cold Spring Harb. Perspect. Med., **8**(4), p. a024224.
42. Lee, J., Hyeon, S. J., Im, H., Ryu, H., Kim, Y., and Ryu, H., 2016, "Astrocytes and Microglia as Non-Cell Autonomous Players in the Pathogenesis of ALS," Exp. Neurobiol., **25**(5), pp. 233–240.
43. Rostalski, H., Leskelä, S., Huber, N., Katisko, K., Cajanus, A., Solje, E., Marttinen, M., Natunen, T., Remes, A. M., Hiltunen, M., and Haapasalo, A., 2019, "Astrocytes and Microglia as Potential Contributors to the Pathogenesis of C9orf72 Repeat Expansion-Associated FTLD and ALS," Front. Neurosci., **13**, p. 486.
44. MacDonald, M. E., Ambrose, C. M., Duyao, M. P., Myers, R. H., Lin, C., Srinidhi, L., Barnes, G., Taylor, S. A., James, M., Groot, N., MacFarlane, H., Jenkins, B., Anderson, M. A., Wexler, N. S., Gusella, J. F., Bates, G. P., Baxendale, S., Hummerich, H., Kirby, S., North, M., Youngman, S., Mott, R., Zehetner, G., Sedlacek, Z., Poustka, A., Frischauf, A.-M., Lehrach, H., Buckler, A. J., Church, D., Doucette-Stamm, L., O'Donovan, M. C., Riba-Ramirez, L., Shah, M., Stanton, V. P., Strobel, S. A., Draths, K. M., Wales, J. L., Dervan, P., Housman, D. E., Altherr, M., Shiang, R., Thompson, L., Fielder, T., Wasmuth, J. J., Tagle, D., Valdes, J., Elmer, L., Allard, M., Castilla, L., Swaroop, M., Blanchard, K., Collins, F. S., Snell, R., Holloway, T., Gillespie, K., Datson, N., Shaw, D., and Harper, P. S., 1993, "A Novel Gene Containing a Trinucleotide Repeat That Is Expanded and Unstable on Huntington's Disease Chromosomes," Cell, **72**(6), pp. 971–983.
45. Gusella, J. F., MacDonald, M. E., and Lee, J.-M., 2014, "Genetic Modifiers of Huntington's Disease," Mov. Disord. Off. J. Mov. Disord. Soc., **29**(11), pp. 1359–1365.
46. Rojas, N. G., Cesarini, M. E., Peker, G., Prat, G. A. D., Etcheverry, J. L., and Gatto, E. M., 2022, "Review of Huntington's Disease: From Basics to Advances in Diagnosis and Treatment," J. Neurol. Res., **12**(3), pp. 93–113.
47. Saft, C., 2017, "Identifying Modifiers of Huntington's Disease Progression," Lancet Neurol., **16**(9), pp. 679–680.
48. Schulte, J., and Littleton, J. T., 2011, "The Biological Function of the Huntingtin Protein and Its Relevance to Huntington's Disease Pathology," Curr. Trends Neurol., **5**, pp. 65–78.
49. Saudou, F., and Humbert, S., 2016, "The Biology of Huntingtin," Neuron, **89**(5), pp. 910–926.
50. DeGuire, S. M., Ruggeri, F. S., Fares, M.-B., Chiki, A., Cendrowska, U., Dietler, G., and Lashuel, H. A., 2018, "N-Terminal Huntingtin (Htt) Phosphorylation Is a Molecular Switch Regulating Htt Aggregation, Helical Conformation, Internalization, and Nuclear Targeting," J. Biol. Chem., **293**(48), pp. 18540–18558.
51. Zuccato, C., Ciammola, A., Rigamonti, D., Leavitt, B. R., Goffredo, D., Conti, L., MacDonald, M. E., Friedlander, R. M., Silani, V., Hayden, M. R., Timmusk, T., Sipione, S., and Cattaneo, E., 2001, "Loss of Huntingtin-Mediated BDNF Gene Transcription in Huntington's Disease," Science, **293**(5529), pp. 493–498.
52. Zuccato, C., and Cattaneo, E., 2007, "Role of Brain-Derived Neurotrophic Factor in Huntington's Disease," Prog. Neurobiol., **81**(5), pp. 294–330.
53. Bersuker, K., Hipp, M. S., Calamini, B., Morimoto, R. I., and Kopito, R. R., 2013, "Heat Shock Response Activation Exacerbates Inclusion Body Formation in a Cellular Model of Huntington Disease," J. Biol. Chem., **288**(33), pp. 23633–23638.
54. Labbadia, J., Cunliffe, H., Weiss, A., Katsyuba, E., Sathasivam, K., Seredenina, T., Woodman, B., Moussaoui, S., Frentzel, S., Luthi-Carter, R., Paganetti, P., and Bates, G. P., 2011, "Altered Chromatin Architecture Underlies Progressive Impairment of the Heat Shock Response in Mouse Models of Huntington Disease," J. Clin. Invest., **121**(8), pp. 3306–3319.
55. Kumar, A., Sidhu, J., Goyal, A., and Tsao, J. W., 2023, "Alzheimer Disease," *StatPearls*, StatPearls Publishing, Treasure Island (FL), PMID: 33760564.
56. Calabrò, M., Rinaldi, C., Santoro, G., and Crisafulli, C., 2021, "The Biological Pathways of Alzheimer Disease: A Review," AIMS Neurosci., **8**(1), pp. 86–132.
57. Andrade-Guerrero, J., Santiago-Balmaseda, A., Jeronimo-Aguilar, P., Vargas-Rodríguez, I., Cadena-Suárez, A. R., Sánchez-Garibay, C., Pozo-Molina, G., Méndez-Catalá, C. F., Cardenas-Aguayo, M.-C., Diaz-Cintra, S., Pacheco-Herrero, M., Luna-Muñoz, J., and Soto-Rojas, L. O., 2023, "Alzheimer's Disease: An Updated Overview of Its Genetics," Int. J. Mol. Sci., **24**(4), p. 3754.
58. Cobos, I., and Palop, J. J., 2022, "Swollen Axons Impair Neuronal Circuits in Alzheimer's Disease," Nature, **612**(7939), pp. 218–220.

59. Zhang, W., Xu, C., Sun, J., Shen, H.-M., Wang, J., and Yang, C., 2022, "Impairment of the Autophagy–Lysosomal Pathway in Alzheimer's Diseases: Pathogenic Mechanisms and Therapeutic Potential," Acta Pharm. Sin. B, **12**(3), pp. 1019–1040.
60. Beck, T. N., Nicolas, E., Kopp, M. C., and Golemis, E. A., 2014, "Adaptors for Disorders of the Brain? The Cancer Signaling Proteins NEDD9, CASS4, and PTK2B in Alzheimer's Disease," Oncoscience, **1**(7), pp. 486–503.
61. Chapuis, J., Flaig, A., Grenier-Boley, B., Eysert, F., Pottiez, V., Deloison, G., Vandeputte, A., Ayral, A.-M., Mendes, T., Desai, S., Goate, A. M., Kauwe, J. S. K., Leroux, F., Herledan, A., Demiautte, F., Bauer, C., Checler, F., Petersen, R. C., Blennow, K., Zetterberg, H., Minthon, L., Van Deerlin, V. M., Lee, V. M.-Y., Shaw, L. M., Trojanowski, J. Q., Albert, M., Moghekar, A., O'Brien, R., Peskind, E. R., Malmanche, N., Schellenberg, G. D., Dourlen, P., Song, O.-R., Cruchaga, C., Amouyel, P., Deprez, B., Brodin, P., and Lambert, J.-C., 2017, "Genome-Wide, High-Content siRNA Screening Identifies the Alzheimer's Genetic Risk Factor FERMT2 as a Major Modulator of APP Metabolism," Acta Neuropathol. (Berl.), **133**(6), pp. 955–966.
62. Mez, J., Chung, J., Jun, G., Kriegel, J., Bourlas, A. P., Sherva, R., Logue, M. W., Barnes, L. L., Bennett, D. A., Buxbaum, J. D., Byrd, G. S., Crane, P. K., Ertekin-Taner, N., Evans, D., Fallin, M. D., Foroud, T., Goate, A., Graff-Radford, N. R., Hall, K. S., Kamboh, M. I., Kukull, W. A., Larson, E. B., Manly, J. J., Alzheimer's Disease Genetics Consortium, Haines, J. L., Mayeux, R., Pericak-Vance, M. A., Schellenberg, G. D., Lunetta, K. L., and Farrer, L. A., 2017, "Two Novel Loci, COBL and SLC10A2, for Alzheimer's Disease in African Americans," Alzheimers Dement. J. Alzheimers Assoc., **13**(2), pp. 119–129.
63. Yin, F., 2023, "Lipid Metabolism and Alzheimer's Disease: Clinical Evidence, Mechanistic Link and Therapeutic Promise," FEBS J., **290**(6), pp. 1420–1453.
64. Zhu, Z., Yang, Y., Xiao, Z., Zhao, Q., Wu, W., Liang, X., Luo, J., Cao, Y., Shao, M., Guo, Q., and Ding, D., 2021, "TOMM40 and APOE Variants Synergistically Increase the Risk of Alzheimer's Disease in a Chinese Population," Aging Clin. Exp. Res., **33**(6), pp. 1667–1675.
65. Gella, A., and Durany, N., 2009, "Oxidative Stress in Alzheimer Disease," Cell Adhes. Migr., **3**(1), pp. 88–93.
66. Sobue, A., Komine, O., and Yamanaka, K., 2023, "Neuroinflammation in Alzheimer's Disease: Microglial Signature and Their Relevance to Disease," Inflamm. Regen., **43**(1), p. 26.
67. Cuervo, A. M., Stefanis, L., Fredenburg, R., Lansbury, P. T., and Sulzer, D., 2004, "Impaired Degradation of Mutant Alpha-Synuclein by Chaperone-Mediated Autophagy," Science, **305**(5688), pp. 1292–1295.
68. Di Maio, R., Barrett, P. J., Hoffman, E. K., Barrett, C. W., Zharikov, A., Borah, A., Hu, X., McCoy, J., Chu, C. T., Burton, E. A., Hastings, T. G., and Greenamyre, J. T., 2016, "α-Synuclein Binds to TOM20 and Inhibits Mitochondrial Protein Import in Parkinson's Disease," Sci. Transl. Med., **8**(342), p. 342ra78.
69. Winslow, A. R., Chen, C.-W., Corrochano, S., Acevedo-Arozena, A., Gordon, D. E., Peden, A. A., Lichtenberg, M., Menzies, F. M., Ravikumar, B., Imarisio, S., Brown, S., O'Kane, C. J., and Rubinsztein, D. C., 2010, "α-Synuclein Impairs Macroautophagy: Implications for Parkinson's Disease," J. Cell Biol., **190**(6), pp. 1023–1037.
70. Xilouri, M., Vogiatzi, T., Vekrellis, K., Park, D., and Stefanis, L., 2009, "Abberant Alpha-Synuclein Confers Toxicity to Neurons in Part through Inhibition of Chaperone-Mediated Autophagy," PloS One, **4**(5), p. e5515.
71. Dorszewska, J., Kowalska, M., Prendecki, M., Piekut, T., Kozłowska, J., and Kozubski, W., 2020, "Oxidative Stress Factors in Parkinson's Disease," Neural Regen. Res., **16**(7), pp. 1383–1391.
72. Kitada, T., Asakawa, S., Hattori, N., Matsumine, H., Yamamura, Y., Minoshima, S., Yokochi, M., Mizuno, Y., and Shimizu, N., 1998, "Mutations in the Parkin Gene Cause Autosomal Recessive Juvenile Parkinsonism," Nature, **392**(6676), pp. 605–608.
73. Nishikawa, K., Li, H., Kawamura, R., Osaka, H., Wang, Y.-L., Hara, Y., Hirokawa, T., Manago, Y., Amano, T., Noda, M., Aoki, S., and Wada, K., 2003, "Alterations of Structure and Hydrolase Activity of Parkinsonism-Associated Human Ubiquitin Carboxyl-Terminal Hydrolase L1 Variants," Biochem. Biophys. Res. Commun., **304**(1), pp. 176–183.
74. Panicker, N., Ge, P., Dawson, V. L., and Dawson, T. M., 2021, "The Cell Biology of Parkinson's Disease," J. Cell Biol., **220**(4), p. e202012095.
75. Um, J. W., Im, E., Lee, H. J., Min, B., Yoo, L., Yoo, J., Lübbert, H., Stichel-Gunkel, C., Cho, H.-S., Yoon, J. B., and Chung, K. C., 2010, "Parkin Directly Modulates 26S Proteasome Activity," J. Neurosci. Off. J. Soc. Neurosci., **30**(35), pp. 11805–11814.
76. Valente, E. M., Bentivoglio, A. R., Dixon, P. H., Ferraris, A., Ialongo, T., Frontali, M., Albanese, A., and Wood, N. W., 2001, "Localization of a Novel Locus for Autosomal Recessive Early-Onset Parkinsonism, PARK6, on Human Chromosome 1p35-P36," Am. J. Hum. Genet., **68**(4), pp. 895–900.

77. Youle, R. J., and Narendra, D. P., 2011, "Mechanisms of Mitophagy," Nat. Rev. Mol. Cell Biol., **12**(1), pp. 9–14.
78. Gómez-Suaga, P., Fdez, E., Blanca Ramírez, M., and Hilfiker, S., 2012, "A Link between Autophagy and the Pathophysiology of LRRK2 in Parkinson's Disease," Park. Dis., **2012**, p. e324521.
79. Ramirez, A., Heimbach, A., Gründemann, J., Stiller, B., Hampshire, D., Cid, L. P., Goebel, I., Mubaidin, A. F., Wriekat, A.-L., Roeper, J., Al-Din, A., Hillmer, A. M., Karsak, M., Liss, B., Woods, C. G., Behrens, M. I., and Kubisch, C., 2006, "Hereditary Parkinsonism with Dementia Is Caused by Mutations in ATP13A2, Encoding a Lysosomal Type 5 P-Type ATPase," Nat. Genet., **38**(10), pp. 1184–1191.
80. Tong, Y., Yamaguchi, H., Giaime, E., Boyle, S., Kopan, R., Kelleher, R. J., and Shen, J., 2010, "Loss of Leucine-Rich Repeat Kinase 2 Causes Impairment of Protein Degradation Pathways, Accumulation of Alpha-Synuclein, and Apoptotic Cell Death in Aged Mice," Proc. Natl. Acad. Sci. U. S. A., **107**(21), pp. 9879–9884.
81. Rahman, A. A., and Morrison, B. E., 2019, "Contributions of VPS35 Mutations to Parkinson's Disease," Neuroscience, **401**, pp. 1–10.
82. Sassone, J., Reale, C., Dati, G., Regoni, M., Pellecchia, M. T., and Garavaglia, B., 2021, "The Role of VPS35 in the Pathobiology of Parkinson's Disease," Cell. Mol. Neurobiol., **41**(2), pp. 199–227.
83. Williams, E. T., Chen, X., and Moore, D. J., 2017, "VPS35, the Retromer Complex and Parkinson's Disease," J. Park. Dis., **7**(2), pp. 219–233.
84. Ayton, S., and Lei, P., 2014, "Nigral Iron Elevation Is an Invariable Feature of Parkinson's Disease and Is a Sufficient Cause of Neurodegeneration," BioMed Res. Int., **2014**, p. 581256.
85. Golts, N., Snyder, H., Frasier, M., Theisler, C., Choi, P., and Wolozin, B., 2002, "Magnesium Inhibits Spontaneous and Iron-Induced Aggregation of Alpha-Synuclein," J. Biol. Chem., **277**(18), pp. 16116–16123.
86. Wang, Z. L., Yuan, L., Li, W., and Li, J. Y., 2022, "Ferroptosis in Parkinson's Disease: Glia–Neuron Crosstalk," Trends Mol. Med., **28**(4), pp. 258–269.
87. Amin, A., Perera, N. D., Beart, P. M., Turner, B. J., and Shabanpoor, F., 2020, "Amyotrophic Lateral Sclerosis and Autophagy: Dysfunction and Therapeutic Targeting," Cells, **9**(11), p. 2413.
88. Mead, R. J., Shan, N., Reiser, H. J., Marshall, F., and Shaw, P. J., 2023, "Amyotrophic Lateral Sclerosis: A Neurodegenerative Disorder Poised for Successful Therapeutic Translation," Nat. Rev. Drug Discov., **22**(3), pp. 185–212.
89. Neumann, M., Sampathu, D. M., Kwong, L. K., Truax, A. C., Micsenyi, M. C., Chou, T. T., Bruce, J., Schuck, T., Grossman, M., Clark, C. M., McCluskey, L. F., Miller, B. L., Masliah, E., Mackenzie, I. R., Feldman, H., Feiden, W., Kretzschmar, H. A., Trojanowski, J. Q., and Lee, V. M.-Y., 2006, "Ubiquitinated TDP-43 in Frontotemporal Lobar Degeneration and Amyotrophic Lateral Sclerosis," Science, **314**(5796), pp. 130–133.
90. Hooten, K. G., Beers, D. R., Zhao, W., and Appel, S. H., 2015, "Protective and Toxic Neuroinflammation in Amyotrophic Lateral Sclerosis," Neurother. J. Am. Soc. Exp. Neurother., **12**(2), pp. 364–375.
91. Moll, T., Graves, E., Urbanek, A., Soni, N., Ranganathan, R., Higginbottom, A., Wang, S., Head, B. P., Cooper-Knock, J., and Shaw, P. J., 2022, "GLT8D1 Mutations Cause Amyotrophic Lateral Sclerosis via Disruption of Neurotrophin Signalling within Membrane Lipid Rafts," bioRxiv. https://doi.org/10.1101/2022.06.28.497990
92. Peggion, C., Scalcon, V., Massimino, M. L., Nies, K., Lopreiato, R., Rigobello, M. P., and Bertoli, A., 2022, "SOD1 in ALS: Taking Stock in Pathogenic Mechanisms and the Role of Glial and Muscle Cells," Antioxid. Basel Switz., **11**(4), p. 614.
93. Weydt, P., Yuen, E. C., Ransom, B. R., and Möller, T., 2004, "Increased Cytotoxic Potential of Microglia from ALS-Transgenic Mice," Glia, **48**(2), pp. 179–182.
94. Dobson-Stone, C., Hallupp, M., Shahheydari, H., Ragagnin, A. M. G., Chatterton, Z., Carew-Jones, F., Shepherd, C. E., Stefen, H., Paric, E., Fath, T., Thompson, E. M., Blumbergs, P., Short, C. L., Field, C. D., Panegyres, P. K., Hecker, J., Nicholson, G., Shaw, A. D., Fullerton, J. M., Luty, A. A., Schofield, P. R., Brooks, W. S., Rajan, N., Bennett, M. F., Bahlo, M., Landers, J. E., Piguet, O., Hodges, J. R., Halliday, G. M., Topp, S. D., Smith, B. N., Shaw, C. E., McCann, E., Fifita, J. A., Williams, K. L., Atkin, J. D., Blair, I. P., and Kwok, J. B., 2020, "CYLD Is a Causative Gene for Frontotemporal Dementia - Amyotrophic Lateral Sclerosis," Brain, **143**(3), pp. 783–799.
95. Freischmidt, A., Wieland, T., Richter, B., Ruf, W., Schaeffer, V., Müller, K., Marroquin, N., Nordin, F., Hübers, A., Weydt, P., Pinto, S., Press, R., Millecamps, S., Molko, N., Bernard, E., Desnuelle, C., Soriani, M.-H., Dorst, J., Graf, E., Nordström, U., Feiler, M. S., Putz, S., Boeckers, T. M., Meyer, T., Winkler, A. S., Winkelman, J., de Carvalho, M., Thal, D. R., Otto, M., Brännström, T., Volk, A. E., Kursula, P., Danzer, K. M., Lichtner, P., Dikic, I., Meitinger, T., Ludolph, A. C., Strom, T. M., Andersen,

P. M., and Weishaupt, J. H., 2015, "Haploinsufficiency of TBK1 Causes Familial ALS and Fronto-Temporal Dementia," Nat. Neurosci., **18**(5), pp. 631–636.

96. Jiang, Z., Wang, Z., Wei, X., and Yu, X.-F., 2022, "Inflammatory Checkpoints in Amyotrophic Lateral Sclerosis: From Biomarkers to Therapeutic Targets," Front. Immunol., **13**, p. 1059994.
97. Markovinovic, A., Ljutic, T., Béland, L.-C., and Munitic, I., 2018, "Optineurin Insufficiency Disbalances Proinflammatory and Anti-Inflammatory Factors by Reducing Microglial IFN-β Responses," Neuroscience, **388**, pp. 139–151.
98. Maruyama, H., Morino, H., Ito, H., Izumi, Y., Kato, H., Watanabe, Y., Kinoshita, Y., Kamada, M., Nodera, H., Suzuki, H., Komure, O., Matsuura, S., Kobatake, K., Morimoto, N., Abe, K., Suzuki, N., Aoki, M., Kawata, A., Hirai, T., Kato, T., Ogasawara, K., Hirano, A., Takumi, T., Kusaka, H., Hagiwara, K., Kaji, R., and Kawakami, H., 2010, "Mutations of Optineurin in Amyotrophic Lateral Sclerosis," Nature, **465**(7295), pp. 223–226.
99. Toth, R. P., and Atkin, J. D., 2018, "Dysfunction of Optineurin in Amyotrophic Lateral Sclerosis and Glaucoma," Front. Immunol., **9**, p. 1017.
100. Cirulli, E. T., Lasseigne, B. N., Petrovski, S., Sapp, P. C., Dion, P. A., Leblond, C. S., Couthouis, J., Lu, Y.-F., Wang, Q., Krueger, B. J., Ren, Z., Keebler, J., Han, Y., Levy, S. E., Boone, B. E., Wimbish, J. R., Waite, L. L., Jones, A. L., Carulli, J. P., Day-Williams, A. G., Staropoli, J. F., Xin, W. W., Chesi, A., Raphael, A. R., McKenna-Yasek, D., Cady, J., Vianney de Jong, J. M. B., Kenna, K. P., Smith, B. N., Topp, S., Miller, J., Gkazi, A., FALS Sequencing Consortium, Al-Chalabi, A., van den Berg, L. H., Veldink, J., Silani, V., Ticozzi, N., Shaw, C. E., Baloh, R. H., Appel, S., Simpson, E., Lagier-Tourenne, C., Pulst, S. M., Gibson, S., Trojanowski, J. Q., Elman, L., McCluskey, L., Grossman, M., Shneider, N. A., Chung, W. K., Ravits, J. M., Glass, J. D., Sims, K. B., Van Deerlin, V. M., Maniatis, T., Hayes, S. D., Ordureau, A., Swarup, S., Landers, J., Baas, F., Allen, A. S., Bedlack, R. S., Harper, J. W., Gitler, A. D., Rouleau, G. A., Brown, R., Harms, M. B., Cooper, G. M., Harris, T., Myers, R. M., and Goldstein, D. B., 2015, "Exome Sequencing in Amyotrophic Lateral Sclerosis Identifies Risk Genes and Pathways," Science, **347**(6229), pp. 1436–1441.
101. Foster, A. D., Downing, P., Figredo, E., Polain, N., Stott, A., Layfield, R., and Rea, S. L., 2020, "ALS-Associated TBK1 Variant p.G175S Is Defective in Phosphorylation of P62 and Impacts TBK1-Mediated Signalling and TDP-43 Autophagic Degradation," Mol. Cell. Neurosci., **108**, p. 103539.
102. Butti, Z., and Patten, S. A., 2018, "RNA Dysregulation in Amyotrophic Lateral Sclerosis," Front. Genet., **9**, p. 712.
103. Weskamp, K., and Barmada, S. J., 2018, "TDP43 and RNA Instability in Amyotrophic Lateral Sclerosis," Brain Res., **1693**(Pt A), pp. 67–74.
104. Tak, Y. J., Park, J.-H., Rhim, H., and Kang, S., 2020, "ALS-Related Mutant SOD1 Aggregates Interfere with Mitophagy by Sequestering the Autophagy Receptor Optineurin," Int. J. Mol. Sci., **21**(20), p. 7525.
105. Todd, T. W., Shao, W., Zhang, Y.-J., and Petrucelli, L., 2023, "The Endolysosomal Pathway and ALS/FTD," Trends Neurosci., **46**(12), pp. 1025–1041.
106. Baron, D. M., Fenton, A. R., Saez-Atienzar, S., Giampetruzzi, A., Sreeram, A., Shankaracharya, Keagle, P. J., Doocy, V. R., Smith, N. J., Danielson, E. W., Andresano, M., McCormack, M. C., Garcia, J., Bercier, V., Van Den Bosch, L., Brent, J. R., Fallini, C., Traynor, B. J., Holzbaur, E. L. F., and Landers, J. E., 2022, "ALS-Associated KIF5A Mutations Abolish Autoinhibition Resulting in a Toxic Gain of Function," Cell Rep., **39**(1), p. 110598.
107. Brenner, D., Yilmaz, R., Müller, K., Grehl, T., Petri, S., Meyer, T., Grosskreutz, J., Weydt, P., Ruf, W., Neuwirth, C., Weber, M., Pinto, S., Claeys, K. G., Schrank, B., Jordan, B., Knehr, A., Günther, K., Hübers, A., Zeller, D., Kubisch, C., Jablonka, S., Sendtner, M., Klopstock, T., de Carvalho, M., Sperfeld, A., Borck, G., Volk, A. E., Dorst, J., Weis, J., Otto, M., Schuster, J., Del Tredici, K., Braak, H., Danzer, K. M., Freischmidt, A., Meitinger, T., Strom, T. M., Ludolph, A. C., Andersen, P. M., Weishaupt, J. H., and The German ALS network MND-NET, 2018, "Hot-Spot KIF5A Mutations Cause Familial ALS," Brain, **141**(3), pp. 688–697.
108. Ganne, A., Balasubramaniam, M., Ayyadevara, H., Kiaei, L., Shmookler Reis, R. J., Varughese, K. I., and Kiaei, M., 2023, "In Silico Analysis of TUBA4A Mutations in Amyotrophic Lateral Sclerosis to Define Mechanisms of Microtubule Disintegration," Sci. Rep., **13**(1), p. 2096.
109. Pant, D. C., Parameswaran, J., Rao, L., Loss, I., Chilukuri, G., Parlato, R., Shi, L., Glass, J. D., Bassell, G. J., Koch, P., Yilmaz, R., Weishaupt, J. H., Gennerich, A., and Jiang, J., 2022, "ALS-Linked KIF5A ΔExon27 Mutant Causes Neuronal Toxicity through Gain-of-Function," EMBO Rep., **23**(8), p. e54234.
110. Jimenez-Sanchez, M., Licitra, F., Underwood, B. R., and Rubinsztein, D. C., 2017, "Huntington's Disease: Mechanisms of Pathogenesis and Therapeutic Strategies," Cold Spring Harb. Perspect. Med., **7**(7), p. a024240.

111. Harding, R. J., and Tong, Y.-F., 2018, "Proteostasis in Huntington's Disease: Disease Mechanisms and Therapeutic Opportunities," Acta Pharmacol. Sin., **39**(5), pp. 754–769.
112. Labbadia, J., and Morimoto, R. I., 2013, "Huntington's Disease: Underlying Molecular Mechanisms and Emerging Concepts," Trends Biochem. Sci., **38**(8), pp. 378–385.
113. Cui, L., Jeong, H., Borovecki, F., Parkhurst, C. N., Tanese, N., and Krainc, D., 2006, "Transcriptional Repression of PGC-1alpha by Mutant Huntingtin Leads to Mitochondrial Dysfunction and Neurodegeneration," Cell, **127**(1), pp. 59–69.
114. Gatto, E. M., Rojas, N. G., Persi, G., Etcheverry, J. L., Cesarini, M. E., and Perandones, C., 2020, "Huntington Disease: Advances in the Understanding of Its Mechanisms," Clin. Park. Relat. Disord., **3**, p. 100056.
115. Intihar, T. A., Martinez, E. A., and Gomez-Pastor, R., 2019, "Mitochondrial Dysfunction in Huntington's Disease; Interplay Between HSF1, P53 and PGC-1α Transcription Factors," Front. Cell. Neurosci., **13**, p. 103.
116. Saavedra, A., García-Díaz Barriga, G., Pérez-Navarro, E., and Alberch, J., 2018, "Huntington's Disease: Novel Therapeutic Perspectives Hanging in the Balance," Expert Opin. Ther. Targets, **22**(5), pp. 385–399.
117. Tyebji, S., and Hannan, A. J., 2017, "Synaptopathic Mechanisms of Neurodegeneration and Dementia: Insights from Huntington's Disease," Prog. Neurobiol., **153**, pp. 18–45.
118. Ratovitski, T., Chighladze, E., Arbez, N., Boronina, T., Herbrich, S., Cole, R. N., and Ross, C. A., 2012, "Huntingtin Protein Interactions Altered by Polyglutamine Expansion as Determined by Quantitative Proteomic Analysis," Cell Cycle, **11**(10), pp. 2006–2021.
119. Savas, J. N., Ma, B., Deinhardt, K., Culver, B. P., Restituito, S., Wu, L., Belasco, J. G., Chao, M. V., and Tanese, N., 2010, "A Role for Huntington Disease Protein in Dendritic RNA Granules," J. Biol. Chem., **285**(17), pp. 13142–13153.
120. Savas, J. N., Makusky, A., Ottosen, S., Baillat, D., Then, F., Krainc, D., Shiekhattar, R., Markey, S. P., and Tanese, N., 2008, "Huntington's Disease Protein Contributes to RNA-Mediated Gene Silencing through Association with Argonaute and P Bodies," Proc. Natl. Acad. Sci. U. S. A., **105**(31), pp. 10820–10825.
121. Gunawardena, S., Her, L.-S., Brusch, R. G., Laymon, R. A., Niesman, I. R., Gordesky-Gold, B., Sintasath, L., Bonini, N. M., and Goldstein, L. S. B., 2003, "Disruption of Axonal Transport by Loss of Huntingtin or Expression of Pathogenic polyQ Proteins in Drosophila," Neuron, **40**(1), pp. 25–40.
122. Lee, W.-C. M., Yoshihara, M., and Littleton, J. T., 2004, "Cytoplasmic Aggregates Trap Polyglutamine-Containing Proteins and Block Axonal Transport in a Drosophila Model of Huntington's Disease," Proc. Natl. Acad. Sci. U. S. A., **101**(9), pp. 3224–3229.
123. Mandal, M., Wei, J., Zhong, P., Cheng, J., Duffney, L. J., Liu, W., Yuen, E. Y., Twelvetrees, A. E., Li, S., Li, X.-J., Kittler, J. T., and Yan, Z., 2011, "Impaired α-Amino-3-Hydroxy-5-Methyl-4-Isoxazolepropionic Acid (AMPA) Receptor Trafficking and Function by Mutant Huntingtin," J. Biol. Chem., **286**(39), pp. 33719–33728.
124. Szebenyi, G., Morfini, G. A., Babcock, A., Gould, M., Selkoe, K., Stenoien, D. L., Young, M., Faber, P. W., MacDonald, M. E., McPhaul, M. J., and Brady, S. T., 2003, "Neuropathogenic Forms of Huntingtin and Androgen Receptor Inhibit Fast Axonal Transport," Neuron, **40**(1), pp. 41–52.
125. Twelvetrees, A. E., Yuen, E. Y., Arancibia-Carcamo, I. L., MacAskill, A. F., Rostaing, P., Lumb, M. J., Humbert, S., Triller, A., Saudou, F., Yan, Z., and Kittler, J. T., 2010, "Delivery of GABAARs to Synapses Is Mediated by HAP1-KIF5 and Disrupted by Mutant Huntingtin," Neuron, **65**(1), pp. 53–65.
126. Wilton, D. K., and Stevens, B., 2020, "The Contribution of Glial Cells to Huntington's Disease Pathogenesis," Neurobiol. Dis., **143**, p. 104963.
127. Costanzo, M., Abounit, S., Marzo, L., Danckaert, A., Chamoun, Z., Roux, P., and Zurzolo, C., 2013, "Transfer of Polyglutamine Aggregates in Neuronal Cells Occurs in Tunneling Nanotubes," J. Cell Sci., **126**(Pt 16), pp. 3678–3685.
128. Kanehisa, M., 1997, "Toward Pathway Engineering: A New Database of Genetic and Molecular Pathways."
129. Anderson, K. M., and Lee, M., 2016, "Therapeutic Strategies in Neurodegenerative Diseases," *Neuroimmune Pharmacology*, Springer, Cham. https://doi.org/10.1007/978-3-319-44022-4_42.
130. Birks, J. S., Chong, L. Y., and Grimley Evans, J., 2015, "Rivastigmine for Alzheimer's Disease," Cochrane Database Syst. Rev., **9**(9), p. CD001191.
131. Patel, P. H., and Gupta, V., 2023, "Rivastigmine," *StatPearls*, StatPearls Publishing, Treasure Island (FL), PMID: 32491370.
132. Scott, L. J., and Goa, K. L., 2000, "Galantamine: A Review of Its Use in Alzheimer's Disease," Drugs, **60**(5), pp. 1095–1122.
133. Kuns, B., Rosani, A., and Varghese, D., 2024, "Memantine," *StatPearls*, StatPearls Publishing, Treasure Island (FL), PMID: 29763201

134. Gandhi, K. R., and Saadabadi, A., 2023, "Levodopa (L-Dopa)," *StatPearls*, StatPearls Publishing, Treasure Island (FL), PMID: 29489269
135. Leyden, E., and Tadi, P., 2023, "Carbidopa," *StatPearls*, StatPearls Publishing, Treasure Island (FL), PMID: 32119439.
136. Habet, S., 2022, "Clinical Pharmacology of Entacapone (Comtan) From the FDA Reviewer," Int. J. Neuropsychopharmacol., **25**(7), pp. 567–575.
137. Willman, C., and Tadi, P., 2023, "Tolcapone," *StatPearls*, StatPearls Publishing, Treasure Island (FL), PMID: 32809428.
138. Katzenschlager, R., Sampaio, C., Costa, J., and Lees, A., 2003, "Anticholinergics for Symptomatic Management of Parkinson's Disease," Cochrane Database Syst. Rev., **2002**(2), p. CD003735.
139. Ahuja, A., and Abdijadid, S., 2023, "Benztropine," *StatPearls*, StatPearls Publishing, Treasure Island (FL), PMID: 32809468.
140. Jilani, T. N., Sabir, S., and Sharma, S., 2023, "Trihexyphenidyl," *StatPearls*, StatPearls Publishing, Treasure Island (FL), PMID: 30137772.
141. Auffret, M., Drapier, S., and Vérin, M., 2018, "Pharmacological Insights into the Use of Apomorphine in Parkinson's Disease: Clinical Relevance," Clin. Drug Investig., **38**(4), pp. 287–312.
142. Carbone, F., Djamshidian, A., Seppi, K., and Poewe, W., 2019, "Apomorphine for Parkinson's Disease: Efficacy and Safety of Current and New Formulations," CNS Drugs, **33**(9), pp. 905–918.
143. Ernst, A. M., and Smelik, P. G., 1966, "Site of Action of Dopamine and Apomorphine on Compulsive Gnawing Behaviour in Rats," Experientia, **22**(12), pp. 837–838.
144. Hisahara, S., and Shimohama, S., 2011, "Dopamine Receptors and Parkinson's Disease," Int. J. Med. Chem., **2011**, p. e403039.
145. Lieberman, A. N., and Goldstein, M., 1985, "Bromocriptine in Parkinson Disease," Pharmacol. Rev., **37**(2), pp. 217–227.
146. Sivanandy, P., Leey, T. C., Xiang, T. C., Ling, T. C., Wey Han, S. A., Semilan, S. L. A., and Hong, P. K., 2021, "Systematic Review on Parkinson's Disease Medications, Emphasizing on Three Recently Approved Drugs to Control Parkinson's Symptoms," Int. J. Environ. Res. Public. Health, **19**(1), p. 364.
147. Brooks, D. J., 2000, "Dopamine Agonists: Their Role in the Treatment of Parkinson's Disease," J. Neurol. Neurosurg. Psychiatry, **68**(6), pp. 685–689.
148. Rajendran, A., Reddy, A. J., Bisaga, K., Sommer, D. A., Prakash, N., Pokala, V. T., Yu, Z., Bachir, M., Nawathey, N., Brahmbhatt, T., and Patel, R., 2023, "A Systematic Review of the Usage of Rotigotine During Early and Advanced Stage Parkinson's," Cureus, **15**(3), p. e36211.
149. Moore, J. J., and Saadabadi, A., 2023, "Selegiline," *StatPearls*, StatPearls Publishing, Treasure Island (FL), PMID: 30252350.
150. M. Finberg, J. P., Takeshima, T., Johnston, J. M., and Commissiong, J. W., 1998, "Increased Survival of Dopaminergic Neurons by Rasagiline, a Monoamine Oxidase B Inhibitor," NeuroReport, **9**(4), p. 703.
151. Mandel, S., Weinreb, O., Amit, T., and Youdim, M. B. H., 2005, "Mechanism of Neuroprotective Action of the Anti-Parkinson Drug Rasagiline and Its Derivatives," Brain Res. Brain Res. Rev., **48**(2), pp. 379–387.
152. Nayak, L., and Henchcliffe, C., 2008, "Rasagiline in Treatment of Parkinson's Disease," Neuropsychiatr. Dis. Treat., **4**(1), pp. 23–32.
153. Youdim, M. B., Wadia, A., Tatton, W., and Weinstock, M., 2001, "The Anti-Parkinson Drug Rasagiline and Its Cholinesterase Inhibitor Derivatives Exert Neuroprotection Unrelated to MAO Inhibition in Cell Culture and in Vivo," Ann. N. Y. Acad. Sci., **939**, pp. 450–458.
154. Lu, H., Le, W. D., Xie, Y.-Y., and Wang, X.-P., 2016, "Current Therapy of Drugs in Amyotrophic Lateral Sclerosis," Curr. Neuropharmacol., **14**(4), pp. 314–321.
155. Nikitin, D., Makam, A. N., Suh, K., McKenna, A., Carlson, J. J., Richardson, M., Rind, D. M., and Pearson, S. D., 2023, "The Effectiveness and Value of AMX0035 and Oral Edaravone for Amyotrophic Lateral Sclerosis: A Summary from the Institute for Clinical and Economic Review's Midwest Comparative Effectiveness Public Advisory Council," J. Manag. Care Spec. Pharm., **29**(2), pp. 216–221.
156. Paganoni, S., Macklin, E. A., Hendrix, S., Berry, J. D., Elliott, M. A., Maiser, S., Karam, C., Caress, J. B., Owegi, M. A., Quick, A., Wymer, J., Goutman, S. A., Heitzman, D., Heiman-Patterson, T., Jackson, C. E., Quinn, C., Rothstein, J. D., Kasarskis, E. J., Katz, J., Jenkins, L., Ladha, S., Miller, T. M., Scelsa, S. N., Vu, T. H., Fournier, C. N., Glass, J. D., Johnson, K. M., Swenson, A., Goyal, N. A., Pattee, G. L., Andres, P. L., Babu, S., Chase, M., Dagostino, D., Dickson, S. P., Ellison, N., Hall, M., Hendrix, K., Kittle, G., McGovern, M., Ostrow, J., Pothier, L., Randall, R., Shefner, J. M., Sherman, A. V., Tustison, E., Vigneswaran, P., Walker, J., Yu, H., Chan, J., Wittes, J., Cohen, J., Klee, J., Leslie, K., Tanzi, R. E., Gilbert, W., Yeramian, P. D., Schoenfeld, D., and Cudkowicz, M. E., 2020, "Trial of Sodium Phenylbutyrate-Taurursodiol for Amyotrophic Lateral Sclerosis," N. Engl. J. Med., **383**(10), pp. 919–930.

157. Cho, H., and Shukla, S., 2020, "Role of Edaravone as a Treatment Option for Patients with Amyotrophic Lateral Sclerosis," Pharmaceuticals, **14**(1), p. 29.
158. Ferguson, M. W., Kennedy, C. J., Palpagama, T. H., Waldvogel, H. J., Faull, R. L. M., and Kwakowsky, A., 2022, "Current and Possible Future Therapeutic Options for Huntington's Disease," J. Cent. Nerv. Syst. Dis., **14**. https://doi.org/10.1177/11795735221092517
159. Zheng, G., Dwoskin, L. P., and Crooks, P. A., 2006, "Vesicular Monoamine Transporter 2: Role as a Novel Target for Drug Development," AAPS J., **8**(4), pp. E682–E692.
160. Valera, E., and Masliah, E., 2013, "Immunotherapy for Neurodegenerative Diseases: Focus on α-Synucleinopathies," Pharmacol. Ther., **138**(3), pp. 311–322.
161. Vellas, B., Black, R., Thal, L. J., Fox, N. C., Daniels, M., McLennan, G., Tompkins, C., Leibman, C., Pomfret, M., Grundman, M., and AN1792 (QS-21)-251 Study Team, 2009, "Long-Term Follow-up of Patients Immunized with AN1792: Reduced Functional Decline in Antibody Responders," Curr. Alzheimer Res., **6**(2), pp. 144–151.
162. Winblad, B., Andreasen, N., Minthon, L., Floesser, A., Imbert, G., Dumortier, T., Maguire, R. P., Blennow, K., Lundmark, J., Staufenbiel, M., Orgogozo, J.-M., and Graf, A., 2012, "Safety, Tolerability, and Antibody Response of Active Aβ Immunotherapy with CAD106 in Patients with Alzheimer's Disease: Randomised, Double-Blind, Placebo-Controlled, First-in-Human Study," Lancet Neurol., **11**(7), pp. 597–604.
163. Yu, H. J., Dickson, S. P., Wang, P.-N., Chiu, M.-J., Huang, C.-C., Chang, C.-C., Liu, H., Hendrix, S. B., Dodart, J.-C., Verma, A., Wang, C. Y., and Cummings, J., 2023, "Safety, Tolerability, Immunogenicity, and Efficacy of UB-311 in Participants with Mild Alzheimer's Disease: A Randomised, Double-Blind, Placebo-Controlled, Phase 2a Study," eBioMedicine, **94**, p. 104665.
164. Novak, P., Schmidt, R., Kontsekova, E., Kovacech, B., Smolek, T., Katina, S., Fialova, L., Prcina, M., Parrak, V., Dal-Bianco, P., Brunner, M., Staffen, W., Rainer, M., Ondrus, M., Ropele, S., Smisek, M., Sivak, R., Zilka, N., Winblad, B., and Novak, M., 2018, "FUNDAMANT: An Interventional 72-Week Phase 1 Follow-up Study of AADvac1, an Active Immunotherapy against Tau Protein Pathology in Alzheimer's Disease," Alzheimers Res. Ther., **10**(1), p. 108.
165. Allen Reish, H. E., and Standaert, D. G., 2015, "Role of α-Synuclein in Inducing Innate and Adaptive Immunity in Parkinson Disease," J. Park. Dis., **5**(1), pp. 1–19.
166. Bhattacharya, P., Budnick, I., Singh, M., Thiruppathi, M., Alharshawi, K., Elshabrawy, H., Holterman, M. J., and Prabhakar, B. S., 2015, "Dual Role of GM-CSF as a Pro-Inflammatory and a Regulatory Cytokine: Implications for Immune Therapy," J. Interferon Cytokine Res. Off. J. Int. Soc. Interferon Cytokine Res., **35**(8), pp. 585–599.
167. Schutt, C. R., Gendelman, H. E., and Mosley, R. L., 2018, "Tolerogenic Bone Marrow-Derived Dendritic Cells Induce Neuroprotective Regulatory T Cells in a Model of Parkinson's Disease," Mol. Neurodegener., **13**(1), p. 26.
168. Brundin, P., Dave, K. D., and Kordower, J. H., 2017, "Therapeutic Approaches to Target Alpha-Synuclein Pathology," Exp. Neurol., **298**(Pt B), pp. 225–235.
169. Takeuchi, S., Fujiwara, N., Ido, A., Oono, M., Takeuchi, Y., Tateno, M., Suzuki, K., Takahashi, R., Tooyama, I., Taniguchi, N., Julien, J.-P., and Urushitani, M., 2010, "Induction of Protective Immunity by Vaccination with Wild-Type Apo Superoxide Dismutase 1 in Mutant SOD1 Transgenic Mice," J. Neuropathol. Exp. Neurol., **69**(10), pp. 1044–1056.
170. Dong, Q.-X., Zhu, J., Liu, S.-Y., Yu, X.-L., and Liu, R.-T., 2018, "An Oligomer-Specific Antibody Improved Motor Function and Attenuated Neuropathology in the SOD1-G93A Transgenic Mouse Model of ALS," Int. Immunopharmacol., **65**, pp. 413–421.
171. Fatoba, O., Ohtake, Y., Itokazu, T., and Yamashita, T., 2020, "Immunotherapies in Huntington's Disease and α-Synucleinopathies," Front. Immunol., **11**, p. 337.
172. Miller, T. W., Shirley, T. L., Wolfgang, W. J., Kang, X., and Messer, A., 2003, "DNA Vaccination against Mutant Huntingtin Ameliorates the HDR6/2 Diabetic Phenotype," Mol. Ther. J. Am. Soc. Gene Ther., **7**(5 Pt 1), pp. 572–579.
173. Ramsingh, A. I., Manley, K., Rong, Y., Reilly, A., and Messer, A., 2015, "Transcriptional Dysregulation of Inflammatory/Immune Pathways after Active Vaccination against Huntington's Disease," Hum. Mol. Genet., **24**(21), pp. 6186–6197.
174. Snyder-Keller, A., McLear, J. A., Hathorn, T., and Messer, A., 2010, "Early or Late-Stage Anti-N-Terminal Huntingtin Intrabody Gene Therapy Reduces Pathological Features in B6.HDR6/1 Mice," J. Neuropathol. Exp. Neurol., **69**(10), pp. 1078–1085.
175. Tuszynski, M. H., Thal, L., Pay, M., Salmon, D. P., U, H. S., Bakay, R., Patel, P., Blesch, A., Vahlsing, H. L., Ho, G., Tong, G., Potkin, S. G., Fallon, J., Hansen, L., Mufson, E. J., Kordower, J. H., Gall, C.,

and Conner, J., 2005, "A Phase 1 Clinical Trial of Nerve Growth Factor Gene Therapy for Alzheimer Disease," Nat. Med., **11**(5), pp. 551–555.
176. Rafii, M. S., Tuszynski, M. H., Thomas, R. G., Barba, D., Brewer, J. B., Rissman, R. A., Siffert, J., Aisen, P. S., and AAV2-NGF Study Team, 2018, "Adeno-Associated Viral Vector (Serotype 2)-Nerve Growth Factor for Patients With Alzheimer Disease: A Randomized Clinical Trial," JAMA Neurol., **75**(7), pp. 834–841.
177. Nagahara, A. H., Wilson, B. R., Ivasyk, I., Kovacs, I., Rawalji, S., Bringas, J. R., Pivirotto, P. J., Sebastian, W. S., Samaranch, L., Bankiewicz, K. S., and Tuszynski, M. H., 2018, "MR-Guided Delivery of AAV2-BDNF into the Entorhinal Cortex of Non-Human Primates," Gene Ther., **25**(2), pp. 104–114.
178. Bu, G., 2022, "APOE Targeting Strategy in Alzheimer's Disease: Lessons Learned from Protective Variants," Mol. Neurodegener., **17**(1), p. 51.
179. Lin, Y.-T., Seo, J., Gao, F., Feldman, H. M., Wen, H.-L., Penney, J., Cam, H. P., Gjoneska, E., Raja, W. K., Cheng, J., Rueda, R., Kritskiy, O., Abdurrob, F., Peng, Z., Milo, B., Yu, C. J., Elmsaouri, S., Dey, D., Ko, T., Yankner, B. A., and Tsai, L.-H., 2018, "APOE4 Causes Widespread Molecular and Cellular Alterations Associated with Alzheimer's Disease Phenotypes in Human iPSC-Derived Brain Cell Types," Neuron, **98**(6), pp. 1141–1154.e7.
180. Chang, J. L., Hinrich, A. J., Roman, B., Norrbom, M., Rigo, F., Marr, R. A., Norstrom, E. M., and Hastings, M. L., 2018, "Targeting Amyloid-β Precursor Protein, APP, Splicing with Antisense Oligonucleotides Reduces Toxic Amyloid-β Production," Mol. Ther. J. Am. Soc. Gene Ther., **26**(6), pp. 1539–1551.
181. György, B., Lööv, C., Zaborowski, M. P., Takeda, S., Kleinstiver, B. P., Commins, C., Kastanenka, K., Mu, D., Volak, A., Giedraitis, V., Lannfelt, L., Maguire, C. A., Joung, J. K., Hyman, B. T., Breakefield, X. O., and Ingelsson, M., 2018, "CRISPR/Cas9 Mediated Disruption of the Swedish APP Allele as a Therapeutic Approach for Early-Onset Alzheimer's Disease," Mol. Ther. Nucleic Acids, **11**, pp. 429–440.
182. Fol, R., Braudeau, J., Ludewig, S., Abel, T., Weyer, S. W., Roederer, J.-P., Brod, F., Audrain, M., Bemelmans, A.-P., Buchholz, C. J., Korte, M., Cartier, N., and Müller, U. C., 2016, "Viral Gene Transfer of APPsα Rescues Synaptic Failure in an Alzheimer's Disease Mouse Model," Acta Neuropathol. (Berl.), **131**(2), pp. 247–266.
183. Golde, T. E., Petrucelli, L., and Lewis, J., 2010, "Targeting Aβ and Tau in Alzheimer's Disease, an Early Interim Report," Exp. Neurol., **223**(2), pp. 252–266.
184. DeVos, S. L., Miller, R. L., Schoch, K. M., Holmes, B. B., Kebodeaux, C. S., Wegener, A. J., Chen, G., Shen, T., Tran, H., Nichols, B., Zanardi, T. A., Kordasiewicz, H. B., Swayze, E. E., Bennett, C. F., Diamond, M. I., and Miller, T. M., 2017, "Tau Reduction Prevents Neuronal Loss and Reverses Pathological Tau Deposition and Seeding in Mice with Tauopathy," Sci. Transl. Med., **9**(374), p. eaag0481.
185. Mummery, C. J., Börjesson-Hanson, A., Blackburn, D. J., Vijverberg, E. G. B., De Deyn, P. P., Ducharme, S., Jonsson, M., Schneider, A., Rinne, J. O., Ludolph, A. C., Bodenschatz, R., Kordasiewicz, H., Swayze, E. E., Fitzsimmons, B., Mignon, L., Moore, K. M., Yun, C., Baumann, T., Li, D., Norris, D. A., Crean, R., Graham, D. L., Huang, E., Ratti, E., Bennett, C. F., Junge, C., and Lane, R. M., 2023, "Tau-Targeting Antisense Oligonucleotide MAPTRx in Mild Alzheimer's Disease: A Phase 1b, Randomized, Placebo-Controlled Trial," Nat. Med., **29**(6), pp. 1437–1447.
186. Zhang, W., Zhao, H., Wu, Q., Xu, W., and Xia, M., 2018, "Knockdown of BACE1-AS by siRNA Improves Memory and Learning Behaviors in Alzheimer's Disease Animal Model," Exp. Ther. Med., **16**(3), pp. 2080–2086.
187. Zhang, X., Li, Y., Xu, H., and Zhang, Y.-W., 2014, "The γ-Secretase Complex: From Structure to Function," Front. Cell. Neurosci., **8**, p. 427.
188. Kandimalla, R. J. L., Wani, W. Y., Binukumar, B. K., and Gill, K. D., 2012, "siRNA against Presenilin 1 (PS1) down Regulates Amyloid B42 Production in IMR-32 Cells," J. Biomed. Sci., **19**(1), p. 2.
189. Luo, J., Kaplitt, M. G., Fitzsimons, H. L., Zuzga, D. S., Liu, Y., Oshinsky, M. L., and During, M. J., 2002, "Subthalamic GAD Gene Therapy in a Parkinson's Disease Rat Model," Science, **298**(5592), pp. 425–429.
190. Axelsen, T. M., and Woldbye, D. P. D., 2018, "Gene Therapy for Parkinson's Disease, An Update," J. Park. Dis., **8**(2), pp. 195–215.
191. Kantor, B., Tagliafierro, L., Gu, J., Zamora, M. E., Ilich, E., Grenier, C., Huang, Z. Y., Murphy, S., and Chiba-Falek, O., 2018, "Downregulation of SNCA Expression by Targeted Editing of DNA Methylation: A Potential Strategy for Precision Therapy in PD," Mol. Ther. J. Am. Soc. Gene Ther., **26**(11), pp. 2638–2649.
192. Gorbatyuk, O. S., Li, S., Nash, K., Gorbatyuk, M., Lewin, A. S., Sullivan, L. F., Mandel, R. J., Chen, W., Meyers, C., Manfredsson, F. P., and Muzyczka, N., 2010, "In Vivo RNAi-Mediated α-Synuclein Silencing Induces Nigrostriatal Degeneration," Mol. Ther., **18**(8), pp. 1450–1457.

193. Gandhi, P. N., Chen, S. G., and Wilson-Delfosse, A. L., 2009, "Leucine-Rich Repeat Kinase 2 (LRRK2): A Key Player in the Pathogenesis of Parkinson's Disease," J. Neurosci. Res., **87**(6), pp. 1283–1295.
194. Lewis, J., Melrose, H., Bumcrot, D., Hope, A., Zehr, C., Lincoln, S., Braithwaite, A., He, Z., Ogholikhan, S., Hinkle, K., Kent, C., Toudjarska, I., Charisse, K., Braich, R., Pandey, R. K., Heckman, M., Maraganore, D. M., Crook, J., and Farrer, M. J., 2008, "In Vivo Silencing of Alpha-Synuclein Using Naked siRNA," Mol. Neurodegener., **3**(1), p. 19.
195. McCormack, A. L., Mak, S. K., Henderson, J. M., Bumcrot, D., Farrer, M. J., and Monte, D. A. D., 2010, "α-Synuclein Suppression by Targeted Small Interfering RNA in the Primate Substantia Nigra," PLoS One, **5**(8), p. e12122.
196. Takahashi, M., Suzuki, M., Fukuoka, M., Fujikake, N., Watanabe, S., Murata, M., Wada, K., Nagai, Y., and Hohjoh, H., 2015, "Normalization of Overexpressed α-Synuclein Causing Parkinson's Disease By a Moderate Gene Silencing With RNA Interference," Mol. Ther. - Nucleic Acids, **4**, p. e241.
197. Evers, M. M., and Konstantinova, P., 2020, "AAV5-miHTT Gene Therapy for Huntington Disease: Lowering Both Huntingtins," Expert Opin. Biol. Ther., **20**(10), pp. 1121–1124.
198. Sun, J., and Roy, S., 2021, "Gene-Based Therapies for Neurodegenerative Diseases," Nat. Neurosci., **24**(3), pp. 297–311.
199. Spronck, E. A., Brouwers, C. C., Vallès, A., de Haan, M., Petry, H., van Deventer, S. J., Konstantinova, P., and Evers, M. M., 2019, "AAV5-miHTT Gene Therapy Demonstrates Sustained Huntingtin Lowering and Functional Improvement in Huntington Disease Mouse Models," Mol. Ther. Methods Clin. Dev., **13**, pp. 334–343.
200. Miniarikova, J., Zanella, I., Huseinovic, A., van der Zon, T., Hanemaaijer, E., Martier, R., Koornneef, A., Southwell, A. L., Hayden, M. R., van Deventer, S. J., Petry, H., and Konstantinova, P., 2016, "Design, Characterization, and Lead Selection of Therapeutic miRNAs Targeting Huntingtin for Development of Gene Therapy for Huntington's Disease," Mol. Ther. Nucleic Acids, **5**(3), p. e297.
201. Fang, T., Je, G., Pacut, P., Keyhanian, K., Gao, J., and Ghasemi, M., 2022, "Gene Therapy in Amyotrophic Lateral Sclerosis," Cells, **11**(13), p. 2066.
202. Gittings, L. M., and Sattler, R., 2020, "Recent Advances in Understanding Amyotrophic Lateral Sclerosis and Emerging Therapies," Fac. Rev., **9**, p. 12.
203. Kim, G., Gautier, O., Tassoni-Tsuchida, E., Ma, X. R., and Gitler, A. D., 2020, "ALS Genetics: Gains, Losses, and Implications for Future Therapies," Neuron, **108**(5), pp. 822–842.
204. Lin, H., Hu, H., Duan, W., Liu, Y., Tan, G., Li, Z., Liu, Y., Deng, B., Song, X., Wang, W., Wen, D., Wang, Y., and Li, C., 2018, "Intramuscular Delivery of scAAV9-hIGF1 Prolongs Survival in the hSOD1G93A ALS Mouse Model via Upregulation of D-Amino Acid Oxidase," Mol. Neurobiol., **55**(1), pp. 682–695.
205. Wang, Y., Duan, W., Wang, W., Di Wen, null, Liu, Y., Liu, Y., Li, Z., Hu, H., Lin, H., Cui, C., Li, D., Dong, H., and Li, C., 2016, "scAAV9-VEGF Prolongs the Survival of Transgenic ALS Mice by Promoting Activation of M2 Microglia and the PI3K/Akt Pathway," Brain Res., **1648**(Pt A), pp. 1–10.
206. Smith, R. A., Miller, T. M., Yamanaka, K., Monia, B. P., Condon, T. P., Hung, G., Lobsiger, C. S., Ward, C. M., McAlonis-Downes, M., Wei, H., Wancewicz, E. V., Bennett, C. F., and Cleveland, D. W., 2006, "Antisense Oligonucleotide Therapy for Neurodegenerative Disease," J. Clin. Invest., **116**(8), pp. 2290–2296.
207. Jiang, J., Zhu, Q., Gendron, T. F., Saberi, S., McAlonis-Downes, M., Seelman, A., Stauffer, J. E., Jafar-Nejad, P., Drenner, K., Schulte, D., Chun, S., Sun, S., Ling, S.-C., Myers, B., Engelhardt, J., Katz, M., Baughn, M., Platoshyn, O., Marsala, M., Watt, A., Heyser, C. J., Ard, M. C., De Muynck, L., Daughrity, L. M., Swing, D. A., Tessarollo, L., Jung, C. J., Delpoux, A., Utzschneider, D. T., Hedrick, S. M., de Jong, P. J., Edbauer, D., Van Damme, P., Petrucelli, L., Shaw, C. E., Bennett, C. F., Da Cruz, S., Ravits, J., Rigo, F., Cleveland, D. W., and Lagier-Tourenne, C., 2016, "Gain of Toxicity from ALS/FTD-Linked Repeat Expansions in C9ORF72 Is Alleviated by Antisense Oligonucleotides Targeting GGGGCC-Containing RNAs," Neuron, **90**(3), pp. 535–550.
208. Foust, K. D., Salazar, D. L., Likhite, S., Ferraiuolo, L., Ditsworth, D., Ilieva, H., Meyer, K., Schmelzer, L., Braun, L., Cleveland, D. W., and Kaspar, B. K., 2013, "Therapeutic AAV9-Mediated Suppression of Mutant SOD1 Slows Disease Progression and Extends Survival in Models of Inherited ALS," Mol. Ther. J. Am. Soc. Gene Ther., **21**(12), pp. 2148–2159.
209. Gaj, T., Ojala, D. S., Ekman, F. K., Byrne, L. C., Limsirichai, P., and Schaffer, D. V., 2017, "In Vivo Genome Editing Improves Motor Function and Extends Survival in a Mouse Model of ALS," Sci. Adv., **3**(12), p. eaar3952.
210. Wu, K., Meyers, C. A., Guerra, N. K., King, M. A., and Meyer, E. M., 2004, "The Effects of rAAV2-Mediated NGF Gene Delivery in Adult and Aged Rats," Mol. Ther., **9**(2), pp. 262–269.
211. Iwata, N., Mizukami, H., Shirotani, K., Takaki, Y., Muramatsu, S., Lu, B., Gerard, N. P., Gerard, C., Ozawa, K., and Saido, T. C., 2004, "Presynaptic Localization of Neprilysin Contributes to Efficient Clearance of Amyloid-Beta Peptide in Mouse Brain," J. Neurosci. Off. J. Soc. Neurosci., **24**(4), pp. 991–998.

212. Hemming, M. L., Patterson, M., Reske-Nielsen, C., Lin, L., Isacson, O., and Selkoe, D. J., 2007, "Reducing Amyloid Plaque Burden via Ex Vivo Gene Delivery of an Abeta-Degrading Protease: A Novel Therapeutic Approach to Alzheimer Disease," PLoS Med., **4**(8), p. e262.
213. Wood, L. S., Pickering, E. H., McHale, D., and Dechairo, B. M., 2007, "Association between Neprilysin Polymorphisms and Sporadic Alzheimer's Disease," Neurosci. Lett., **427**(2), pp. 103–106.
214. Carty, N. C., Nash, K., Lee, D., Mercer, M., Gottschall, P. E., Meyers, C., Muzyczka, N., Gordon, M. N., and Morgan, D., 2008, "Adeno-Associated Viral (AAV) Serotype 5 Vector Mediated Gene Delivery of Endothelin-Converting Enzyme Reduces Aβ Deposits in APP + PS1 Transgenic Mice," Mol. Ther. J. Am. Soc. Gene Ther., **16**(9), pp. 1580–1586.
215. Dodart, J.-C., Marr, R. A., Koistinaho, M., Gregersen, B. M., Malkani, S., Verma, I. M., and Paul, S. M., 2005, "Gene Delivery of Human Apolipoprotein E Alters Brain Abeta Burden in a Mouse Model of Alzheimer's Disease," Proc. Natl. Acad. Sci. U. S. A., **102**(4), pp. 1211–1216.
216. Hong, C.-S., Goins, W. F., Goss, J. R., Burton, E. A., and Glorioso, J. C., 2006, "Herpes Simplex Virus RNAi and Neprilysin Gene Transfer Vectors Reduce Accumulation of Alzheimer's Disease-Related Amyloid-Beta Peptide in Vivo," Gene Ther., **13**(14), pp. 1068–1079.
217. Singer, O., Marr, R. A., Rockenstein, E., Crews, L., Coufal, N. G., Gage, F. H., Verma, I. M., and Masliah, E., 2005, "Targeting BACE1 with siRNAs Ameliorates Alzheimer Disease Neuropathology in a Transgenic Model," Nat. Neurosci., **8**(10), pp. 1343–1349.
218. Azevedo, M. D., Prince, N., Humbert-Claude, M., Mesa-Infante, V., Jeanneret, C., Golzne, V., De Matos, K., Jamot, B. B., Magara, F., Gonzalez-Hernandez, T., and Tenenbaum, L., 2023, "Oxidative Stress Induced by Sustained Supraphysiological Intrastriatal GDNF Delivery Is Prevented by Dose Regulation," Mol. Ther. Methods Clin. Dev., **31**, p. 101106.
219. Palasz, E., Wysocka, A., Gasiorowska, A., Chalimoniuk, M., Niewiadomski, W., and Niewiadomska, G., 2020, "BDNF as a Promising Therapeutic Agent in Parkinson's Disease," Int. J. Mol. Sci., **21**(3), p. 1170.
220. Huttunen, H. J., and Saarma, M., 2019, "CDNF Protein Therapy in Parkinson's Disease," Cell Transplant., **28**(4), pp. 349–366.
221. Tian, Y., Tang, C.-J., Wang, J., Feng, Y., Chen, X., Wang, L., Qiao, X., and Sun, S., 2007, "Favorable Effects of VEGF Gene Transfer on a Rat Model of Parkinson Disease Using Adeno-Associated Viral Vectors," Neurosci. Lett., **421**(3), pp. 239–244.
222. Sullivan, A. M., and O'Keeffe, G. W., 2016, "Neurotrophic Factor Therapy for Parkinson's Disease: Past, Present and Future," Neural Regen. Res., **11**(2), pp. 205–207.
223. Nutt, J. G., Curtze, C., Hiller, A., Anderson, S., Larson, P. S., Van Laar, A. D., Richardson, R. M., Thompson, M. E., Sedkov, A., Leinonen, M., Ravina, B., Bankiewicz, K. S., and Christine, C. W., 2020, "Aromatic L-Amino Acid Decarboxylase Gene Therapy Enhances Levodopa Response in Parkinson's Disease," Mov. Disord. Off. J. Mov. Disord. Soc., **35**(5), pp. 851–858.
224. Miller, T., Cudkowicz, M., Shaw, P. J., Andersen, P. M., Atassi, N., Bucelli, R. C., Genge, A., Glass, J., Ladha, S., Ludolph, A. L., Maragakis, N. J., McDermott, C. J., Pestronk, A., Ravits, J., Salachas, F., Trudell, R., Van Damme, P., Zinman, L., Bennett, C. F., Lane, R., Sandrock, A., Runz, H., Graham, D., Houshyar, H., McCampbell, A., Nestorov, I., Chang, I., McNeill, M., Fanning, L., Fradette, S., and Ferguson, T. A., 2020, "Phase 1–2 Trial of Antisense Oligonucleotide Tofersen for SOD1 ALS," N. Engl. J. Med., **383**(2), pp. 109–119.
225. Miller, T. M., Cudkowicz, M. E., Genge, A., Shaw, P. J., Sobue, G., Bucelli, R. C., Chiò, A., Van Damme, P., Ludolph, A. C., Glass, J. D., Andrews, J. A., Babu, S., Benatar, M., McDermott, C. J., Cochrane, T., Chary, S., Chew, S., Zhu, H., Wu, F., Nestorov, I., Graham, D., Sun, P., McNeill, M., Fanning, L., Ferguson, T. A., Fradette, S., and VALOR and OLE Working Group, 2022, "Trial of Antisense Oligonucleotide Tofersen for SOD1 ALS," N. Engl. J. Med., **387**(12), pp. 1099–1110.
226. Mueller, C., Berry, J. D., McKenna-Yasek, D. M., Gernoux, G., Owegi, M. A., Pothier, L. M., Douthwright, C. L., Gelevski, D., Luppino, S. D., Blackwood, M., Wightman, N. S., Oakley, D. H., Frosch, M. P., Flotte, T. R., Cudkowicz, M. E., and Brown, R. H., 2020, "SOD1 Suppression with Adeno-Associated Virus and MicroRNA in Familial ALS," N. Engl. J. Med., **383**(2), pp. 151–158.
227. McEachin, Z. T., Parameswaran, J., Raj, N., Bassell, G. J., and Jiang, J., 2020, "RNA-Mediated Toxicity in C9orf72 ALS and FTD," Neurobiol. Dis., **145**, p. 105055.
228. Korobeynikov, V. A., Lyashchenko, A. K., Blanco-Redondo, B., Jafar-Nejad, P., and Shneider, N. A., 2022, "Antisense Oligonucleotide Silencing of FUS Expression as a Therapeutic Approach in Amyotrophic Lateral Sclerosis," Nat. Med., **28**(1), pp. 104–116.
229. Keskin, S., Brouwers, C. C., Sogorb-Gonzalez, M., Martier, R., Depla, J. A., Vallès, A., van Deventer, S. J., Konstantinova, P., and Evers, M. M., 2019, "AAV5-miHTT Lowers Huntingtin mRNA and Protein without Off-Target Effects in Patient-Derived Neuronal Cultures and Astrocytes," Mol. Ther. Methods Clin. Dev., **15**, pp. 275–284.

230. Chen, B. K., Staff, N. P., Knight, A. M., Nesbitt, J. J., Butler, G. W., Padley, D. J., Parisi, J. E., Dietz, A. B., and Windebank, A. J., 2015, "A Safety Study on Intrathecal Delivery of Autologous Mesenchymal Stromal Cells in Rabbits Directly Supporting Phase I Human Trials," Transfusion (Paris), **55**(5), pp. 1013–1020.
231. Nicaise, C., Mitrecic, D., Falnikar, A., and Lepore, A. C., 2015, "Transplantation of Stem Cell-Derived Astrocytes for the Treatment of Amyotrophic Lateral Sclerosis and Spinal Cord Injury," World J. Stem Cells, **7**(2), pp. 380–398.
232. Wen, M. M., El-Salamouni, N. S., El-Refaie, W. M., Hazzah, H. A., Ali, M. M., Tosi, G., Farid, R. M., Blanco-Prieto, M. J., Billa, N., and Hanafy, A. S., 2017, "Nanotechnology-Based Drug Delivery Systems for Alzheimer's Disease Management: Technical, Industrial, and Clinical Challenges," J. Controlled Release, **245**, pp. 95–107.
233. Blurton-Jones, M., Kitazawa, M., Martinez-Coria, H., Castello, N. A., Müller, F.-J., Loring, J. F., Yamasaki, T. R., Poon, W. W., Green, K. N., and LaFerla, F. M., 2009, "Neural Stem Cells Improve Cognition via BDNF in a Transgenic Model of Alzheimer Disease," Proc. Natl. Acad. Sci. U. S. A., **106**(32), pp. 13594–13599.
234. Park, D., Yang, Y.-H., Bae, D. K., Lee, S. H., Yang, G., Kyung, J., Kim, D., Choi, E.-K., Lee, S. W., Kim, G. H., Hong, J. T., Choi, K.-C., Lee, H. J., Kim, S. U., and Kim, Y.-B., 2013, "Improvement of Cognitive Function and Physical Activity of Aging Mice by Human Neural Stem Cells Over-Expressing Choline Acetyltransferase," Neurobiol. Aging, **34**(11), pp. 2639–2646.
235. Xuan, A. G., Luo, M., Ji, W. D., and Long, D. H., 2009, "Effects of Engrafted Neural Stem Cells in Alzheimer's Disease Rats," Neurosci. Lett., **450**(2), pp. 167–171.
236. Sakthiswary, R., and Raymond, A. A., 2012, "Stem Cell Therapy in Neurodegenerative Diseases," Neural Regen. Res., **7**(23), pp. 1822–1831.
237. Guo, Z., Zhang, L., Wu, Z., Chen, Y., Wang, F., and Chen, G., 2014, "In Vivo Direct Reprogramming of Reactive Glial Cells into Functional Neurons after Brain Injury and in an Alzheimer's Disease Model," Cell Stem Cell, **14**(2), pp. 188–202.
238. Liu, Y., Weick, J. P., Liu, H., Krencik, R., Zhang, X., Ma, L., Zhou, G., Ayala, M., and Zhang, S.-C., 2013, "Medial Ganglionic Eminence-like Cells Derived from Human Embryonic Stem Cells Correct Learning and Memory Deficits," Nat. Biotechnol., **31**(5), pp. 440–447.
239. Fujiwara, N., Shimizu, J., Takai, K., Arimitsu, N., Saito, A., Kono, T., Umehara, T., Ueda, Y., Wakisaka, S., Suzuki, T., and Suzuki, N., 2013, "Restoration of Spatial Memory Dysfunction of Human APP Transgenic Mice by Transplantation of Neuronal Precursors Derived from Human iPS Cells," Neurosci. Lett., **557**(Pt B), pp. 129–134.
240. Takamatsu, K., Ikeda, T., Haruta, M., Matsumura, K., Ogi, Y., Nakagata, N., Uchino, M., Ando, Y., Nishimura, Y., and Senju, S., 2014, "Degradation of Amyloid Beta by Human Induced Pluripotent Stem Cell-Derived Macrophages Expressing Neprilysin-2," Stem Cell Res., **13**(3 Pt A), pp. 442–453.
241. Kim, D. H., Lim, H., Lee, D., Choi, S. J., Oh, W., Yang, Y. S., Oh, J. S., Hwang, H. H., and Jeon, H. B., 2018, "Thrombospondin-1 Secreted by Human Umbilical Cord Blood-Derived Mesenchymal Stem Cells Rescues Neurons from Synaptic Dysfunction in Alzheimer's Disease Model," Sci. Rep., **8**(1), p. 354.
242. Garcia, K. O., Ornellas, F. L. M., Martin, P. K. M., Patti, C. L., Mello, L. E., Frussa-Filho, R., Han, S. W., and Longo, B. M., 2014, "Therapeutic Effects of the Transplantation of VEGF Overexpressing Bone Marrow Mesenchymal Stem Cells in the Hippocampus of Murine Model of Alzheimer's Disease," Front. Aging Neurosci., **6**, p. 30.
243. Hunsberger, J. G., Rao, M., Kurtzberg, J., Bulte, J. W. M., Atala, A., LaFerla, F. M., Greely, H. T., Sawa, A., Gandy, S., Schneider, L. S., and Doraiswamy, P. M., 2016, "Accelerating Stem Cell Trials for Alzheimer's Disease," Lancet Neurol., **15**(2), pp. 219–230.
244. Kan, I., Barhum, Y., Melamed, E., and Offen, D., 2011, "Mesenchymal Stem Cells Stimulate Endogenous Neurogenesis in the Subventricular Zone of Adult Mice," Stem Cell Rev. Rep., **7**(2), pp. 404–412.
245. Kim, H. J., Seo, S. W., Chang, J. W., Lee, J. I., Kim, C. H., Chin, J., Choi, S. J., Kwon, H., Yun, H. J., Lee, J. M., Kim, S. T., Choe, Y. S., Lee, K.-H., and Na, D. L., 2015, "Stereotactic Brain Injection of Human Umbilical Cord Blood Mesenchymal Stem Cells in Patients with Alzheimer's Disease Dementia: A Phase 1 Clinical Trial," Alzheimers Dement. N. Y. N, **1**(2), pp. 95–102.
246. Kim, H.-J., 2011, "Stem Cell Potential in Parkinson's Disease and Molecular Factors for the Generation of Dopamine Neurons," Biochim. Biophys. Acta, **1812**(1), pp. 1–11.
247. Takagi, Y., Takahashi, J., Saiki, H., Morizane, A., Hayashi, T., Kishi, Y., Fukuda, H., Okamoto, Y., Koyanagi, M., Ideguchi, M., Hayashi, H., Imazato, T., Kawasaki, H., Suemori, H., Omachi, S., Iida, H., Itoh, N., Nakatsuji, N., Sasai, Y., and Hashimoto, N., 2005, "Dopaminergic Neurons Generated from Monkey Embryonic Stem Cells Function in a Parkinson Primate Model," J. Clin. Invest., **115**(1), pp. 102–109.

248. Wang, Y.-K., Zhu, W.-W., Wu, M.-H., Wu, Y.-H., Liu, Z.-X., Liang, L.-M., Sheng, C., Hao, J., Wang, L., Li, W., Zhou, Q., and Hu, B.-Y., 2018, "Human Clinical-Grade Parthenogenetic ESC-Derived Dopaminergic Neurons Recover Locomotive Defects of Nonhuman Primate Models of Parkinson's Disease," Stem Cell Rep., **11**(1), pp. 171–182.
249. Nishino, H., Hida, H., Takei, N., Kumazaki, M., Nakajima, K., and Baba, H., 2000, "Mesencephalic Neural Stem (Progenitor) Cells Develop to Dopaminergic Neurons More Strongly in Dopamine-Depleted Striatum than in Intact Striatum," Exp. Neurol., **164**(1), pp. 209–214.
250. Yasuhara, T., Matsukawa, N., Hara, K., Yu, G., Xu, L., Maki, M., Kim, S. U., and Borlongan, C. V., 2006, "Transplantation of Human Neural Stem Cells Exerts Neuroprotection in a Rat Model of Parkinson's Disease," J. Neurosci. Off. J. Soc. Neurosci., **26**(48), pp. 12497–12511.
251. Soler, R., Füllhase, C., Hanson, A., Campeau, L., Santos, C., and Andersson, K.-E., 2012, "Stem Cell Therapy Ameliorates Bladder Dysfunction in an Animal Model of Parkinson Disease," J. Urol., **187**(4), pp. 1491–1497.
252. Sivandzade, F., and Cucullo, L., 2021, "Regenerative Stem Cell Therapy for Neurodegenerative Diseases: An Overview," Int. J. Mol. Sci., **22**(4), p. 2153.
253. Venkataramana, N. K., Kumar, S. K. V., Balaraju, S., Radhakrishnan, R. C., Bansal, A., Dixit, A., Rao, D. K., Das, M., Jan, M., Gupta, P. K., and Totey, S. M., 2010, "Open-Labeled Study of Unilateral Autologous Bone-Marrow-Derived Mesenchymal Stem Cell Transplantation in Parkinson's Disease," Transl. Res. J. Lab. Clin. Med., **155**(2), pp. 62–70.
254. Wernig, M., Zhao, J.-P., Pruszak, J., Hedlund, E., Fu, D., Soldner, F., Broccoli, V., Constantine-Paton, M., Isacson, O., and Jaenisch, R., 2008, "Neurons Derived from Reprogrammed Fibroblasts Functionally Integrate into the Fetal Brain and Improve Symptoms of Rats with Parkinson's Disease," Proc. Natl. Acad. Sci. U. S. A., **105**(15), pp. 5856–5861.
255. Barker, R. A., Drouin-Ouellet, J., and Parmar, M., 2015, "Cell-Based Therapies for Parkinson Disease—Past Insights and Future Potential," Nat. Rev. Neurol., **11**(9), pp. 492–503.
256. Han, F., Baremberg, D., Gao, J., Duan, J., Lu, X., Zhang, N., and Chen, Q., 2015, "Development of Stem Cell-Based Therapy for Parkinson's Disease," Transl. Neurodegener., **4**(1), p. 16.
257. Fu, M.-H., Li, C.-L., Lin, H.-L., Chen, P.-C., Calkins, M. J., Chang, Y.-F., Cheng, P.-H., and Yang, S.-H., 2015, "Stem Cell Transplantation Therapy in Parkinson's Disease," SpringerPlus, **4**(1), p. 597.
258. Karussis, D., Petrou, P., and Kassis, I., 2013, "Clinical Experience with Stem Cells and Other Cell Therapies in Neurological Diseases," J. Neurol. Sci., **324**(1–2), pp. 1–9.
259. Chiò, A., Mora, G., La Bella, V., Caponnetto, C., Mancardi, G., Sabatelli, M., Siciliano, G., Silani, V., Corbo, M., Moglia, C., Calvo, A., Mutani, R., Rutella, S., Gualandi, F., Melazzini, M., Scimè, R., Petrini, M., Bondesan, P., Garbelli, S., Mantovani, S., Bendotti, C., Tarella, C., and STEMALS Study Group, 2011, "Repeated Courses of Granulocyte Colony-Stimulating Factor in Amyotrophic Lateral Sclerosis: Clinical and Biological Results from a Prospective Multicenter Study," Muscle Nerve, **43**(2), pp. 189–195.
260. Martínez, H. R., Molina-Lopez, J. F., González-Garza, M. T., Moreno-Cuevas, J. E., Caro-Osorio, E., Gil-Valadez, A., Gutierrez-Jimenez, E., Zazueta-Fierro, O. E., Meza, J. A., Couret-Alcaraz, P., and Hernandez-Torre, M., 2012, "Stem Cell Transplantation in Amyotrophic Lateral Sclerosis Patients: Methodological Approach, Safety, and Feasibility," Cell Transplant., **21**(9), pp. 1899–1907.
261. Wyatt, T. J., Rossi, S. L., Siegenthaler, M. M., Frame, J., Robles, R., Nistor, G., and Keirstead, H. S., 2011, "Human Motor Neuron Progenitor Transplantation Leads to Endogenous Neuronal Sparing in 3 Models of Motor Neuron Loss," Stem Cells Int., **2011**, p. 207230.
262. Xu, L., Ryugo, D. K., Pongstaporn, T., Johe, K., and Koliatsos, V. E., 2009, "Human Neural Stem Cell Grafts in the Spinal Cord of SOD1 Transgenic Rats: Differentiation and Structural Integration into the Segmental Motor Circuitry," J. Comp. Neurol., **514**(4), pp. 297–309.
263. Yan, J., Xu, L., Welsh, A. M., Hatfield, G., Hazel, T., Johe, K., and Koliatsos, V. E., 2007, "Extensive Neuronal Differentiation of Human Neural Stem Cell Grafts in Adult Rat Spinal Cord," PLoS Med., **4**(2), p. e39.
264. Blanquer, M., Iniesta, F., Gómez Espuch, J., Ramón, V., Pérez Espejo, M. Á., Ruíz López, F. J., García Santos, J. M., Bleda, P., Izura, V., Sáez, M. V., de Mingo, P., Vivancos, L., Carles, R., Majado, M. J., Sánchez Salinas, A., Martínez-Lage, J. F., Martínez, S., and Moraleda, J. M., 2011, "Neurotrophic Bone Marrow Cellular Nests Prevent Spinal Motoneuron Degeneration in Amyotrophic Lateral Sclerosis Patients," Blood, **118**(21), p. 4393.
265. Popescu, I. R., Nicaise, C., Liu, S., Bisch, G., Knippenberg, S., Daubie, V., Bohl, D., and Pochet, R., 2013, "Neural Progenitors Derived From Human Induced Pluripotent Stem Cells Survive and Differentiate Upon Transplantation Into a Rat Model of Amyotrophic Lateral Sclerosis," Stem Cells Transl. Med., **2**(3), pp. 167–174.

266. Mazzini, L., Mareschi, K., Ferrero, I., Miglioretti, M., Stecco, A., Servo, S., Carriero, A., Monaco, F., and Fagioli, F., 2012, "Mesenchymal Stromal Cell Transplantation in Amyotrophic Lateral Sclerosis: A Long-Term Safety Study," Cytotherapy, **14**(1), pp. 56–60.
267. Sironi, F., De Marchi, F., Mazzini, L., and Bendotti, C., 2023, "Cell Therapy in ALS: An Update on Preclinical and Clinical Studies," Brain Res. Bull., **194**, pp. 64–81.
268. Uccelli, A., Milanese, M., Principato, M. C., Morando, S., Bonifacino, T., Vergani, L., Giunti, D., Voci, A., Carminati, E., Giribaldi, F., Caponnetto, C., and Bonanno, G., 2012, "Intravenous Mesenchymal Stem Cells Improve Survival and Motor Function in Experimental Amyotrophic Lateral Sclerosis," Mol. Med., **18**(1), pp. 794–804.
269. Connor, B., 2018, "Concise Review: The Use of Stem Cells for Understanding and Treating Huntington's Disease," Stem Cells Dayt. Ohio, **36**(2), pp. 146–160.
270. McBride, J. L., Behrstock, S. P., Chen, E.-Y., Jakel, R. J., Siegel, I., Svendsen, C. N., and Kordower, J. H., 2004, "Human Neural Stem Cell Transplants Improve Motor Function in a Rat Model of Huntington's Disease," J. Comp. Neurol., **475**(2), pp. 211–219.
271. Olson, S. D., Pollock, K., Kambal, A., Cary, W., Mitchell, G.-M., Tempkin, J., Stewart, H., McGee, J., Bauer, G., Kim, H. S., Tempkin, T., Wheelock, V., Annett, G., Dunbar, G., and Nolta, J. A., 2012, "Genetically Engineered Mesenchymal Stem Cells as a Proposed Therapeutic for Huntington's Disease," Mol. Neurobiol., **45**(1), pp. 87–98.
272. Ebert, A. D., Barber, A. E., Heins, B. M., and Svendsen, C. N., 2010, "Ex Vivo Delivery of GDNF Maintains Motor Function and Prevents Neuronal Loss in a Transgenic Mouse Model of Huntington's Disease," Exp. Neurol., **224**(1), pp. 155–162.
273. Dey, N. D., Bombard, M. C., Roland, B. P., Davidson, S., Lu, M., Rossignol, J., Sandstrom, M. I., Skeel, R. L., Lescaudron, L., and Dunbar, G. L., 2010, "Genetically Engineered Mesenchymal Stem Cells Reduce Behavioral Deficits in the YAC 128 Mouse Model of Huntington's Disease," Behav. Brain Res., **214**(2), pp. 193–200.
274. Zimmermann, T., Remmers, F., Lutz, B., and Leschik, J., 2016, "ESC-Derived BDNF-Overexpressing Neural Progenitors Differentially Promote Recovery in Huntington's Disease Models by Enhanced Striatal Differentiation," Stem Cell Rep., **7**(4), pp. 693–706.
275. Snyder, B. R., Chiu, A. M., Prockop, D. J., and Chan, A. W. S., 2010, "Human Multipotent Stromal Cells (MSCs) Increase Neurogenesis and Decrease Atrophy of the Striatum in a Transgenic Mouse Model for Huntington's Disease," PLoS One, **5**(2), p. e9347.
276. Snyder, B. R., Cheng, P.-H., Yang, J., Yang, S.-H., Huang, A. H., and Chan, A. W., 2011, "Characterization of Dental Pulp Stem/Stromal Cells of Huntington Monkey Tooth Germs," BMC Cell Biol., **12**, p. 39.
277. Jeon, I., Lee, N., Li, J.-Y., Park, I.-H., Park, K. S., Moon, J., Shim, S. H., Choi, C., Chang, D.-J., Kwon, J., Oh, S.-H., Shin, D. A., Kim, H. S., Do, J. T., Lee, D. R., Kim, M., Kang, K.-S., Daley, G. Q., Brundin, P., and Song, J., 2012, "Neuronal Properties, in Vivo Effects, and Pathology of a Huntington's Disease Patient-Derived Induced Pluripotent Stem Cells," Stem Cells Dayt. Ohio, **30**(9), pp. 2054–2062.

Part II

Synaptic Plasticity and Computational Neuroscience

8 Computational Neuroscience

Recent Advancement

*Mohammad Asif**
Indian Institute of Information Technology Allahabad
Prayagraj, Uttar Pradesh, India

Princy Choudhary
National Centre for Cell Science
Ganeshkhind, Pune, Maharashtra, India
and
National Institute of Mental Health, and Neurosciences
Bengaluru, Karnataka, India

Anushka Ajit Dandawate
Indian Institute of Information Technology Allahabad
Prayagraj, Uttar Pradesh, India

The chapter also focuses on the exploration of neurodegenerative diseases, such as Alzheimer's and Parkinson's. Recent advancements in modeling techniques offer a nuanced perspective on the altered information flow within neural circuits associated with these conditions. This provides a bridge between theoretical constructs and empirical observations, shedding light on potential avenues for therapeutic interventions and a deeper understanding of these debilitating disorders.

In summary, this chapter serves as a comprehensive compendium of current knowledge and recent advancements in computational neuroscience. The integration of theoretical frameworks with cutting-edge technologies shapes the landscape of understanding neural intricacies, offering promising opportunities for future research, applications in neurological disorders, and the continued evolution of computational neuroscience.

8.1 INTRODUCTION

Computational neuroscience is a fascinating field that merges biology, computer science, and mathematics to gain insights into how the brain processes information and generates behaviors. It's amazing how far this field has come in recent years, and this chapter explores the latest advancements that have propelled the field forward. By using mathematical models, computer simulations, and analytical tools, researchers are able to understand the underlying principles of neural processes and bridge the gap between the brain's complexity and modern technology's computational capabilities. It's a multidisciplinary field that's breaking traditional boundaries and opening up new possibilities to unravel the mysteries of the brain [1,2].

Computational neuroscience is a fascinating field that has its roots in the mid-20th century. It was founded by brilliant researchers like Warren McCulloch and Walter Pitts, who created the foundation for artificial neural networks. Since then, the field has undergone significant changes, driven by advances in computing power, neuroscience techniques, and interdisciplinary collaborations [3].

DOI: 10.1201/9781003464648-10

What's exciting about this field is the ongoing interplay between theoretical models and empirical discoveries. This creates a feedback loop that propels the entire field forward, allowing researchers to gain a deeper understanding of the brain and its workings. It's amazing to see how far we've come, and it's even more exciting to think about where we're headed in the future!

The brain is a complex organ with billions of neurons and intricate. Understanding how it works is one of the biggest challenges in science [4]. Computational neuroscience offers a systematic approach to unraveling the mysteries of neural function. It's important because it helps us understand the fundamental mechanisms of the brain and find practical applications. These range from developing artificial intelligence to creating innovative treatments for neurological disorders.

As we explore "Computational Neuroscience: Recent Advancements," we aim to navigate the landscapes of neural networks, modeling techniques, and emerging technologies that have propelled the field into a new era. Through this journey, we will witness how recent breakthroughs are reshaping our comprehension of the brain, opening doors to unprecedented possibilities, and paving the way for deeper integration of computational approaches with the complexities of neural systems.

8.2 BRIEF HISTORY AND EVOLUTION OF COMPUTATIONAL NEUROSCIENCE

The history of computational neuroscience is a fascinating story of brilliant minds coming together, ideas evolving, and technology advancing. This interdisciplinary field, which combines biology and computer science, has a rich history spanning several decades. It is a tapestry of innovation and progress that has transformed the way we understand the brain and the mind.

8.2.1 Early Sparks: 1940s–1950s

Computational neuroscience has its roots in the 1940s and 1950s, when Warren McCulloch and Walter Pitts pioneered the development of artificial neural networks [5]. Their work aimed to replicate the electrical impulses of the brain, leading to the idea that computers could imitate the complexities of biological neurons. This groundbreaking research paved the way for further advances in the field of neuroscience and artificial intelligence.

8.2.2 From Neurons to Networks: 1960s–1980s

During the 1960s, there was an emergence of computational models aimed at imitating neural networks. One of the first artificial neurons, known as the perceptron, was developed by Frank Rosenblatt, which opened up doors for exploring machine learning [6,7]. However, the excitement was short-lived as people started to perceive some limitations in the models. This caused a temporary decline in interest towards computational neuroscience.

8.2.3 Revival and Renaissance: 1990s–2000s

In the 1990s, technological advancements led to renewed interest in the field of neural networks. This era saw the emergence of advanced neural network architectures, including recurrent neural networks (RNNs), and the exploration of connectionist models [7]. With the help of these technological advancements, scientists were able to gain a more detailed and nuanced understanding of the workings of the brain.

8.2.4 Deep Learning and the 21st Century: 2010s–Present

During the 21st century, there was a significant advancement in computational neuroscience. The field made a big leap with the introduction of deep learning, a sophisticated form of neural network modeling. This allowed for better simulations and predictions, as the intricate layers of artificial neural networks mimicked the complexity of the brain. The relationship between neuroscience and artificial intelligence also improved as machine learning contributed to a better understanding of biological neural networks [7].

8.2.5 Interdisciplinary Collaborations: Present and Future

In the contemporary landscape, computational neuroscience is a testament to the power of collaboration. Interdisciplinary teams, comprising neuroscientists, computer scientists, and mathematicians join forces to push the boundaries of knowledge. The evolution continues, with the field poised at the intersection of cutting-edge technology, expansive data sets, and the unrelenting quest to unravel the brain's enigmatic code.

As we navigate the pages of history, the story of computational neuroscience unfolds not merely as a chronicle of progress but as a testament to human curiosity and ingenuity. Each chapter reveals a symbiotic dance between theoretical insight and technological advancement, propelling us into a future where the marriage of computation and neuroscience promises to unlock the mind's deepest secrets.

8.3 IMPORTANCE OF COMPUTATIONAL NEUROSCIENCE IN UNDERSTANDING THE BRAIN

Computational neuroscience is a vital field of study that offers a unique pathway to unlock the mysteries of the brain. Its significance lies in the fact that it helps us to delve deeper into the inner workings of the brain than ever before. By providing us with a range of powerful tools, it enables us to explore the complexities of the brain in a way that is not possible with traditional neuroscience methods. Through the use of advanced techniques, such as computer simulations and mathematical models, we are able to gain a deeper understanding of the brain's intricate workings and the processes that underlie cognition.

As we continue to advance our understanding of the brain, the role of this field becomes increasingly pronounced. It promises to provide us with the tools and insights we need to truly comprehend the nature of cognition and the workings of the brain. As we journey into the future, the significance of computational neuroscience in unlocking the brain's secrets will continue to grow, leading us towards a future where we are able to understand and harness the full potential of this extraordinary organ. Here are some points worth mentioning:

8.3.1 Decoding the Brain's Language

The brain, with its intricate network of neurons communicating through electrical impulses, speaks a language that is both intricate and cryptic. Computational neuroscience acts as a linguistic interpreter, translating the electrochemical dance of neurons into the binary code of computers. By creating mathematical models and simulations, scientists can grasp neural communication rules and understand how these interactions give rise to cognition [8,9].

8.3.2 Unraveling Neural Complexity

The brain is an incredibly complex organ, with billions of neurons and trillions of synapses. Although traditional research methods are valuable, they often fail to capture the intricate workings of neural networks. Computational neuroscience, on the other hand, is highly effective in managing this complexity. By utilizing advanced modeling techniques and simulations, scientists are able to investigate how individual neurons connect, communicate, and work together to contribute to cognitive processes.

8.3.3 Prediction and Understanding

Computational models are not just a reflection of the brain's internal processes, but they also provide a means for predicting how the brain will respond to different stimuli and conditions. This predictive ability is vital for understanding the brain's functions and behaviors. By simulating the impact of various factors, researchers can uncover the principles that govern learning, memory, decision-making, and other cognitive functions [10].

8.3.4 Innovations in Medicine and Technology

Computational neuroscience is a field of study that has significant implications beyond research laboratories. It enables us to explore the mechanisms of neurological and psychiatric disorders through computational models, which in turn contributes to the development of new therapies and treatment strategies. Furthermore, the knowledge and insights that we gain from understanding neural processes are being used to fuel advancements in brain-machine interfaces, neuroprosthetics, and artificial intelligence. This is shaping the landscape of medical and technological innovation in ways we have never seen before [11].

8.3.5 Unlocking Personalized Medicine

Every person's brain is unique, which means that what works for one individual may not necessarily work for another. Computational neuroscience is a field that allows for the creation of personalized medicine, where treatments can be customized based on an individual's neural profile. By gaining insights into the variations in neural circuits and responses, scientists can develop precise and effective interventions, which can lead to the development of healthcare that is tailored to each person's unique neurological makeup [12].

8.4 CONTRIBUTIONS OF COMPUTATIONAL NEUROSCIENCE TO UNDERSTANDING NEURODEGENERATIVE DISEASES

The field of neuroscience faces a significant challenge in understanding and combating neurodegenerative diseases. However, computational neuroscience offers a powerful ally in this battle. With its ability to create computational models and simulations, this interdisciplinary field provides hope in unraveling the complexities of disorders that damage the delicate architecture of the brain.

8.4.1 Modeling Disease Progression

Neurodegenerative diseases, like Alzheimer's, Parkinson's, and Huntington's, progress gradually with complex changes at the molecular, cellular, and network levels [13]. Computational neuroscience helps scientists develop dynamic models that can simulate the advancement of these diseases. By replicating the chain reaction of pathological processes, researchers can get a better understanding of the series of events that lead to neuronal degeneration.

8.4.2 Identifying Biomarkers and Risk Factors

The pursuit of understanding neurodegenerative diseases involves the identification of biomarkers and risk factors [14]. Computational models, powered by extensive datasets, assist in identifying molecular signatures and genetic factors linked to disease susceptibility. This not only improves early detection but also uncovers potential targets for therapeutic interventions.

8.4.3 Understanding Synaptic Dysfunction

Synapses act as communication hubs between neurons and have a significant impact on neurodegenerative diseases [15]. Computational neuroscience offers a virtual laboratory for investigating synaptic dysfunction. Researchers can simulate how changes in synaptic plasticity contribute to disease pathology, allowing them to explore new avenues for intervention and possible approaches to maintaining synaptic health.

8.4.4 Drug Discovery and Treatment Optimization

Neurodegenerative diseases are complex and require precise and targeted treatments. Computational models accelerate drug discovery by predicting the effects of potential therapeutic agents on disease-related pathways [16]. It has been reported that when computational neuroscience is used in collaboration with quantitative system pharmacology modeled molecular pathways, it elevates the success rate of drug development for neurodegenerative diseases [17]. Computational models not only expedite the identification of promising drug candidates but also facilitate the optimization of treatment regimens tailored to individual patients. In recent years, the computational approach has been effectively utilized for drug development for neurodegenerative disorders [18–21].

8.4.5 Personalized Medicine and Treatment Response

The variation in how diseases show up and how people respond to treatments is a significant problem when it comes to neurodegenerative diseases. However, computational models that take into account the distinct neural profiles of patients can help create personalized medicine. By customizing treatments based on an individual's neural characteristics, researchers hope to improve the effectiveness of therapy while reducing side effects.

8.4.6 Simulating Cognitive Decline

Many neurodegenerative diseases are characterized by cognitive decline [22]. Computational models that use cognitive architectures can simulate how certain brain regions that are responsible for memory, attention, and decision-making are affected by the disease. This helps in comprehending the cognitive progression of the diseases and provides guidance in developing interventions to reduce cognitive decline.

8.4.7 Integration with Experimental Data

Computational neuroscience serves as a connection between experimental discoveries and theoretical comprehension. By combining computational models with empirical data from neuroimaging, genetics, and clinical studies, researchers are able to gain a comprehensive perspective on neurodegenerative diseases [23]. This collaboration improves the accuracy of predictions and encourages a thorough understanding of the mechanisms behind these diseases.

The field of computational neuroscience has made significant contributions to our understanding of neurodegenerative diseases. It allows for an in-depth exploration of disease processes at various levels and speeds up the translation of findings into tangible clinical applications. As research in computational neuroscience progresses, there is great potential for revolutionary discoveries in the diagnosis, treatment, and prevention of neurodegenerative diseases. This offers hope for a future where these conditions are not just managed but successfully conquered.

8.5 NEURAL NETWORKS AND MODELS

In the vast field of computational neuroscience, understanding the fundamental principles is similar to decoding the language of the brain's computational machinery. This section aims to explore the basics, revealing the complexities of neural networks and the dynamic communication among neurons.

Neural networks are the foundation of computational neuroscience and imitate the intricate connections found in the human brain. In this exploration of the basics, we will break down the components and types of neural network models that support our journey into the computational marvels of the mind.

Neural networks act as digital counterparts to the complex web of neurons within the human brain. This section offers a basic introduction to the concept of neural networks, explaining their structure and functionality, and the inspiration drawn from the biological nervous system.

8.5.1 Neural Networks

Neural networks are computer models that take inspiration from the brain's remarkable network of neurons. Artificial neural networks consist of interconnected nodes (neurons) that work together to perform complex computations, much like the brain's neurons communicate to process information. This introduction lays the foundation for comprehending how these networks can imitate cognitive processes.

8.5.1.1 Basic Components

Understanding the basics of neural networks lays the groundwork for exploring their applications, adaptations, and the nuanced dynamics that emerge when these computational entities engage in information processing. As we delve deeper, the intricate dance of neurons within these networks will come to life, shedding light on the computational wonders that underpin the field of computational neuroscience. Refer to Figure 8.1 for more information.

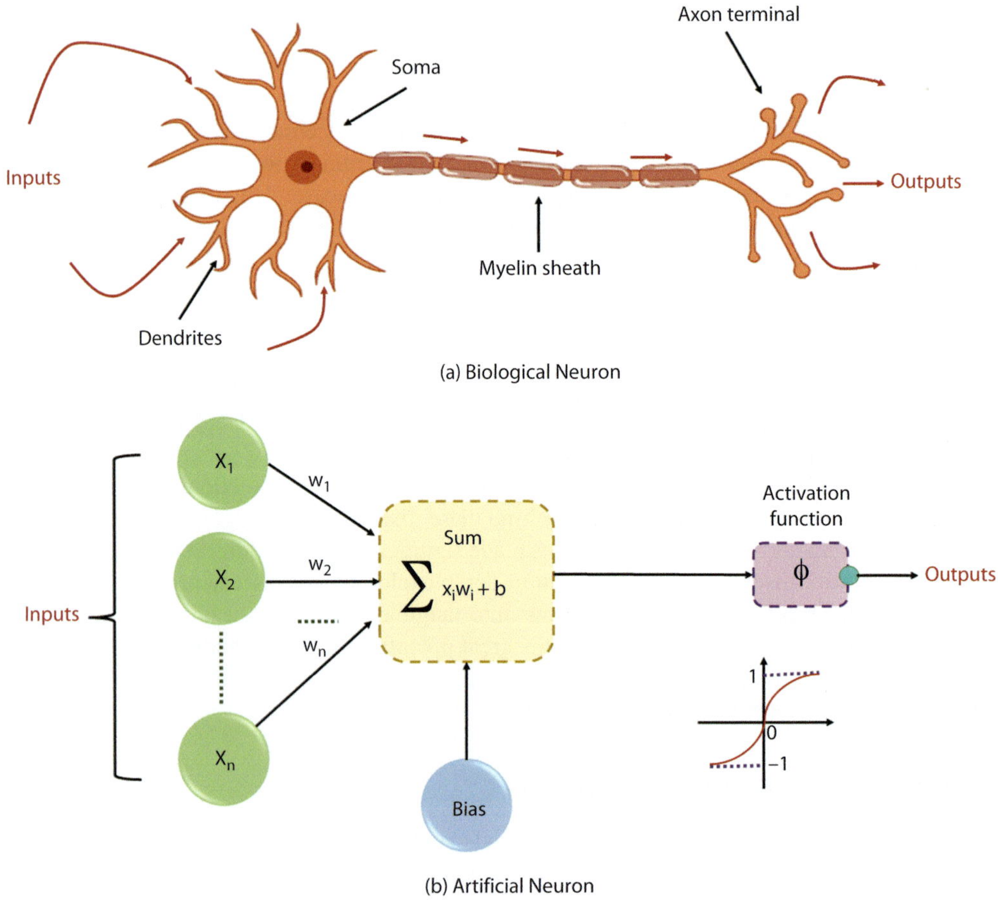

FIGURE 8.1 This diagram shows (a) A biological neuron with its various parts. (b) An artificial neuron and its functions. A neuron takes input from other neurons (X_1, X_2,. . . , Xn) and takes a dot product with their respective weights (w_1, w_2,. . . , wn) and adds a bias to it b. After that, an activation function decides what will be its output based on these calculations.

8.5.1.1.1 Neurons

The building blocks of neural networks, artificial neurons receive and transmit information. They process inputs, apply mathematical operations, and produce outputs that contribute to the network's decision-making.

8.5.1.1.2 Synapses

An artificial neural network's synapses function similarly to biological neurons' connections. They carry weights that determine the strength of the connections and influence the impact of one neuron on another.

8.5.1.1.3 Layers

Neural networks consist of layers, including input, hidden, and output layers. Each layer has a unique function in processing and transforming information as it moves through the network.

8.5.2 Biological and Artificial Parallels

- Neural networks draw inspiration from the biological brain's structure and function. The artificial replication of neurons and their interconnectedness allows computational models to emulate aspects of learning, adaptation, and decision-making observed in living organisms.
- While artificial neural networks simplify the complexity of biological systems, the parallels between the two enable computational models to capture certain aspects of intelligence, making them versatile tools in various applications.

8.5.3 Types of Neural Network Models

A neural network is a type of machine learning model that is designed to mimic the structure and functioning of the human brain. Neural networks consist of a large number of interconnected processing nodes, which are organized into layers. Each layer performs a specific task, such as feature extraction or classification. There are various types of neural network models, each designed for specific tasks. For example, feedforward neural networks are commonly used for classification tasks, while recurrent neural networks are useful for processing sequential data. Convolutional neural networks are often used for image recognition, and deep belief networks are used for unsupervised learning. Please refer to Figure 8.2 for various types of neural architectures. Here are some of the common types.

8.5.3.1 Feedforward Neural Network (FNN)

This is the most basic type of neural network. The flow of information in this model is unidirectional, that is from the input nodes, through the hidden nodes (if any), and to the output nodes, without any cycles or loops [24].

8.5.3.2 Multilayer Perceptron (MLP)

This is a type of neural network known as a feedforward neural network. It is composed of fully connected neurons that use a nonlinear activation function. The network is organized into at least three layers: the input layer, one or more hidden layers, and the output layer. This type of neural network is notable for its ability to distinguish data that is not linearly separable [25].

8.5.3.3 Convolutional Neural Network (CNN)

A convolutional neural network (CNN) is a type of deep neural network that is mainly utilized in image processing and pattern recognition. It is composed of one or more convolutional layers, which are followed by fully connected layers (matching those in typical ANNs). This network type is designed to learn feature engineering by itself via filter (or kernel) optimization, making it a regularized feed-forward neural network [26].

8.5.3.4 Recurrent Neural Network (RNN)

A bi-directional artificial neural network is different from a uni-directional feedforward neural network in that it allows the output from some nodes to affect subsequent input to the same nodes. This makes it possible for them to use internal state (memory) to process arbitrary sequences of inputs. As a result, they can be used for tasks such as unsegmented, connected handwriting recognition or speech recognition [27].

8.5.3.5 Long Short-Term Memory (LSTM) Networks

The long short-term memory (LSTM) network is a type of recurrent neural network (RNN) designed to address the issue of the vanishing gradient problem that is present in traditional RNNs.

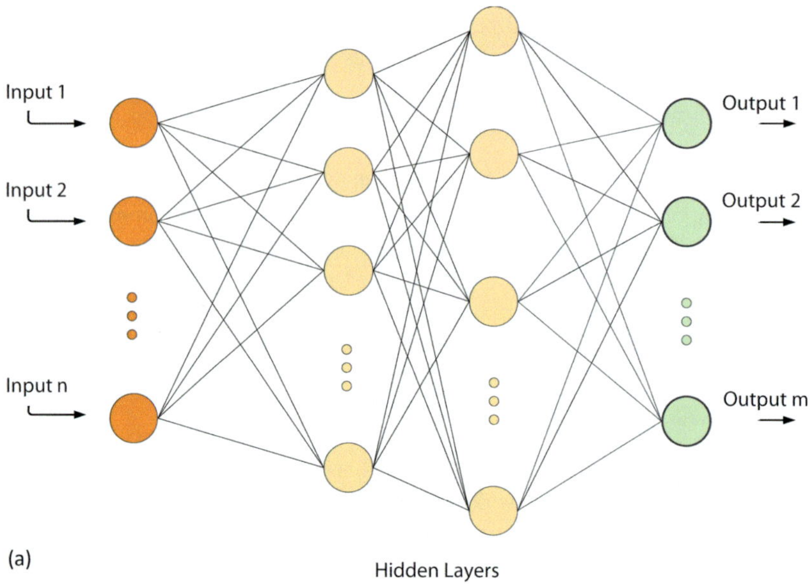

FIGURE 8.2 Types of neural architectures: (a) multilayer perceptron (MLP), (b) convolutional neural network (CNN), (c) long short-term memory (LSTM), and (d) generative adversarial network (GAN) *(Continued)*

Figure 8.2 illustrates the following:

a. Architecture of a multilayer perceptron (MLP) neural network. The network consists of an input layer, one or more hidden layers, and an output layer. Each layer contains multiple nodes (neurons), and connections between nodes are represented by weighted edges. The MLP is designed for complex pattern recognition and can be trained using backpropagation and optimization algorithms.
b. Architecture of a convolutional neural network (CNN). The CNN is designed for image recognition tasks and consists of convolutional layers for feature extraction, pooling layers for down-sampling, and fully connected layers for classification. Convolutional filters capture spatial hierarchies, allowing the network to learn local patterns and global representations. The architecture is effective for tasks such as object detection and image classification.
c. Architecture of a long short-term memory (LSTM) neural network. An LSTM is a type of recurrent neural network designed for processing sequential data. The network includes memory cells and gates that control the flow of information, enabling the model to capture and remember patterns over long sequences. LSTMs are commonly used for tasks such as natural language processing, speech recognition, and time series prediction.
d. Architecture of a generative adversarial network (GAN). GANs consist of a generator network and a discriminator network trained simultaneously through adversarial training. The generator creates synthetic data, while the discriminator distinguishes between real and generated samples. The competition between the generator and discriminator leads to the generation of realistic data. GANs are widely used for image synthesis, style transfer, and generating novel content.

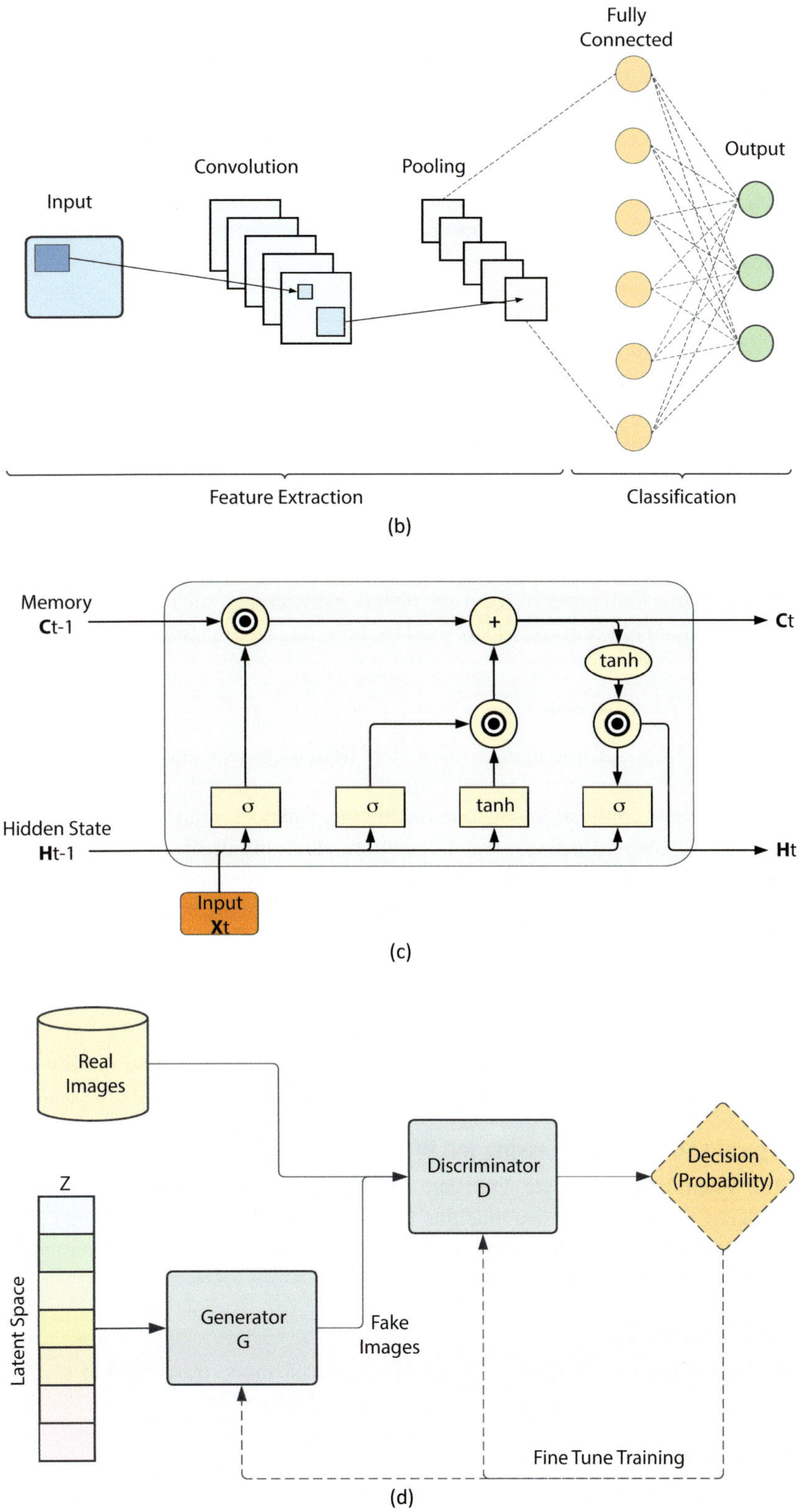

FIGURE 8.2 *(Continued)*

The LSTM network is capable of selectively outputting only the relevant information from the current state, which allows the network to maintain useful long-term dependencies to make predictions both in the present and future time-steps [28].

8.5.3.6 Autoencoders

An autoencoder is a type of artificial neural network that is used to learn efficient codings of unlabeled data, which is a form of unsupervised learning. It consists of two main components: an encoder that maps the input data to a latent space, and a decoder that reconstructs the input from the latent representation [29].

8.5.3.7 Generative Adversarial Networks (GAN)

Generative Adversarial Network (GAN) is a type of machine learning framework and a popular approach to generative AI. The concept was first introduced by Ian Goodfellow and his colleagues in June 2014 [30]. In a GAN, two neural networks, the generator and the discriminator, compete with each other in a zero-sum game where one agent's gain is the other's loss. This technique helps to generate new data that has similar statistics to the training set.

8.5.3.8 Transformers

A transformer is a type of deep learning architecture that uses the multi-head attention mechanism [31]. It requires less training time than previous recurrent neural architectures, such as long short-term memory (LSTM). Its later variation has become widely used for training large language models on big datasets.

8.5.4 Real-world Applications

Neural networks and deep learning models are widely used across various industries. They perform tasks such as image and speech recognition, natural language processing, predictive analytics, and anomaly detection. Their applications include healthcare, finance, retail, transportation, and more. They are used for diagnosing diseases, fraud detection, autonomous driving, personalized product recommendations, and traffic optimization, to name a few. With advancements in technology, we can expect to see even more innovative applications in the future. Here are some real-world applications:

8.5.4.1 Image and Video Recognition

- **Facial Recognition**: Facial recognition technology is widely used in security systems, social media tagging, and authentication to identify human faces [32].
- **Object Detection**: Applied in autonomous vehicles, surveillance, and quality control in manufacturing. Deep learning-trained vehicles now interpret 360° camera views [33].

8.5.4.2 Natural Language Processing (NLP)

- **Machine Translation**: Google Translate and other translation services use neural networks. LSTM helped to improve machine translation and language modeling [34].
- **Chatbots and Virtual Assistants**: Siri, Alexa, and Google Assistant leverage NLP for understanding and generating human-like responses.

8.5.4.3 Speech Recognition

- **Spoken Language Processing**: Utilized in voice assistants, transcription services, and voice-activated systems for spoken language understanding [35].

8.5.4.4 Healthcare

- **Disease Diagnosis**: Medical images are analyzed by neural networks to detect diseases like cancer. Modern deep learning tools show high accuracy in detecting various diseases and the helpfulness of specialists in improving diagnosis efficiency [36].

- **Drug Discovery**: Accelerating drug discovery by predicting potential drug interactions and identifying new compounds. Research has explored the use of deep learning to predict the biomolecular targets, off-targets, and toxic effects of environmental chemicals in nutrients, household products, and drugs [37].

8.5.4.5 Finance

- **Algorithmic Trading**: Predicting market trends and making trading decisions based on historical data.
- **Credit Scoring**: Assessing credit risk by analyzing financial and personal data. Deep learning is being successfully applied to financial fraud detection, tax evasion detection [5], and anti-money laundering.

8.5.4.6 Retail

- **Recommendation Systems**: Suggesting products to customers based on their preferences and behaviors. Recommendation systems have used deep learning to extract meaningful features for a latent factor model for content-based music and journal recommendations [38]. Multi-view deep learning has been applied for learning user preferences from multiple domains [39].
- **Deep Reinforcement Learning in Marketing**: Deep reinforcement learning has been used to approximate the value of possible direct marketing actions, defined in terms of RFM variables. The estimated value function was shown to have a natural interpretation as customer lifetime value [40].

These examples highlight the versatility of neural networks and deep learning in addressing complex problems and improving efficiency across diverse industries. The application of these technologies continues to grow as research advances and new use cases are discovered. Neural networks and deep learning have proven themselves to be highly versatile in solving complex problems and enhancing productivity across various industries. With continuous research and the discovery of new use cases, these technologies are being increasingly adopted, and their application is expanding rapidly. The examples of their successful implementation only add to their reputation and potential for revolutionizing the way we work and live.

8.6 NEURONAL DYNAMICS

Neural dynamics refer to the temporal patterns and changes in the activity of neural networks, particularly the dynamic interactions among neurons that give rise to various cognitive processes and behaviors. It encompasses the complex and ever-changing electrical and chemical activity occurring within the brain's intricate network of neurons. Studying neural dynamics is essential for understanding the mechanisms underlying brain function and dysfunction. Computational models of neural dynamics, combined with experimental approaches such as neuroimaging and electrophysiology, provide a holistic understanding of how the brain processes information and adapts to a changing environment. Additionally, disruptions in neural dynamics are implicated in various neurological and psychiatric disorders, making this field crucial for advancing our understanding of brain health and disease.

8.6.1 Key Aspects of Neuronal Dynamics

8.6.1.1 Temporal Patterns

Neural dynamics involve studying how neural activity unfolds over time. This includes the patterns of firing and communication between neurons, as well as the synchronization and coordination of neural activity in different brain regions.

8.6.1.2 Information Processing

Neural dynamics are closely linked to the brain's information-processing capabilities. The dynamic interactions among neurons allow for the integration and processing of sensory inputs, the formation of memories, decision-making, and the execution of motor functions [41].

8.6.1.3 Synaptic Plasticity

Neural dynamics also encompass the adaptability of neural connections, known as synaptic plasticity. This involves changes in the strength of synapses, allowing the neural network to learn and adapt based on experience [42].

8.6.1.4 Cognitive Functions

The dynamic nature of neural activity underlies various cognitive functions, including perception, attention, learning, and memory [43]. Neural dynamics are fundamental to understanding how the brain generates thoughts, emotions, and behaviors.

8.6.1.5 Brain Oscillations

Neural dynamics often manifest as rhythmic patterns of activity known as brain oscillations. These oscillations are crucial in coordinating communication between brain regions and are associated with different cognitive states [44].

8.6.1.6 Network Resilience

Neural dynamics also contribute to the resilience and stability of neural networks. The ability of the brain to maintain stable patterns of activity despite external perturbations is a key aspect of neural dynamics.

8.6.2 Basis of Neuronal Communication

Neuronal communication is the intricate process through which neurons, the building blocks of the nervous system, transmit information to orchestrate the myriad functions of the brain and body. This communication occurs through a combination of electrical and chemical signals, forming the foundation of neural dynamics. Understanding the basics of neuronal communication is essential for grasping the fundamental principles of neural function. This intricate dance of electrical and chemical signals forms the basis for higher-order cognitive processes, sensory perception, and motor functions, laying the groundwork for the dynamic neural dynamics that characterize the brain's computational power.

8.6.2.1 Neuron Anatomy

Neurons have distinct structural components that enable communication. These include dendrites, which are branch-like extensions responsible for receiving incoming signals from other neurons. The cell body (soma) serves as the central region where signals are integrated and decisions are made. Additionally, neurons feature axons, long and slender projections that transmit signals to other neurons or effector cells.

8.6.2.2 Action Potentials

8.6.2.2.1 Generation

Neuronal communication begins with generating electrical impulses called action potentials. Refer to Figure 8.3 for more information.

8.6.2.2.2 Threshold and Firing

When the membrane potential of a neuron reaches a critical threshold, an action potential is initiated, leading to rapid depolarization and subsequent repolarization.

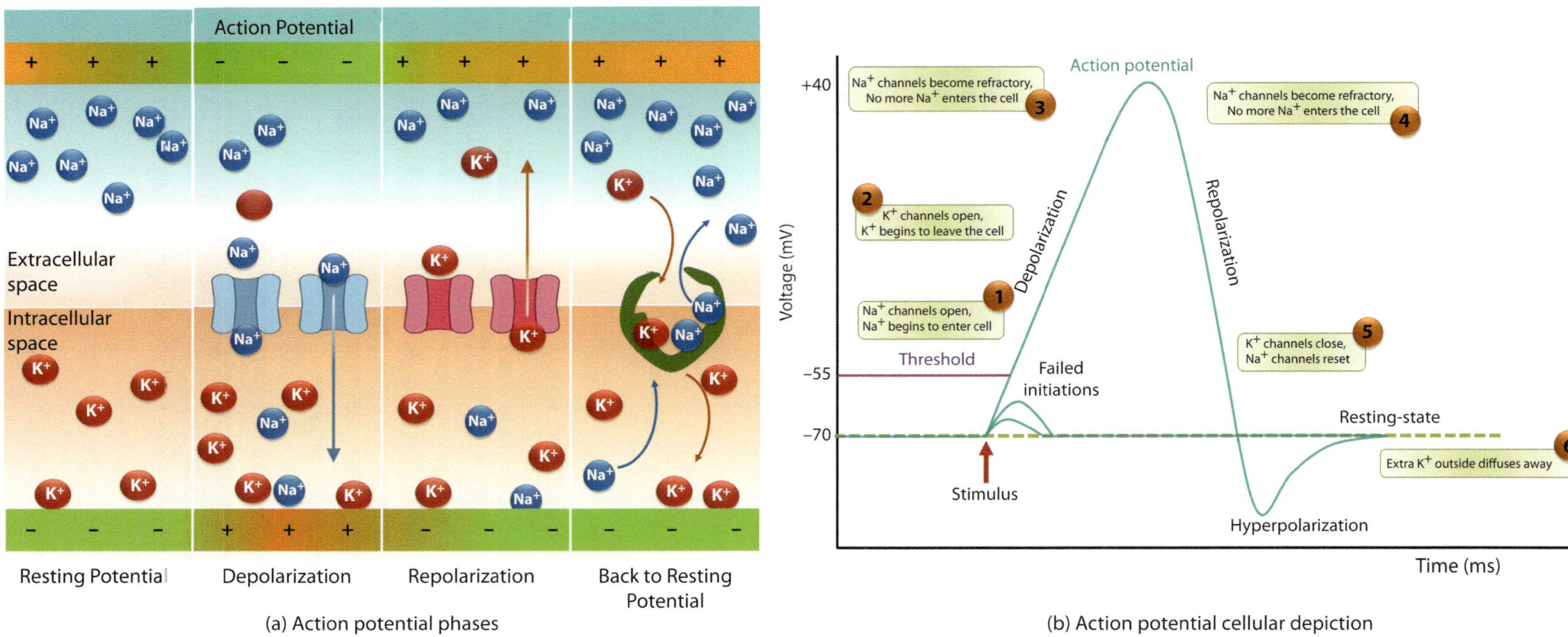

FIGURE 8.3 Diagram illustrating (a) the phases of a standard action potential, showcasing depolarization, repolarization, and hyperpolarization during the transmission of electrical signals in neurons. (b) Depiction of action potential phases emphasizing the transfer of ions between intracellular and extracellular spaces during depolarization, repolarization, and hyperpolarization in neuronal signaling.

8.6.2.2.3 Propagation

Action potentials travel along the axon, ensuring the transmission of signals over long distances.

8.6.2.3 Neurotransmission

8.6.2.3.1 Synaptic Transmission

Neurons communicate at specialized junctions called synapses.

8.6.2.3.2 Neurotransmitters

Chemical messengers are released by the presynaptic neuron into the synaptic cleft.

8.6.2.3.3 Receptor Activation

Neurotransmitters bind to receptors on the postsynaptic neuron, inducing changes in its membrane potential.

8.6.2.4 Postsynaptic Potentials

8.6.2.4.1 Excitatory Postsynaptic Potentials (EPSPs)

Depolarizations that make the postsynaptic neuron more likely to fire an action potential.

8.6.2.4.2 Inhibitory Postsynaptic Potentials (IPSPs)

Hyperpolarizations that decrease the likelihood of the postsynaptic neuron firing.

8.6.2.5 Integration of Signals

- The cell body integrates excitatory and inhibitory signals from multiple synapses.
- A new action potential is generated at the axon hillock if the overall signal surpasses a certain threshold.

8.6.2.6 Communication Between Neurons

- Neurons communicate in networks, with axons transmitting signals to the dendrites of other neurons.
- The sequence of action potentials and neurotransmitter release allows for the relay of information through neural circuits.

8.6.3 Synaptic Transmission and Plasticity

Synaptic transmission and plasticity constitute neuronal communication's dynamic and adaptable aspects, playing a crucial role in shaping the strength and efficacy of connections between neurons. These processes are fundamental to learning, memory, and the overall flexibility of the nervous system. Understanding synaptic transmission and plasticity provides insights into the fundamental mechanisms that underlie learning, memory, and the flexibility of neural circuits. These processes are essential for normal brain function and contribute to our understanding of various neurological disorders and potential therapeutic interventions.

8.6.3.1 Synaptic Transmission Release of Neurotransmitters:

- When an action potential reaches the presynaptic terminal, it triggers the release of neurotransmitters into the synaptic cleft.
- Neurotransmitters are stored in vesicles and are released in response to the influx of calcium ions.

8.6.3.1.1 Binding to Receptors

- Neurotransmitters bind to receptors on the postsynaptic membrane.
- This binding activates the receptors and generates postsynaptic potentials.

8.6.3.1.2 *Excitatory and Inhibitory Signals*

- Excitatory neurotransmitters, such as glutamate, depolarize the postsynaptic membrane, increasing the likelihood of an action potential.
- Inhibitory neurotransmitters, such as GABA, hyperpolarize the postsynaptic membrane, reducing the likelihood of an action potential.

8.6.3.1.3 *Propagation of Signals*

- The postsynaptic potentials, whether excitatory or inhibitory, travel toward the axon hillock, where the decision to generate an action potential is made.

8.6.3.2 Long-Term Potentiation (LTP) and Long-Term Depression (LTD) LTP:

- Long-term potentiation involves the strengthening of synaptic connections.
- Repetitive and persistent stimulation of a synapse leads to an enduring increase in its efficacy.
- NMDA receptors and changes in synaptic strength contribute to the mechanisms of LTP.

8.6.3.2.1 *LTD*

- Long-term depression involves the weakening of synaptic connections.
- Prolonged low-frequency stimulation or low levels of neurotransmitters result in a persistent decrease in synaptic efficacy.
- LTD is crucial for the refinement and flexibility of neural circuits.

8.6.3.3 Plasticity Mechanisms Structural Changes

- Synaptic plasticity involves structural alterations, such as changes in the number and sensitivity of receptors.
- This structural adaptability allows synapses to modify their strength based on experience.

8.6.3.3.1 *Homeostatic Plasticity*

- Neurons and synapses exhibit homeostatic plasticity to maintain overall stability.
- If overall activity is too high, synapses may undergo a weakening process, and vice versa.

8.6.3.3.2 *Adaptation to Experience*

- Synaptic plasticity enables neurons to adapt to experiences, learn from them, and modify their responses accordingly.
- This adaptability is fundamental to memory formation and learning from the environment.

8.7 RECENT ADVANCEMENTS IN COMPUTATIONAL NEUROSCIENCE

Advancements in computational neuroscience reflect a paradigm shift in our understanding of the brain's intricacies and the development of innovative technologies. This section explores cutting-edge techniques and breakthroughs that have propelled the field forward. In recent years, computational neuroscience has witnessed remarkable strides in modeling techniques, pushing the boundaries of our understanding of brain function. This section excavates into two cutting-edge approaches that have garnered significant attention: spiking neural networks and deep learning approaches in neuroscience.

These advanced modeling techniques depart from traditional approaches, offering more biologically faithful and versatile tools for understanding neural processes. The integration of spiking neural networks and deep learning methodologies reflects a convergence of computational prowess and biological realism, propelling the field of computational neuroscience into new realms of exploration and discovery. Understanding how the brain represents information and processes it through neural circuits is central to computational neuroscience. This section explores the intricacies of neural coding and the dynamic flow of information within neural circuits. In essence, the representation of information in the brain and its flow through neural circuits embody a complex dance of coding schemes, dynamic interactions, and adaptive mechanisms. The orchestration of these processes is central to cognitive functions, sensory perception, and the seamless integration of information that defines the remarkable capabilities of the human brain.

The brain is a complex organ that processes information through a series of coordinated neural activities. Understanding how the brain encodes and represents information is crucial to uncovering the secrets of cognition, perception, and behavior. Let's explore the different aspects of information representation in the brain.

8.7.1 Spiking Neural Networks

Spiking neural networks (SNNs) are a new paradigm in computational neuroscience that aim to simulate the complexity of the human brain. Unlike traditional rate-based models, SNNs capture the intricacies of neural communication with a high degree of biological accuracy. This comprehensive study outlines the unique features, applications, and challenges that are associated with spiking neural networks. SNNs are at the forefront of computational neuroscience, providing a promising pathway to bridge the gap between artificial neural networks and the intricacies of biological neural systems. Their focus on temporal dynamics and biological plausibility makes them valuable in understanding complex neural processes and developing advanced brain-inspired computing systems. See Figure 8.4 for more information.

8.7.1.1 Biological Fidelity

8.7.1.1.1 Emulating Neuronal Behavior

At the heart of SNNs lies a commitment to closely mimic the firing patterns of biological neurons. These networks operate on the principle of discrete spikes or action potentials, introducing a level of realism not achievable with traditional artificial neural networks [45].

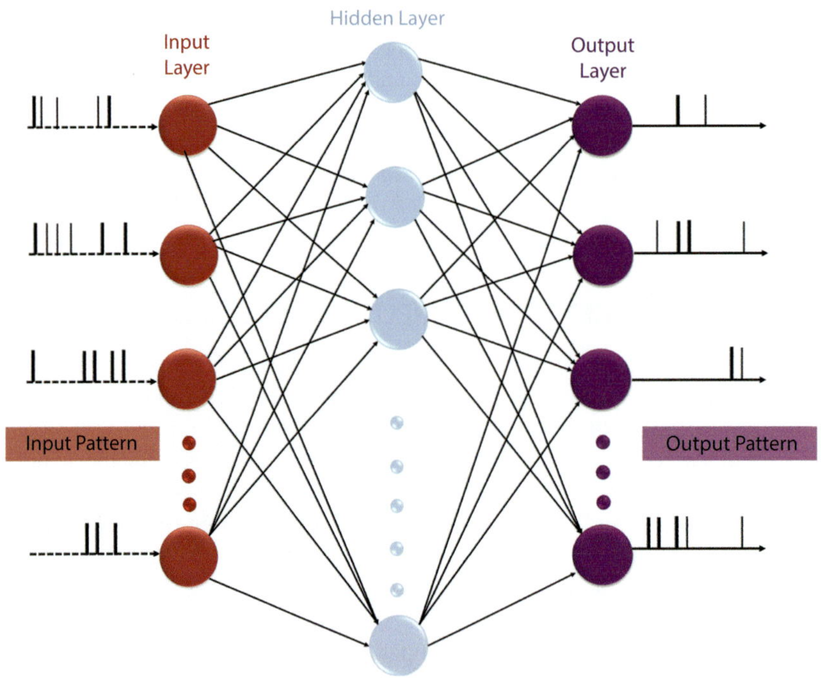

FIGURE 8.4 Schematic representation of a spiking neural network (SNN), showcasing the interconnected neurons and the transmission of information through spikes. Neurons communicate through brief, intense electrical pulses known as spikes, reflecting the dynamic and event-driven nature of SNNs. It consists of an input layer and an output layer with hidden layer(s).

8.7.1.1.2 Temporal Precision

The temporal dimension takes center stage in SNNs. Unlike rate-based models that focus on the firing frequency, SNNs consider the precise timing of spikes. This departure reflects a keen awareness of the temporal intricacies observed in biological neural systems [46].

8.7.1.1.3 Spike-Based Representation

The departure from traditional rate-based models is at the core of biological fidelity in SNNs. Instead, SNNs operate on the premise of discrete spikes or action potentials, mirroring the physiological behavior of real neurons [47].

8.7.1.1.4 Capture of Neural Dynamics

By embracing the spiking nature of neurons, SNNs capture the rich dynamics inherent in biological neural systems. This departure from continuous firing rates allows for a more accurate representation of how neurons transmit information.

8.7.1.2 Temporal Coding

8.7.1.2.1 Information Processing Nuances

SNNs leverage the temporal aspect of neural spiking for information coding and processing. This approach allows for a more nuanced representation of information, where the timing of spikes carries crucial significance.

8.7.1.2.2 Biological Plausibility

The emphasis on temporal coding aligns SNNs with the biological plausibility sought in computational neuroscience. This fidelity to the temporal dynamics of neural communication is instrumental in capturing the essence of information processing in the brain.

8.7.1.3 Neuromorphic Hardware

8.7.1.3.1 Parallelism and Connectivity

SNNs find natural synergy with neuromorphic hardware architectures. These specialized hardware designs mimic the parallelism and connectivity observed in biological neural networks, amplifying the biological fidelity of SNNs.

8.7.1.3.2 Real-Time Simulation

The collaboration between SNNs and neuromorphic hardware goes beyond efficiency; it facilitates real-time simulations. This alignment with the brain's biological structure enhances computational capabilities and brings us closer to emulating the brain's temporal processing.

8.7.1.4 Applications and Real-World Significance

8.7.1.4.1 Sensory Processing Excellence

The commitment to biological fidelity in SNNs positions them as excellent candidates for sensory processing tasks. In applications such as vision and audition, where precise timing is paramount, SNNs excel.

8.7.1.4.2 Neuromorphic Vision Systems

The real-world impact is evident in developing neuromorphic vision systems. SNNs contribute to creating systems that process visual information with efficiency and fidelity reflective of biological vision [48].

8.7.1.5 Learning Mechanisms

8.7.1.5.1 Spike-Timing-Dependent Plasticity (STDP)

Learning in SNNs is orchestrated through mechanisms like spike-timing-dependent plasticity (STDP). This learning rule, inspired by biological synapses, adjusts synaptic weights based on the temporal relationship between pre- and post-synaptic spikes [49].

8.7.1.5.2 Adaptive Synapses

The concept of adaptive synapses further enhances biological fidelity. SNNs equipped with adaptive synapses exhibit learning and adaptation akin to the plasticity observed in biological neural networks.

8.7.1.6 Challenges and Future Directions

8.7.1.6.1 Training Complexity

While SNNs offer unparalleled biological realism, training these networks poses challenges due to the temporal nature of spiking events. Developing efficient algorithms and optimization techniques for training SNNs remains an active area of research.

8.7.1.6.2 Integration with Deep Learning

Efforts are underway to integrate SNNs with deep learning approaches, combining the advantages of both paradigms. This integration aims to harness the power of deep learning while preserving the biological nuances SNNs bring.

8.7.2 Deep Learning Approaches in Neuroscience

Integrating deep learning approaches into neuroscience has marked a transformative juncture in our quest to comprehend the intricate workings of the brain. This discussion explores the multifaceted applications, challenges, and future prospects of deploying deep learning within the realm of neuroscience.

8.7.2.1 Neural Networks in the Era of Big Data

The advent of big data in neuroscience, characterized by vast datasets from neuroimaging, electrophysiology, and genomics, has necessitated advanced analytical tools. Deep learning, particularly convolutional neural networks (CNNs) and recurrent neural networks (RNNs), has emerged as a powerful ally in decoding the complexity embedded in these datasets [50,51]. Deep learning provides the computational muscle required for data-driven discoveries, from analyzing brain scans to deciphering intricate patterns in functional connectivity data.

8.7.2.2 Automated Feature Extraction

One of the distinctive features of deep learning is its capability for automated feature extraction. In traditional machine learning, the burden of manual feature engineering often hampers progress. Deep learning models, however, can autonomously learn hierarchical features from raw data [52]. This capability is harnessed in neuroscience for tasks such as image recognition in neuroimaging and the analysis of intricate neural datasets, freeing researchers from the constraints of labor-intensive feature engineering.

8.7.2.3 Transfer Learning in Neuroscientific Endeavors

Transfer learning, a technique where pre-trained models are fine-tuned for specific tasks, has gained prominence in neuroscience. This approach maximizes the utility of deep learning even when labeled data is limited. The ability to transfer knowledge from models trained on large, diverse datasets to neuroscience applications accelerates the development of models, enhancing their adaptability and broadening their applicability across different domains within the field.

8.7.2.4 Representation of Information in the Brain

In the pursuit of understanding how the brain represents information, deep learning models offer insights into population coding. Deep learning sheds light on the distributed nature of neural representations by investigating how groups of neurons collectively encode information. This aligns with the brain's use of ensemble coding, where groups of neurons work together to encode complex stimuli, memories, and cognitive functions [53].

8.7.2.5 Information Flow in Neural Circuits

The exploration of information flow within neural circuits is greatly enriched by deep learning. Connectomics, the mapping of neural connectivity, is advanced through techniques like diffusion MRI and electron microscopy [54]. Deep learning methods contribute to the construction of detailed connectomes, unraveling the intricate patterns of connectivity that govern information transmission within the brain. Additionally, studying dynamic changes in neural connectivity over time enhances our understanding of how information is dynamically routed through circuits during different cognitive states.

8.7.2.6 Applications in Brain-Machine Interfaces

Deep learning profoundly impacts the development of brain-machine interfaces (BMIs). These interfaces, which facilitate communication between the brain and external devices, leverage the capabilities of deep learning for decoding neural signals. From neuroprosthetics that restore lost functionalities to brain-controlled devices, deep learning enhances the precision and efficiency of brain-machine interactions [55]. Closed-loop systems, which dynamically adapt to user intent based on real-time neural feedback, exemplify the potential of deep learning in optimizing the functionality of BMIs.

8.7.3 Connectomics

Connectomics refers to the comprehensive mapping of neural connections in the brain, essentially creating a "wiring diagram" of the nervous system. This field seeks to understand the intricate network of neurons, their synaptic connections, and the functional pathways that underlie information processing and communication within the brain. The ultimate goal of connectomics is to unravel the structural and functional basis of cognition, behavior, and neurological disorders.

Connectomics is a rapidly evolving field with great promise for advancing our understanding of the brain's intricate wiring and its role in cognition and behavior. As technology and methodologies continue to improve, connectomics is expected to contribute significantly to both basic neuroscience research and clinical applications.

8.7.3.1 Key Concepts in Connectomics

8.7.3.1.1 Neural Circuits

Connectomics focuses on identifying and understanding neural circuits, which are interconnected networks of neurons responsible for specific functions or behaviors.

8.7.3.1.2 Mapping Connectivity

Techniques used in connectomics involve mapping the connections between neurons. This can be done at various scales, from individual synapses to larger brain regions.

8.7.3.1.3 Electron Microscopy (EM)

EM is a crucial tool in connectomics. It allows for high resolution imaging, making it possible to visualize the fine details of synaptic connections.

8.7.3.1.4 Light Microscopy

While not as high in resolution as EM, light microscopy is essential for studying larger-scale connectivity. Fluorescent tracers and markers help identify specific types of neurons and their connections.

8.7.3.1.5 Functional Connectomics

Beyond structural connectivity, functional connectomics examines the dynamic patterns of activity within neural circuits. Techniques like functional MRI (fMRI) and electrophysiological recordings provide insights into the functional aspects of brain networks.

8.7.3.1.6 Graph Theory

Connectomics often employs graph theory to represent and analyze neural networks. Nodes in the graph represent neurons or brain regions, and edges represent the connections between them.

8.7.3.2 Significance and Applications

8.7.3.2.1 Understanding Brain Function

Connectomics provides insights into how information is processed and transmitted in the brain. It helps elucidate the neural basis of various cognitive functions, such as learning, memory, and decision-making [56].

8.7.3.2.2 Neurological Disorders

Studying the connectivity of the brain is crucial for understanding neurological disorders. Aberrant connections or disruptions in neural circuits are often associated with conditions like Alzheimer's, schizophrenia, and epilepsy [57].

8.7.3.2.3 Plasticity and Learning

Connectomics contributes to the understanding of synaptic plasticity, the ability of synaptic connections to change over time. This is fundamental to learning and memory.

8.7.3.2.4 Brain-Computer Interfaces (BCIs)

Connectomics plays a role in the development of BCIs by providing insights into how the brain's neural circuits can be interfaced with external devices.

8.7.3.2.5 Comparative Neuroscience

Comparing the connectomes of different species can shed light on evolutionary aspects of brain development and function.

8.7.4 Brain-Machine Interfaces: Pioneering the Convergence of Mind and Machine

Advancements in brain-machine interfaces (BMIs) herald a new era where the human mind interfaces seamlessly with machines, offering novel solutions for neurodegenerative diseases, disabilities, and beyond [55]. This section explores the multifaceted landscape of BMIs, encompassing neuroprosthetics, brain-controlled devices, and brain-computer interfaces (BCIs).

8.7.4.1 Neuroprosthetics and Brain-Controlled Devices

Neuroprosthetics, which are artificial devices designed to replace or enhance nervous system functionality [58], represent a groundbreaking frontier in addressing neurodegenerative diseases such as ALS or paralysis. These technologies interface directly with the brain, offering potential solutions for restoring lost motor functions.

Brain-controlled devices, such as robotic limbs and exoskeletons, have the ability to be controlled by direct neural signals. This means that individuals can interact with their environment using their thoughts alone. These devices have the potential to significantly enhance autonomy and improve the quality of life for people with conditions such as spinal cord injuries.

Advancements in sensory neuroprosthetics introduce bidirectional connections that enable users to perceive touch or temperature through artificial limbs. Examples include cochlear implants, which restore auditory sensation, and retinal implants, which aim to bring vision back to those with sensory impairments [59].

8.7.4.2 Brain-Computer Interfaces (BCIs)

BCIs establish direct communication channels between the brain and external devices, enabling real-time information exchange without relying on traditional motor outputs. Their potential applications extend beyond neurodegenerative diseases, encompassing communication enhancement, control augmentation, and cognitive empowerment.

8.7.4.2.1 Non-Invasive BCIs

Non-invasive BCIs, utilizing technologies such as EEG (electroencephalogram), fMRI (functional magnetic resonance imaging), and fNIRS (functional near-infrared spectroscopy), offer alternative pathways for interfacing without surgical procedures. These applications span from assistive technology and mental health monitoring to neurofeedback training for cognitive enhancement.

8.7.4.2.2 Invasive BCIs

Invasive BCIs, involving technologies like intracranial electrodes, provide higher precision in decoding neural signals. Their promise lies in restoring complex motor functions, such as precise control of robotic limbs or communication devices, particularly relevant in neurodegenerative conditions like ALS.

8.7.4.2.3 BCIs for Cognitive Enhancement

BCIs for cognitive enhancement explore the amplification of cognitive functions by facilitating direct communication between the brain and external computational devices. Ongoing research aims to boost memory, attention, and information processing, offering potential benefits for individuals facing cognitive decline in neurodegenerative diseases.

8.7.4.3 Ethical Considerations in BMI Development

8.7.4.3.1 Privacy and Security Concerns

Privacy and security concerns are paramount as BMIs collect sensitive neural data. The ethical development of these interfaces requires robust frameworks to safeguard privacy and prevent unauthorized access.

8.7.4.3.2 Informed Consent and Autonomy

Informed consent and autonomy are crucial principles in BMI development. Users must be thoroughly informed about risks, benefits, and potential outcomes, ensuring that their autonomy is respected in adopting these transformative technologies.

8.7.4.3.3 Equitable Access to BMI Technologies

Equitable access to BMI technologies is essential to prevent disparities based on socioeconomic factors. Policymakers play a crucial role in shaping the distribution and accessibility of BMI technologies to ensure fairness and prevent exacerbation of existing inequalities.

In the unfolding narrative of brain-machine interfaces, the convergence of neuroscience and technology is reshaping possibilities for individuals affected by neurodegenerative diseases. From restoring motor function to enhancing cognitive abilities, BMIs stand as beacons of hope, navigating the intersection of human potential and technological innovation.

8.7.5 Neural Encoding Strategies and Representation of Information in the Brain

8.7.5.1 Sparse Coding

Sparse coding is a method where a particular feature or idea is represented by the activation of a small group of neurons [60]. For instance, in visual processing, sparse coding may take place when a specific neuron responds selectively to a particular orientation of a visual stimulus.

8.7.5.2 Rate Coding

Our brain encodes information by the frequency or rate of neural firing. The higher the rate of firing, the more intense the representation of a particular attribute [61]. For example, in motor control, the rate at which motor neurons fire can indicate the strength or intensity of a movement.

8.7.5.3 Temporal Coding

The precise timing of neural spikes carries information. Patterns of temporal spikes encode features or events [62]. For example, in auditory processing, the timing of spikes may encode the frequency or duration of a sound.

8.7.5.4 Population Coding

8.7.5.4.1 Distributed Representation

Information is not confined to individual neurons but distributed across a population. Patterns of activity across multiple neurons collectively represent a complex concept [63]. For example, the memory of a face might be distributed across a network of neurons, each contributing to different aspects of facial features.

8.7.5.4.2 Ensemble Coding

Groups of neurons work together as ensembles to represent specific features or concepts [63]. For example, the concept of a "red apple" might be encoded by an ensemble of neurons representing color, shape, and taste.

8.7.5.5 Representation of Information in the Brain

8.7.5.5.1 Semantic Representations

Neural circuits are engaged in representing abstract concepts and semantic meanings. For example, the concept of "justice" might activate a distributed network of neurons associated with fairness, morality, and legal principles.

8.7.5.5.2 Memory Representations

Memory traces involve the persistent activation of specific neural circuits that represent past experiences. For example, a recall of a childhood event might involve the reactivation of neurons associated with the sights, sounds, and emotions of that particular memory.

8.7.5.5.3 Synaptic Plasticity

Synapses between neurons can change in strength, allowing for the adaptation of neural circuits over time. For example, learning a new skill involves the strengthening of synapses in the corresponding neural circuits.

8.7.5.5.4 Dynamic Adaptation

Neural circuits can adapt to changing circumstances and learning experiences. For instance, repeated exposure to a stimulus can enhance recognition or discrimination.

8.7.6 Information Flow in Neural Circuits

8.7.6.1 Feedforward Processing

8.7.6.1.1 Sensory Pathways

Information flows in a unidirectional manner from sensory organs to the brain's higher processing centers. For example, in vision, sensory information from the retina travels through the optic nerve to the visual cortex, where it undergoes initial processing.

8.7.6.1.2 Motor Commands

Motor commands are generated in higher cortical areas and transmitted to the spinal cord or peripheral muscles. For example, decisions made in the prefrontal cortex to move a hand trigger a cascade of motor commands, ultimately resulting in the intended movement.

8.7.6.2 Feedback Mechanisms

8.7.6.2.1 Top-Down Modulation

Higher cortical areas send feedback signals to lower-level regions, modulating processing based on cognitive states, expectations, or attention. For example, during a visual task, feedback from higher visual areas might enhance the processing of relevant visual stimuli in lower visual regions.

8.7.6.2.2 Reciprocal Connections

Many neural circuits exhibit reciprocal connections, allowing for bidirectional communication between regions [64]. For example, the prefrontal cortex and the hippocampus engage in reciprocal connections, facilitating memory processes and decision-making.

8.7.6.3 Dynamic Information Routing

8.7.6.3.1 Dynamic Synaptic Plasticity

Synapses can undergo changes in strength based on neural activity, enabling dynamic information routing. For example, learning a new skill involves the strengthening of synapses in the corresponding neural circuits, optimizing information flow.

8.7.6.3.2 Reentrant Processing

Information loops back and forth between brain regions, creating reentrant circuits that contribute to the integration of information across distributed networks. For example, in visual processing, feedback loops between higher and lower visual areas refine the interpretation of visual stimuli.

8.7.6.4 Selective Attention and Filtering

8.7.6.4.1 Attentional Mechanisms

Neural circuits implement selective attention, emphasizing relevant information while filtering out distractions. For example, during a complex auditory scene, selective attention might enhance the processing of a specific voice while filtering out background noise [65].

8.7.6.4.2 Filtering in Sensory Pathways

Neural circuits act as filters, emphasizing salient features and discarding less relevant information. For example, in the olfactory system, filtering mechanisms allow the brain to focus on specific odorants amidst myriad scents.

8.7.6.5 Integrative Processing

8.7.6.5.1 Association Areas

Information from different sensory modalities converges in association areas, facilitating integrative processing. For example, multisensory integration in the superior colliculus involves combining visual and auditory information to guide orienting movements.

8.7.6.5.2 Cross-Modal Integration

Neural circuits integrate information from different sensory modalities to construct a unified perceptual experience. For example, integrating visual and tactile information in the somatosensory cortex enhances object recognition through touch.

8.7.6.6 Plasticity and Adaptation

8.7.6.6.1 Learning-Induced Changes

Neural circuits change in response to learning experiences, adapting to new information. For example, long-term potentiation in the hippocampus involves strengthening synapses, facilitating the encoding of new memories.

8.7.6.6.2 Adaptive Neural Circuits

Neural circuits exhibit adaptability, allowing for adjustments in response to changing environmental conditions. For example, the neural circuits involved in motor control adapt to changes in limb dynamics, ensuring precise and coordinated movements.

8.8 NEURODEGENERATIVE DISEASES: RECENT ADVANCEMENT IN MODELING TECHNIQUES AND INFORMATION FLOW IN NEURAL CIRCUITS

Neurodegenerative diseases present a formidable challenge, characterized by intricate pathological processes and multifaceted clinical manifestations. Advanced modeling techniques in computational neuroscience have emerged as invaluable tools to dissect and understand the complexities associated with these diseases. This discussion delves into the sophisticated modeling approaches employed to unravel the mysteries of neurodegenerative disorders. Neurodegenerative diseases inflict a profound impact on the delicate dance of neural coding and information flow within the brain's circuits. The intricate processes of encoding information and the dynamic flow of signals are disrupted, contributing to the manifestation of cognitive decline, motor dysfunction, and other symptoms characteristic of these diseases. Let's explore how neural coding and information flow are altered in the context of neurodegenerative diseases.

8.8.1 Spiking Neural Networks (SNNs) for Neurodegenerative Diseases

8.8.1.1 Biological Realism

SNNs stand out for their emphasis on biological fidelity, mirroring the spiking behavior of real neurons. In the context of neurodegenerative diseases, SNNs provide a platform to simulate the impact of pathological changes on neural activity, offering insights into disease progression at a level of detail not achievable with traditional models.

8.8.1.2 Temporal Coding

The temporal precision embedded in SNNs aligns with the importance of temporal aspects in neurodegenerative diseases. This modeling approach captures the nuanced changes in neural dynamics over time, shedding light on the temporal evolution of pathological processes.

8.8.2 Agent-Based Modeling for Neurodegenerative Diseases

8.8.2.1 Microscale Interactions

Agent-based modeling dredges into the microscale interactions between individual agents, representing cells or molecules. In the context of neurodegenerative diseases, this approach allows researchers to simulate the interactions between neurons, glial cells, and pathological agents (such as misfolded proteins) at a granular level. This microscopic perspective aids in understanding emergent properties and local dynamics contributing to disease progression.

8.8.2.2 Spatial Considerations

Agent-based models incorporate spatial considerations, enabling the exploration of how spatial patterns of cellular interactions contribute to the spread of pathological changes. This is particularly relevant for diseases like Alzheimer's, where the propagation of misfolded proteins occurs in a spatially defined manner.

8.8.3 Network Models and Graph Theory for Neurodegenerative Diseases

8.8.3.1 Connectome-Based Modeling

Network models leverage the connectome, a comprehensive map of neural connections in the brain. Researchers can model the connectivity disruptions associated with neurodegenerative diseases by integrating graph theory. Changes in network properties, such as altered connectivity patterns and network resilience, offer valuable insights into disease-related structural alterations.

8.8.3.2 Propagation Dynamics

Graph-based modeling enables the simulation of disease-related alterations in information propagation across neural networks. This sheds light on how disruptions in connectivity contribute to the spread of pathological changes, a phenomenon observed in diseases like Parkinson's and Huntington's.

8.8.4 Hybrid Models for Neurodegenerative Diseases

8.8.4.1 Integrating Multiple Scales

Neurodegenerative diseases manifest at multiple scales, from molecular interactions to whole-brain dynamics. Hybrid models seamlessly integrate different modeling approaches, combining the strengths of individual techniques. For example, a model may incorporate agent-based components to simulate cellular interactions alongside network models to capture large-scale connectivity patterns.

8.8.4.2 Cross-Scale Validation

Hybrid models facilitate cross-scale validation, allowing researchers to validate predictions made at the microscopic level against observations at the macroscopic level. This enhances the robustness and reliability of the models, providing a more comprehensive understanding of the disease processes.

8.8.5 Machine Learning for Biomarker Discovery for Neurodegenerative Diseases

8.8.5.1 Pattern Recognition

Machine learning techniques, particularly deep learning, excel in pattern recognition within complex datasets. In the context of neurodegenerative diseases, these models analyze multi-modal data, including neuroimaging, genomics, and clinical records, to identify subtle patterns indicative of disease presence or progression.

8.8.5.2 Predictive Modeling

Advanced machine learning models can be trained to predict disease outcomes, aiding in early diagnosis and prognostication. By identifying biomarkers associated with disease progression, these models contribute to developing predictive tools for personalized patient management.

8.8.6 Altered Neural Coding Strategies for Neurodegenerative Diseases

8.8.6.1 Sparse Coding Disruptions

8.8.6.1.1 Dysfunction

Degeneration of specific neuronal populations disrupts sparse coding, impacting the selective activation of neurons for distinct features or concepts.

8.8.6.1.2 Example

In Alzheimer's disease, the loss of pyramidal neurons in the hippocampus disrupts sparse coding related to episodic memories.

8.8.6.2 Rate Coding Aberrations

8.8.6.2.1 Dysfunction

Disruptions in neurotransmitter systems and neuronal loss affect rate coding, altering the frequency-based representation of information.

8.8.6.2.2 Example

Parkinson's disease, characterized by dopaminergic degeneration, disrupts the rate coding necessary for smooth motor control.

8.8.6.3 Temporal Coding Disturbances

8.8.6.3.1 Dysfunction

Changes in neural oscillations and synchrony impact temporal coding, affecting the precise timing of information representation.

8.8.6.3.2 Example

In frontotemporal dementia, alterations in the temporal dynamics of neural activity contribute to language processing impairments.

8.8.7 Compromised Population Coding for Neurodegenerative Diseases

8.8.7.1 Distributed Representation Breakdown

8.8.7.1.1 Dysfunction

Neurodegeneration disrupts the distributed representation of information, leading to fragmented and inefficient coding.

8.8.7.1.2 Example

Huntington's disease, marked by striatal atrophy, disrupts the distributed representation of motor and cognitive functions.

8.8.7.2 Ensemble Coding Disruptions

8.8.7.2.1 Dysfunction

Loss of specific neuronal populations hampers ensemble coding, impacting the coordinated activity of neurons representing complex concepts.

8.8.7.2.2 Example

In amyotrophic lateral sclerosis (ALS), the degeneration of motor neurons disrupts ensemble coding necessary for coordinated muscle movements.

8.8.8 Sensory Modality-Specific Coding Challenges for Neurodegenerative Diseases

8.8.8.1 Visual Coding Impairments

8.8.8.1.1 Dysfunction

Degeneration of visual processing regions leads to visual coding impairments, affecting the representation of visual stimuli.

8.8.8.1.2 Example

Progressive supranuclear palsy (PSP) disrupts visual coding, contributing to gaze control and visual attention deficits.

8.8.8.2 Auditory Coding Distortions

8.8.8.2.1 Dysfunction

Auditory processing deficits in neurodegenerative diseases alter the coding of acoustic features, impacting sound perception.

8.8.8.2.2 Example

Auditory processing disorders in Alzheimer's disease affect the coding of auditory stimuli, contributing to communication difficulties.

8.8.9 Disrupted Conceptualization of Cognitive Representations for Neurodegenerative Diseases

8.8.9.1 Semantic Representation Breakdown

8.8.9.1.1 Dysfunction

Cortical atrophy disrupts semantic representations, impacting abstract thinking and conceptual understanding.

8.8.9.1.2 Example

Semantic dementia, a subtype of frontotemporal dementia, impairs the representation of abstract concepts and semantic knowledge.

8.8.9.2 Memory Representation Impairments

8.8.9.2.1 Dysfunction

Hippocampal and cortical degeneration disrupts memory representations, leading to memory retrieval deficits.

8.8.9.2.2 Example

Alzheimer's disease affects the memory representation of events, contributing to episodic memory impairments.

8.8.10 Synaptic Dysfunction and Altered Information Flow for Neurodegenerative Diseases

8.8.10.1 Synaptic Plasticity Disruptions

8.8.10.1.1 Dysfunction

Accumulation of pathological proteins impairs synaptic plasticity, impacting the adaptability of neural circuits.

8.8.10.1.2 Example

Tau and beta-amyloid pathology in Alzheimer's disease disrupts synaptic plasticity, leading to cognitive decline.

8.8.10.2 Impaired Dynamic Information Routing

8.8.10.2.1 Dysfunction

Synaptic loss and dysfunction compromise dynamic information routing, disrupting the flow of signals between neurons.

8.8.10.2.2 Example

Parkinson's disease disrupts the dynamic routing of information between the basal ganglia and motor cortex, leading to motor dysfunction.

8.8.11 Selective Attention and Filtering Deficits for Neurodegenerative Diseases

8.8.11.1 Attentional Impairments

8.8.11.1.1 Dysfunction

Neurodegenerative diseases impact attentional mechanisms, leading to difficulties in focusing on relevant stimuli.

8.8.11.1.2 Example

Lewy body dementia disrupts attentional processes, contributing to fluctuations in cognitive performance.

8.8.11.2 Filtering Dysfunction in Sensory Pathways

8.8.11.2.1 Dysfunction

Altered neural coding and connectivity impair filtering mechanisms, leading to sensory hypersensitivity.

8.8.11.2.2 Example

Individuals with Parkinson's disease may exhibit heightened sensitivity to sensory stimuli due to disruptions in sensory filtering.

8.8.12 Challenges and Future Directions for Neurodegenerative Diseases

8.8.12.1 Data Integration

Integrating diverse data types, including genetic, imaging, and clinical data, poses a significant challenge. Advanced models need to seamlessly integrate these data sources to provide a holistic understanding of neurodegenerative diseases.

8.8.12.2 Validation and Clinical Translation

Translating computational models into clinically relevant insights requires rigorous validation against empirical data. Ensuring that these models accurately represent the biological reality of neurodegenerative diseases is crucial for their clinical utility.

In conclusion, the landscape of advanced modeling techniques in neurodegenerative diseases is marked by innovation and a quest for a deeper understanding of these complex disorders. As computational neuroscience continues to evolve, integrating diverse modeling approaches holds promise for unraveling the intricacies of disease mechanisms and, ultimately, advancing therapeutic strategies for neurodegenerative conditions.

In the intricate tapestry of neurodegenerative diseases, the disruptions in neural coding and information flow underscore the challenges faced by affected individuals. Deciphering these alterations is essential for developing targeted interventions that may, in the future, mitigate the progression of these debilitating conditions.

8.9 CONCLUSION

In conclusion, this chapter has traversed the landscape of computational neuroscience, encompassing a spectrum of intricacies from foundational neural networks to the dynamic processes of synaptic transmission and plasticity. Recent strides in the field have propelled us towards a paradigm shift, unveiling the potency of spiking neural networks, the application of advanced deep learning techniques in neuroscience, and the confluence of connectomics and brain-machine interfaces. Our exploration into the realms of neural encoding, information representation, and the nuanced flow of information within neural circuits has not only enriched our comprehension of cerebral complexities but also heralded a new era of decoding the intricacies embedded within neural networks. A pivotal investigation within the purview of neurodegenerative diseases illuminated recent advancements in modeling techniques and elucidated aberrations in the information flow within neural circuits associated with conditions such as Alzheimer's and Parkinson's. The intersection of computational neuroscience with the study of neurodegenerative diseases emerges as a promising avenue, holding the potential to unravel the underlying mechanisms of these disorders.

In summary, this chapter underscores computational neuroscience's dynamic and progressive nature. The amalgamation of theoretical constructs with cutting-edge technologies delineates an era where our understanding of the intricate workings of the brain achieves unprecedented depths. As we navigate the labyrinth of neural systems, the pursuit of knowledge endures, beckoning researchers to forge ahead, transcend boundaries, and contribute to a collective endeavor poised to redefine the future of neuroscience and its applications in addressing neurological disorders.

8.10 KEY TAKEAWAYS

8.10.1 Computational Models Unveil Synaptic Dynamics

Computational neuroscience has provided invaluable tools for simulating synaptic plasticity, enabling the mapping of intricate neural dynamics in neurodegenerative diseases like Alzheimer's and Parkinson's.

8.10.2 Impactful Disease Modeling

By employing diverse computational approaches, researchers can model the progression of neurodegenerative diseases, offering insights into early-stage pathology, identifying vulnerable neural subtypes, and guiding potential therapeutic interventions.

8.10.3 Personalized Treatment Strategies

Advancements in computational models allow for the development of personalized treatment strategies. Patient-specific neural network dynamics in conditions like Parkinson's inform individualized medication plans, enhancing therapeutic outcomes.

8.10.4 Emerging Technologies Shape the Future

Quantum computing and advances in neuroimaging techniques stand as transformative forces in computational neuroscience. Quantum computing promises accelerated simulations, while improved neuroimaging provides enhanced spatial and temporal resolution for deeper insights.

8.10.5 Ethical Considerations are Crucial

Responsible use of computational neuroscience demands ethical considerations. Open access to models and data, ethical guidelines for AI integration, and robust data protection protocols ensure transparency, privacy, and responsible research practices.

8.10.6 Impact on Our Understanding of the Brain

Recent advancements in computational neuroscience have significantly advanced our understanding of the brain. From predictive biomarkers to treatment simulations, these breakthroughs contribute to a deeper comprehension of neurodegenerative diseases.

8.10.7 Computational Neuroscience as a Catalyst

Computational neuroscience plays a pivotal role in neurodegenerative research, acting as a catalyst for innovation. Its ability to bridge theory and experimentation accelerates understanding, diagnosis, and treatment progress.

8.10.8 Encouraging Future Exploration

The conclusion emphasizes the need to encourage curiosity and interest in the field. Aspiring researchers are urged to explore the interdisciplinary landscape, where computational models contribute to a collective understanding of the brain's complexities.

8.10.9 Promising Future Possibilities

The synergy between computational neuroscience and neurodegenerative research holds promise for transformative treatments. The hope is that ongoing advancements will lead to improved diagnostic tools, innovative therapeutic interventions, and ultimately better outcomes for individuals affected by these challenging conditions.

REFERENCES

1. Thomas Trappenberg. *Fundamentals of Computational Neuroscience*. OUP Oxford, Oxford, England, 2009.
2. Losiana Nayak, Abhijit Dasgupta, Ritankar Das, Kuntal Ghosh, and Rajat K De. Computational neuroscience and neuroinformatics: Recent progress and resources. *Journal of Biosciences*, 43:1037–1054, 2018.
3. Kathleen A Powers. *The Cybernetic Origins of Life*. University of California, Berkeley, 2020.
4. Eric R Kandel and Larry R Squire. Neuroscience: Breaking down scientific barriers to the study of brain and mind. *Science*, 290(5494):1113–1120, 2000.
5. Christos Kleanthous, and Sotirios Chatzis. Gated mixture variational autoencoders for value added tax audit case selection. *Knowledge-Based Systems*, 188:105048, 2020.
6. Boris Zlotin, Vladimir Proseanic, Boris Farber, Anatoly Guin, Ivan Nehrishnyi Sr, Ivan Nehrishnyi Jr, Vladimir Matsenko, Anton Belyakov, and Gafur Zainiev. A new approach to building artificial neural networks and medicine. *Annals of Mechnikov's Institute*, (4):101–107, 2021.
7. Hamit Taner Ünal, and Fatih Başçiftçi. Evolutionary design of neural network architectures: A review of three decades of research. *Artificial Intelligence Review*, 55, 1723–1802, 2022.
8. RM Awangga, TLR Mengko, and NP Utama. A literature review of brain decoding research. IOP Conference Series: Materials Science and Engineering, 830:032049, 2020.
9. Célia Loriette, Julian L Amengual, and Suliann Ben Hamed. Beyond the braincomputer interface: Decoding brain activity as a tool to understand neuronal mechanisms subtending cognition and behavior. *Frontiers in Neuroscience*, 16:811736, 2022.
10. Sudhakar Mishra, Mohammad Asif, Narayanan Srinivasan, and Uma Shanker Tiwary. Dataset on emotion with naturalistic stimuli (dens) on indian samples. bioRxiv, pp. 2021–08, 2021.
11. Zhijin Wang, Philippe Fournier-Viger, Yu-Ping Wang, and Yonggang Fu. Crosstalk between computational medicine and neuroscience in healthcare. *Frontiers in Neuroscience*, 17:1333227, 2023.
12. Catherine Bjerre Collin, Tom Gebhardt, Martin Golebiewski, Tugce Karaderi, Maximilian Hillemanns, Faiz Muhammad Khan, Ali Salehzadeh-Yazdi, Marc Kirschner, Sylvia Krobitsch, EU-STANDS4PM consortium, et al. Computational models for clinical applications in personalized medicine—Guidelines and recommendations for data integration and model validation. *Journal of Personalized Medicine*, 12(2):166, 2022.
13. Kurt A Jellinger. Basic mechanisms of neurodegeneration: A critical update. *Journal of Cellular and Molecular Medicine*, 14(3):457–487, 2010.
14. Jose A Santiago, Virginie Bottero, and Judith A Potashkin. Dissecting the molecular mechanisms of neurodegenerative diseases through network biology. *Frontiers in Aging Neuroscience*, 9:166, 2017.
15. Jae Ryul Bae, and Sung Hyun Kim. Synapses in neurodegenerative diseases. *BMB Reports*, 50(5):237, 2017.
16. Princy Choudhary, Rameshwar Sankhala, and Sangeeta Singh. Integration of significant signaling pathways involved in neural-fate differentiation of stem cells. In *2023 10th International Conference on Computing for Sustainable Global Development (INDIACom)*, pp. 694–701. IEEE, 2023.
17. Hugo Geerts, Silke Bergeler, William W Lytton, and Piet H van der Graaf. Computational neurosciences and quantitative systems pharmacology: A powerful combination for supporting drug development in neurodegenerative diseases. *Journal of Pharmacokinetics and Pharmacodynamics*, 2023. https://doi.org/10.1007/s10928-023-09876-6
18. Sonu Pahal, Ayushi Gupta, Princy Choudhary, Amit Chaudhary, and Sangeeta Singh. Network pharmacological evaluation of withania somnifera bioactive phytochemicals for identifying novel potential inhibitors against neurodegenerative disorder. *Journal of Biomolecular Structure and Dynamics*, 40(21):10887–10898, 2022.
19. Avik Atta, Ayushi Gupta, Princy Choudhary, Shrey Dwivedi, and Sangeeta Singh. Inhibition of lingo1 as a therapeutic target to promote axonal regeneration and repair for neurological disorders. *3 Biotech*, 13(11):372, 2023.

20. Richa Shukla, Sonu Pahal, Ayushi Gupta, Princy Choudhary, Krishna Misra, and Sangeeta Singh. Modulation of gpcr receptors common to gut inflammatory diseases and neuronal disorders, alzheimer's and parkinson's diseases as druggable targets through withania somnifera bioactives: An in silico study. *Journal of Biomolecular Structure and Dynamics*, 41(10):4485–4503, 2023.
21. Princy Choudhary, Shivani Gupta, Richa Shukla, Ayushi Gupta, Sonu Pahal, and Sangeeta Singh. Regulation of neuronal repair and regeneration through inhibition of oligodendrocyte myelin glycoprotein (OMgp). *Journal of Biomolecular Structure and Dynamics*, 40(24):13936–13952, 2022.
22. Richard NL Lamptey, Bivek Chaulagain, Riddhi Trivedi, Avinash Gothwal, Buddhadev Layek, and Jagdish Singh. A review of the common neurodegenerative disorders: Current therapeutic approaches and the potential role of nanotherapeutics. *International Journal of Molecular Sciences*, 23(3):1851, 2022.
23. Sepehr Golriz Khatami, Sarah Mubeen, and Martin Hofmann-Apitius. Data science in neurodegenerative disease: Its capabilities, limitations, and perspectives. *Current Opinion in Neurology*, 33(2):249, 2020.
24. Andreas Zell. *Simulation Neuronaler Netze*, vol. 1. Addison-Wesley Bonn, 1994.
25. George Cybenko. Approximation by superpositions of a sigmoidal function. *Mathematics of Control, Signals and Systems*, 2(4):303–314, 1989.
26. Rikiya Yamashita, Mizuho Nishio, Richard Kinh Gian Do, and Kaori Togashi. Convolutional neural networks: An overview and application in radiology. *Insights into Imaging*, 9:611–629, 2018.
27. Oludare Isaac Abiodun, Aman Jantan, Abiodun Esther Omolara, Kemi Victoria Dada, Nachaat AbdElatif Mohamed, and Humaira Arshad. State-of-the-art in artificial neural network applications: A survey. *Heliyon*, 4(11), 2018.
28. Sepp Hochreiter, and Jürgen Schmidhuber. Long short-term memory. *Neural Computation*, 9(8): 1735–1780, 1997.
29. Mark A Kramer. Nonlinear principal component analysis using autoassociative neural networks. *AIChE Journal*, 37(2):233–243, 1991.
30. Ian Goodfellow, Jean Pouget-Abadie, Mehdi Mirza, Bing Xu, David Warde-Farley, Sherjil Ozair, Aaron Courville, and Yoshua Bengio. Generative adversarial nets. *Advances in Neural Information Processing Systems*, 27, 2014.
31. Ashish Vaswani, Noam Shazeer, Niki Parmar, Jakob Uszkoreit, Llion Jones, Aidan N Gomez, Łukasz Kaiser, and Illia Polosukhin. Attention is all you need. *Advances in Neural Information Processing Systems*, 30, 2017.
32. Chaochao Lu, and Xiaoou Tang. Surpassing human-level face verification performance on lfw with gaussianface. In *Proceedings of the AAAI conference on artificial intelligence*, volume 29, PKP Publishing Services Network,, pp. 3811–3819, 2015.
33. Sampo Kuutti, Richard Bowden, Yaochu Jin, Phil Barber, and Saber Fallah. A survey of deep learning applications to autonomous vehicle control. *IEEE Transactions on Intelligent Transportation Systems*, 22(2):712–733, 2020.
34. Ilya Sutskever, Oriol Vinyals, and Quoc V Le. Sequence to sequence learning with neural networks. *Advances in Neural Information Processing Systems*, 27, 2014.
35. Grégoire Mesnil, Yann Dauphin, Kaisheng Yao, Yoshua Bengio, Li Deng, Dilek Hakkani-Tur, Xiaodong He, Larry Heck, Gokhan Tur, Dong Yu, et al. Using recurrent neural networks for slot filling in spoken language understanding. *IEEE/ACM Transactions on Audio, Speech, and Language Processing*, 23(3):530–539, 2014.
36. Xin Dong, Yizhao Zhou, Lantian Wang, Jingfeng Peng, Yanbo Lou, and Yiqun Fan. Liver cancer detection using hybridized fully convolutional neural network based on deep learning framework. *IEEE Access*, 8:129889–129898, 2020.
37. Bie Verbist, Günter Klambauer, Liesbet Vervoort, Willem Talloen, Ziv Shkedy, Olivier Thas, Andreas Bender, Hinrich WH Göhlmann, Sepp Hochreiter, QSTAR Consortium, et al. Using transcriptomics to guide lead optimization in drug discovery projects: Lessons learned from the qstar project. *Drug Discovery Today*, 20(5):505–513, 2015.
38. Aaron Van den Oord, Sander Dieleman, and Benjamin Schrauwen. Deep contentbased music recommendation. *Advances in Neural Information Processing Systems*, 26, 2013.
39. Ali Mamdouh Elkahky, Yang Song, and Xiaodong He. A multi-view deep learning approach for cross domain user modeling in recommendation systems. In *Proceedings of the 24th international conference on world wide web*, International World Wide Web Conferences Steering Committee, Republic and Canton of Geneva, Switzerland, pp. 278–288, 2015.
40. Yegor Tkachenko. Autonomous crm control via clv approximation with deep reinforcement learning in discrete and continuous action space. *arXiv preprint arXiv:1504.01840*, 2015.

41. JunHyuk Woo, Kiri Choi, Soon Ho Kim, Kyungreem Han, and MooYoung Choi. The structural aspects of neural dynamics and information flow. *Frontiers in Bioscience- Landmark*, 27(1):15, 2022.
42. John R Hughes. Post-tetanic potentiation. *Physiological Reviews*, 38(1):91–113, 1958.
43. M James Nichols, and William T Newsome. The neurobiology of cognition. *Nature*, 402(Suppl 6761): C35–C38, 1999.
44. Pascal Fries. A mechanism for cognitive dynamics: Neuronal communication through Neuronal coherence. *Trends in Cognitive Sciences*, 9(10):474–480, 2005.
45. Wolfgang Maass. Networks of spiking neurons: The third generation of neural network models. *Neural Networks*, 10(9):1659–1671, 1997.
46. Robert Gütig, and Haim Sompolinsky. The tempotron: A neuron that learns spike timing–based decisions. *Nature Neuroscience*, 9(3):420–428, 2006.
47. Kashu Yamazaki, Viet-Khoa Vo-Ho, Darshan Bulsara, and Ngan Le. Spiking neural networks and their applications: A review. *Brain Sciences*, 12(7):863, 2022.
48. Marc Osswald, Sio-Hoi Ieng, Ryad Benosman, and Giacomo Indiveri. A spiking neural network model of 3d perception for event-based neuromorphic stereo vision systems. *Scientific Reports*, 7(1):40703, 2017.
49. DV Buonomano, and TP Carvalho. Spike-timing-dependent plasticity (STDP), pp. 265–268, Elsevier. 2009.
50. Mohammad Asif, and Sudhakar Mishra, Majithia Tejas Vinodbhai, and Uma Shanker Tiwary. Emotion recognition using temporally localized emotional events in EEG with naturalistic context: Dens# dataset. *IEEE Access*, 11:39913–39925, 2023.
51. Mohammad Asif, Majithia Tejas Vinodbhai, Sudhakar Mishra, Aditya Gupta, and Uma Shanker Tiwary. Emotion recognition in vad space during emotional events using cnn-gru hybrid model on eeg signals. In *International Conference on Intelligent Human Computer Interaction*, pp. 75–84. Springer, 2022.
52. Mohammad Asif, Diya Srivastava, Aditya Gupta, and Uma Shanker Tiwary. Inter subject emotion recognition using spatio-temporal features from EEG signal. In *2023 27th International Computer Science and Engineering Conference (ICSEC)*, pp. 1–4. IEEE, 2023.
53. Haiguang Wen, Junxing Shi, Yizhen Zhang, Kun-Han Lu, Jiayue Cao, and Zhongming Liu. Neural encoding and decoding with deep learning for dynamic natural vision. *Cerebral Cortex*, 28(12): 4136–4160, 2018.
54. Stamatios N Sotiropoulos, and Andrew Zalesky. Building connectomes using diffusion MRI: Why, how and but. *NMR in Biomedicine*, 32(4):e3752, 2019.
55. Michael J Young, David J Lin, and Leigh R Hochberg. Brain–computer interfaces in neurorecovery and neurorehabilitation. Seminars in Neurology, 41:206–216, 2021.
56. R Cameron Craddock, Rosalia L Tungaraza, and Michael P Milham. Connectomics and new approaches for analyzing human brain functional connectivity. *GigaScience*, 4(1):13, 2015.
57. Alex Fornito, Andrew Zalesky, and Michael Breakspear. The connectomics of brain disorders. *Nature Reviews Neuroscience*, 16(3):159–172, 2015.
58. Justin C Sanchez. *Neuroprosthetics: Principles and Applications*. CRC Press, 2018.
59. Kenji Kansaku. Neuroprosthetics in systems neuroscience and medicine. *Scientific Reports*, 11(1):5404, 2021.
60. Michael Beyeler, Emily L Rounds, Kristofor D Carlson, Nikil Dutt, and Jeffrey L Krichmar. Neural correlates of sparse coding and dimensionality reduction. *PLoS Computational Biology*, 15(6):e1006908, 2019.
61. Edmund T Rolls, and Alessandro Treves. The neuronal encoding of information in the brain. *Progress in Neurobiology*, 95(3):448–490, 2011.
62. G Foffani, ML Morales-Botello, and J Aguilar. Spike timing, spike count, and temporal information for the discrimination of tactile stimuli in the rat ventrobasal complex. *Journal of Neuroscience*, 29(18):5964–5973, 2009.
63. Stefano Panzeri, Jakob H Macke, Joachim Gross, and Christoph Kayser. Neural population coding: Combining insights from microscopic and mass signals. *Trends in Cognitive Sciences*, 19(3):162–172, 2015.
64. Gary G Berntson, and Sahib S Khalsa. Neural circuits of interoception. *Trends in Neurosciences*, 44(1):17–28, 2021.
65. Eric I Knudsen. Neural circuits that mediate selective attention: A comparative perspective. *Trends in Neurosciences*, 41(11):789–805, 2018.

9 Tools for Improving Synaptic Plasticity from Computational Neuroscience

Mohammad Asif and Diya Srivastava
Indian Institute of Information Technology Allahabad
Prayagraj, Uttar Pradesh, India

*Princy Choudhary**
National Centre for Cell Science
Ganeshkhind, Pune, Maharashtra, India
and
National Institute of Mental Health and Neurosciences
Bengaluru, Karnataka, India

9.1 INTRODUCTION

Synaptic plasticity, the dynamic modulation of synaptic strength, stands as a cornerstone in the neurobiological substrate of learning, memory, and adaptive behavior [1]. The intricate orchestration of molecular events, cellular processes, and network dynamics underscores the complexity of this phenomenon, demanding advanced tools and methodologies for its comprehensive exploration. At the nexus of neuroscience and computational science, this chapter delineates an exhaustive investigation into the diverse computational tools instrumental in unraveling the subtleties of synaptic plasticity and, concomitantly, in propelling strategies for its deliberate modulation [2].

The overarching goal of this exploration is to navigate through a repertoire of computational tools, each tailored to scrutinize synaptic plasticity from a distinct perspective [3]. By synthesizing empirical observations with computational models, researchers can unravel the intricacies of synaptic plasticity with unprecedented granularity and foster a nuanced understanding of its implications for cognitive function and neurological disorders.

The journey commences with an examination of neural simulations, notably NEURON and Brian, which serve as virtual laboratories for modeling complex neural networks. These simulations, rooted in biophysical realism, provide a dynamic platform for the exploration of temporal and spatial aspects of synaptic plasticity mechanisms [3,4].

Transitioning into the domain of spiking neural networks (SNNs), exemplified by NEST and BindsNET, the exploration probes the departure from conventional neural network architectures [5]. The adoption of spiking dynamics facilitates a more faithful representation of biological neural processes, enriching our understanding of synaptic plasticity at the network level.

Machine learning paradigms, particularly reinforcement learning (RL), are dissected for their utility in optimizing synaptic connections. Drawing inspiration from the reward-based learning observed in biological systems, RL algorithms offer insights into the adaptive nature of synaptic plasticity [6].

The chapter further scrutinizes the integration of computational models with neuromorphic hardware, exemplified by SpiNNaker and TrueNorth. This marriage of simulation and real-time implementation bridges the gap between theoretical constructs and physical hardware, paving the way for advances in neuromorphic computing.

DOI: 10.1201/9781003464648-11

Optogenetics, a revolutionary technique employing light-sensitive proteins such as Channelrhodopsin and Halorhodopsin, is explored for its precision in manipulating neuronal activity [7]. The integration of optogenetic tools with computational models offers a unique avenue to causally dissect synaptic plasticity mechanisms and their behavioral consequences.

Electrophysiological recording techniques, prominently featuring patch-clamp recordings, provide a lens into the intricacies of individual neurons. High-resolution data from these recordings serve as a critical validation ground for computational models of synaptic plasticity.

Fluorescent calcium indicators and calcium imaging techniques form a focal point, offering non-invasive insights into intracellular dynamics correlated with synaptic activity. The confluence of computational models and empirical data from calcium imaging enriches our understanding of synaptic plasticity.

Brain-machine interfaces (BMIs) and closed-loop systems emerge as transformative tools for bidirectional communication between neural circuits and external devices. Integration with computational algorithms allows for real-time modulation of synaptic plasticity, holding promise for therapeutic interventions in neurological disorders [7].

The chapter culminates with an exploration of foundational computational tools, including MATLAB and Python (utilizing libraries like NumPy, SciPy, and TensorFlow), which form the bedrock for processing and interpreting vast datasets generated by both computational and experimental investigations.

As we traverse this scientific odyssey through the myriad computational tools, the synergy between theoretical models and empirical observations not only deepens our understanding of synaptic plasticity but also lays the groundwork for transformative advancements in the realms of cognitive enhancement and therapeutic interventions for neurological disorders. This chapter, thus endeavors to unravel the multifaceted dimensions of synaptic plasticity, poised at the intersection of computational precision and empirical richness.

9.1.1 Importance of Computational Neuroscience in Understanding and Enhancing Synaptic Plasticity

The field of computational neuroscience is essential to improving our knowledge of synaptic plasticity, a fundamental brain function that underpins memory and learning. This multidisciplinary discipline provides unique insights into the intricate workings of neural networks and their plasticity by integrating parts of computer science, physics, mathematics, and neuroscience [8]. Some major contributions made by computational neuroscience are discussed here. Please refer to Figure 9.1 for more information.

9.1.1.1 Modeling Synaptic Processes

Scientists are able to simulate and investigate the complex mechanisms involved in synaptic plasticity through computational models. These models allow for a better understanding of how synaptic alterations occur and are regulated over time. They can represent different levels of complexity, ranging from single synapses to networks of neurons [9].

9.1.1.2 Decoding Neural Codes

A basic aspect of neuroscience is comprehending the encoding and processing of information in brain networks. Deciphering these neural codes is made easier by computational techniques, which are especially useful in comprehending how variations in synaptic strength might impact the general performance of neural networks.

9.1.1.3 Bridging Scales

Different processes are involved in synaptic plasticity, ranging from chemical changes within a single neuron to modifications in large-scale brain networks. By offering instruments to

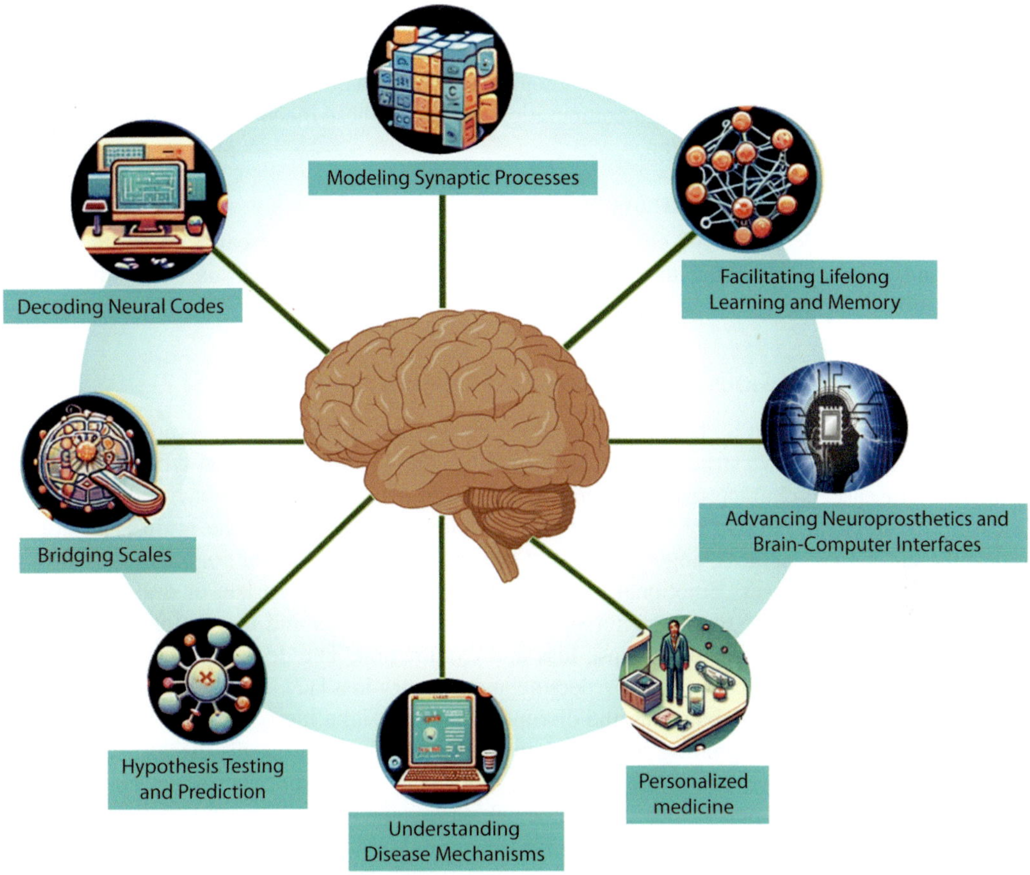

FIGURE 9.1 Importance of computational neuroscience in understanding and enhancing synaptic plasticity.

connect various scales, computational neuroscience contributes to our understanding of how modifications at the molecular and cellular levels might affect network behavior and cognitive processes [9].

9.1.1.4 Hypothesis Testing and Prediction

Researchers can test theories on synaptic plasticity in a safe, virtual setting by using computational models. Researchers can direct laboratory research and lessen the need for considerable trial and error by modifying variables in silico, which allows researchers to predict the results of real-world trials.

9.1.1.5 Understanding Disease Mechanisms

A wide range of neurological conditions include abnormalities in synaptic plasticity. In addition to providing insights into the mechanisms underlying disorders like epilepsy, Parkinson's disease, and Alzheimer's disease, computational neuroscience can assist in mimicking these pathological situations and contribute to the development of novel remedies [10].

9.1.1.6 Personalized Medicine

Computational neuroscience may help personalized medicine methods in neurology and psychiatry by modeling individual changes in synaptic plasticity. This may result in more tailored and efficient treatment regimens for individuals suffering from neurological conditions.

9.1.1.7 Advancing Neuroprosthetics and Brain-Computer Interfaces

Understanding synaptic plasticity computationally is essential for creating brain-computer interfaces and neuroprosthetics [11]. These technologies are dependent on neuroplasticity, the brain's capacity for adaptation, and a thorough computational grasp of these mechanisms can substantially increase their efficacy and integration.

9.1.1.8 Facilitating Lifelong Learning and Memory

Understanding synaptic plasticity's computational foundations may help researchers create lifelong learning and memory-enhancing techniques that could improve education and slow down age-related cognitive decline.

9.1.2 Overview of Various Neurodegenerative Diseases

A class of illnesses known as neurodegenerative diseases are defined by the gradual deterioration of the nervous system's structure and functionality [12]. The most common type of dementia, Alzheimer's disease, is characterized by memory loss, cognitive decline, and behavioral abnormalities. It is linked to the buildup of tau tangles and amyloid-beta plaques in the brain [7]. Patients frequently suffer from sadness, anger, and mood swings, in part because of cognitive impairment and the frustration it causes. Emotional reactions may become inappropriate or dull as the condition worsens. Even before the characteristic amyloid-beta plaques and tangles of tau build-up, Alzheimer's disease is typified by synapse loss and dysfunction. The degradation of synapses in important brain locations causes cognitive decline and memory loss [12]. The disease progresses as a result of neurotransmitter imbalances and disrupted synaptic signaling.

Reduced dopaminergic signaling in the striatum is caused by Parkinson's disease-related loss of dopamine-producing neurons in the substantia nigra [13]. The distinctive motor symptoms of Parkinson's disease are caused by this imbalance, which impairs synaptic communication required for regulated motor movements. Parkinson's disease can cause emotional changes such as melancholy, anxiety, and apathy, in addition to motor symptoms [7]. These may result from the disease's altered neurobiology as well as the psychological effects of living with a chronic, progressive illness.

Huntington's disease is a hereditary condition that results in the progressive degeneration of brain nerve cells, impairing movement, decreasing cognitive function, and causing mental health problems [7]. Significant emotional disruptions brought on by this illness can include despair, impatience, and a diminished capacity to identify the emotional states of others. The psychological burden of knowing that one's offspring may inherit the condition might also be exacerbated by its genetic basis. The huntingtin protein, which is involved in synaptic function, functions abnormally due to this genetic disease. The ensuing synaptic disruption exacerbates Huntington's disease symptoms in the areas of motor, cognitive, and mental health.

9.1.3 A Brief Overview of Synaptic Plasticity in Neurodegenerative Diseases

Synaptic plasticity, a fundamental characteristic of the brain, is the potential of synapses, or the connections between neurons, to become stronger or weaker with time. The dynamic properties of synapses are essential for memory, learning, and the nervous system's general adaptability. The group of diseases known as neurodegenerative diseases are defined by the progressive loss of neuronal structure and function. Huntington's disease, Parkinson's disease, and Alzheimer's disease are examples of common neurodegenerative illnesses [14]. Memory, cognition, and motor function impairments are common outcomes of these disorders.

A significant impact on synaptic plasticity is seen in neurodegenerative disorders. Neural network function declines when there is a reduction in the efficient communication between neurons. One of the main causes of the cognitive and functional deficits linked to these illnesses is this impairment. The most prevalent type of dementia, Alzheimer's disease (AD), is typified by the build-up of tau tangles and amyloid-beta plaques in the brain. Dysfunction and loss of synapses are associated with these pathogenic characteristics. Synaptic plasticity is significantly reduced in AD, especially in areas of the brain related to memory and learning, such as the hippocampus. Parkinson's disease (PD) is generally characterized by impaired motor function as a result of dopaminergic neurons in the substantia nigra dying off. The loss causes changes in the associative and motor circuits of the brain's synaptic plasticity, which in turn causes the motor symptoms and cognitive decline associated with Parkinson's disease (PD). A genetic condition known as Huntington's disease (HD) results in the gradual degeneration of brain nerve cells [14]. A mutation in the huntingtin protein, which is involved in neural communication, is what identifies it. Both motor and cognitive skills are impacted by this mutation's defective synaptic plasticity [15].

To develop viable treatments, it is imperative to comprehend the alterations in synaptic plasticity associated with these disorders. Some of the symptoms of these disorders may be slowed down or even reversed by treatments that work to improve compensatory plasticity or restore synaptic function.

9.2 FUNDAMENTALS OF SYNAPTIC PLASTICITY

The process by which synapses, the points of contact between neurons become stronger or weaker in response to fluctuations in activity is known as synaptic plasticity. This shift in synaptic strength may occur over a short period of time, minutes to hours, or a longer period of time, days to years. The brain's capacity to encode, store, and retrieve information depends on synaptic plasticity. It enables the brain circuits that underpin learning to adjust in response to experiences. For instance, the synaptic connections in the appropriate brain circuits get stronger when you acquire a new skill or memorize information [16].

9.2.1 Biological Mechanisms Enabling Synaptic Plasticity

Five major biological events—induction, signal transduction, expression, maintenance, and modulation—together enable synaptic plasticity. They are discussed below.

9.2.1.1 Induction

The first stage of synaptic plasticity is induction, triggered by specific circumstances or stimuli. This might entail:

Neural Activity: Variations in synaptic strength are caused by specific neuronal firing patterns. For instance, long-term depression (LTD) may be induced by low-frequency stimulation, while long-term potentiation (LTP) is frequently induced by high-frequency stimulation [17].

Biochemical Signals: Important biochemical signals include the activation of receptors on the postsynaptic neuron and the release of neurotransmitters [18]. Glutamate, the main neurotransmitter in excitatory synapses, plays a crucial role.

9.2.1.2 Signal Transduction

Following the receipt of the induction stimuli, a series of intracellular signaling processes take place:

Receptor Activation: Neurotransmitters attach to receptors on postsynaptic neurons, activating them. For example, glutamate triggers responses on NMDA and AMPA receptors.

Calcium Influx: A crucial stage in LTP, the activation of NMDA receptors permits calcium ions to enter the neuron.

Second Messenger Systems: Calcium ions initiate various intracellular processes, acting as second messengers. Important proteins like kinases (e.g., CaMKII) and calmodulin play vital roles [19].

9.2.1.3 Expression

This stage involves the actual changes in synaptic strength:

9.2.1.3.1 For LTP

- **Insertion of AMPA Receptors**: In response to signal transduction pathways, more AMPA receptors are inserted into the postsynaptic membrane, increasing its sensitivity to glutamate.
- **Cytoskeletal Changes**: The structure of the dendritic spine, where the synapse is located, can change, often becoming larger or forming new spines.

9.2.1.3.2 For LTD

- **Removal of AMPA Receptors**: AMPA receptors are internalized from the post-synaptic membrane, decreasing its sensitivity.
- **Spine Shrinkage or Loss**: Dendritic spines may shrink or be eliminated.

9.2.1.4 Maintenance

This phase involves the long-term preservation of changes in synaptic strength:

Protein Synthesis: New proteins are synthesized to consolidate structural changes in the synapse, including locally synthesized proteins at the dendrite and proteins transported from the cell body.

Genetic Changes: Altered gene expression can occur, leading to the long-term stabilization of synaptic modifications [20].

9.2.1.5 Modulation

Synaptic plasticity is modulated by various factors:

Neuromodulators: Substances like dopamine and serotonin can modulate synaptic plasticity, influencing learning and mood.

Neurotrophins: Proteins like brain-derived neurotrophic factor (BDNF) play a crucial role in modulating synaptic plasticity.

Feedback Mechanisms: The neural circuit itself can provide feedback that influences the extent and duration of synaptic changes.

9.2.2 Types of Synaptic Plasticity

Synaptic plasticity exhibits diverse forms based on various criteria, encompassing duration, mechanism, and function [21]. The following classification provides an in-depth understanding of the major types of synaptic plasticity:

9.2.2.1 Long-Term Potentiation (LTP)

Long-term potentiation is characterized by a sustained increase in synaptic strength, typically following high-frequency synapse stimulation. The mechanisms involved in LTP encompass heightened neurotransmitter release, the insertion of additional receptors, notably AMPA receptors, into the postsynaptic membrane, and structural alterations like the development of dendritic spines. LTP plays a pivotal role in learning and memory formation [22].

9.2.2.2 Long-Term Depression (LTD)

In contrast to LTP, long-term depression involves a sustained reduction in synaptic strength, commonly triggered by low-frequency stimulation. The mechanisms underlying LTD include modifications in neurotransmitter release and the removal of receptors from the postsynaptic membrane. LTD contributes to synaptic pruning, memory erasure, and the prevention of neuronal circuit overexcitation.

9.2.2.3 Short-Term Plasticity

Short-term plasticity encompasses two distinctive types:

Facilitation: This involves the transient enhancement of neurotransmitter release, typically induced by the lingering presence of calcium ions in the presynaptic cell.

Depression: A transient loss of synaptic strength, often initiated by the depletion of readily available neurotransmitter vesicles in the presynaptic neuron.

9.2.2.4 Homeostatic Synaptic Plasticity

Homeostatic synaptic plasticity adjusts synaptic strengths in response to broad fluctuations in brain activity, contributing to the overall stability of the neural network. This form of plasticity often necessitates global alterations in synaptic effectiveness to maintain neural activity within a functional range [23].

9.2.2.5 Spike-Timing-Dependent Plasticity (STDP)

In the form of plasticity known as STDP, the timing of action potentials, or spikes, in both postsynaptic and presynaptic neurons controls the occurrence of LTD or LTP. LTP is more likely if the presynaptic neuron fires just prior to the postsynaptic neuron; LTD is more likely if the order is reversed.

9.2.2.6 Metaplasticity

Metaplasticity represents the plasticity of synaptic plasticity itself. The concept of "metaplasticity" elucidates how past synaptic activity can alter the threshold for triggering LTD or LTP. This regulatory mechanism prevents excessive synaptic strengthening or weakening, maintaining a delicate balance in neural circuitry.

9.2.3 Role of Synaptic Plasticity in Maintaining Cognitive Functioning

The preservation and control of cognitive functioning depend heavily on synaptic plasticity. It serves as the molecular foundation for memory and learning and is necessary for many different facets of cognitive processing. In response to new experiences, synaptic connections can become stronger thanks to synaptic plasticity. It is thought that procedures like long-term potentiation (LTP) are essential for encoding and retaining new information. Synaptic alterations contribute to the stabilization and consolidation of memories, converting transient memories into permanent ones [13,23].

Cognitive flexibility is greatly aided by the brain's ability to adapt to novel circumstances and surroundings thanks to synaptic plasticity. Modifying synaptic connections facilitates more effective information processing and integration, which is essential for sophisticated cognitive functions like decision-making and problem-solving. The brain is more resilient against age-related cognitive loss because of synaptic plasticity. Even in the face of aging-related neuronal alterations, cognitive functions are preserved through ongoing synaptic remodeling [24]. Synaptic plasticity in brain damage situations permits neural network reorganization, which aids in compensating processes that aid in cognitive function recovery.

Fine-tuning sensory perceptions, responding to changes in the sensory environment, and focusing on pertinent stimuli are made possible by synaptic alterations in the brain's sensory areas. Coherent perception requires the integration of information from several sensory modalities, which plasticity facilitates. Language learning and development, particularly in the early years of life, depend on the establishment and maturation of synaptic connections. The brain networks responsible for comprehending and generating language, which is a crucial aspect of human communication, are supported by synaptic plasticity. Emotional learning and memory are facilitated by the synaptic connections found in limbic regions, such as the hippocampus and amygdala. These connections influence the development and modification of emotional responses. Certain brain regions have plasticity that influences emotional and psychological resilience by facilitating stress adaptation and coping [24–26].

9.2.4 Link Between Synaptic Plasticity and Neurodegenerative Disease Progression

It is becoming more well-acknowledged that a key component in comprehending neurodegenerative illnesses is the relationship between synaptic plasticity and their progression [12]. Synaptic plasticity, or the capacity of synapses to become stronger or weaker over time, is essential for memory, learning, and the nervous system's general adaptability. Disruptions in synaptic plasticity are considered a major factor in the development and course of neurodegenerative disorders [2]. For example, in Alzheimer's disease, amyloid plaques and neurofibrillary tangles frequently develop before synapse dysfunction, indicating that early alterations in synaptic health may be a major cause of the disease [7]. Similar to this, Parkinson's disease is characterized by motor symptoms resulting from abnormal basal ganglia synaptic activity brought on by the death of dopaminergic neurons [13].

The cognitive and motor deficiencies associated with these disorders can be attributed to a disturbance in brain network function caused by these synaptic alterations, which include the loss of synaptic connections and the impairment of synaptic signaling [2]. Thus, the increasing loss of synaptic plasticity is intimately associated with the advancement of neurodegenerative disorders. This insight has caused a shift in the focus of both research and treatment approaches from the conventional strategy of focusing on neuron loss to one of maintaining or restoring synaptic health. By comprehending and addressing the synaptic components of these illnesses, a new avenue in the therapy of neurodegenerative diseases may be opened up for the purpose of reducing or even stopping their course.

9.3 COMPUTATIONAL MODELS OF SYNAPTIC PLASTICITY

9.3.1 An Overview of Computational Neuroscience and Its Relevance to Synaptic Plasticity

It is due to computational models of neuroplasticity that we are able to comprehend the brain's capacity for self-organization and adaptation. These models provide a system-level knowledge that connects the dots between small-scale cellular changes and large-scale behavioral consequences by simulating the intricate molecular and cellular mechanisms underpinning synaptic changes. Computational models provide an extensive understanding of how synaptic adjustments translate into alterations in neural circuits and brain function overall by enabling precise modeling of neurotransmitter interactions, receptor dynamics, intracellular signaling, and gene expression. The capacity of computational neuroplasticity models to provide prediction and hypothesis testing is one of its main advantages. By adjusting model parameters and seeing the results, researchers can investigate a variety of "what-if" situations and forecast the consequences of particular synapse alterations [27]. This ability to forecast outcomes informs experimental research and speeds up the process of uncovering new information about how the brain works [28]. These models play a crucial role in connecting alterations at the molecular and cellular levels to behavioral and cognitive outcomes. They aid in the comprehension of the ways in which bigger brain networks and cognitive functions are affected by molecular changes. Given the prevalence of neuroplasticity disruptions in neurological and psychiatric illnesses, this is especially pertinent to research in these areas. These disorders can be simulated by computational models, which can provide insight into their mechanisms and provide possible targets for treatment.

Computational models have the potential to be customized for each patient in the field of personalized medicine, providing unique insights into brain illnesses. Furthermore, these models are useful teaching aids that assist researchers and students in understanding the intricate workings of neuroplasticity. Computational models improve the focus and efficiency of laboratory work by forecasting experimental results and directing experimental design [29].

9.3.2 Simulation Models of Synaptic Plasticity

Synaptic plasticity simulation models are computer programs that simulate the physiological activities taking place at nervous system synapses. From intricate biophysical models of individual synapses to network-level simulations that depict the collective behavior of huge neuronal populations, these models range widely in complexity and breadth. Numerous simulation models are available for synaptic plasticity, each accommodating varying degrees of biological detail and scale. They are essential tools in neuroscience, assisting researchers in their exploration of how complex behaviors and cognitive functions might result from alterations in synapses, as well as the fundamental principles of learning and memory [30]. The most prominent simulation models are discussed in the text below.

9.3.2.1 Hodgkin-Huxley Model

The Hodgkin-Huxley model is not a synaptic plasticity model; rather, it does offer a vital foundation for comprehending and simulating the electrical components of synaptic transmission and plasticity due to its thorough explanation of neuronal action potential mechanisms. Based on Alan Hodgkin and Andrew Huxley's groundbreaking research on the giant squid axon, this model was developed in 1952 and depicts the electrical properties of excitable cells, such as neurons. Figure 9.2 represents the Hodgkin-Huxley model.

To explain how these ion channels open and close in response to changes in the membrane potential and how this impacts the flow of ions across the membrane, resulting in the formation of action potentials, the model makes use of a set of four nonlinear differential equations. Not only did Hodgkin and Huxley discover the functions of Na^+ and K^+ ions in brain activity, but they also painstakingly measured their motions and how they affected the electrical state of the neuron. A neuron's internal membrane potential remains negative when it is at rest, mostly because of K^+ ion leakage out of the cell, which is somewhat offset by Na^+ ion leakage inside and sodium-potassium pump activity. Voltage-gated Na^+ channels open to let Na^+ ions enter a neuron in response to a stimulation potent enough to depolarize the membrane. More depolarization results from this inward flow of positive ions, which opens more Na^+ channels in a positive feedback loop. This quickly raises the membrane potential and produces the action potential's rising phase.

When K^+ channels open and N^+ channels begin to inactivate, the action potential peaks and the repolarization process begins. Hyperpolarization is the outcome of the membrane potential being

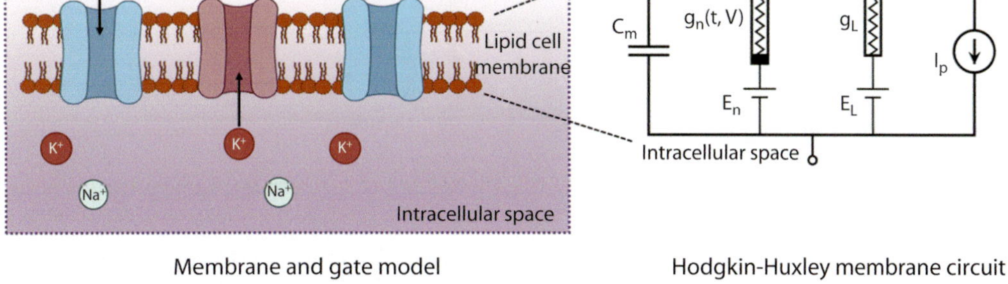

FIGURE 9.2 Hodgkin-Huxley model.

even more negative than its resting state due to the outflow of K^+ ions, which persists even after the membrane potential crosses back to a negative value. The neuron is either less sensitive or incapable of firing another action potential during this time and the refractory phase that follows [31].

The significance of the Hodgkin-Huxley model goes much beyond its direct results. It provided a framework for comprehending how neurons communicate and process information, laying the foundation for the entire science of neurophysiology. This model has played a key role in directing decades of neuroscience research, offering insights into the functions of different kinds of neurons and ion channels, and serving as the foundation for more sophisticated computational models in the field. The Hodgkin-Huxley model is a fundamental tool in the study of neuronal excitability and signaling because its principles have been modified and expanded to examine a broad variety of neurons and their behaviors. With the help of this model, we are better able to comprehend not only the electrical activity of neurons but also how this activity underlies sophisticated mental processes.

9.3.2.2 Integrate-and-Fire Models

A mainstay of computational neuroscience, integrate-and-fire models provide a straightforward but effective framework for comprehending synaptic plasticity and neuronal dynamics. Remove the intricate biophysical specifics of ion channel behavior that define models such as the Hodgkin-Huxley model, and integrate-and-fire models concentrate on the basic mechanism by which neurons combine synaptic inputs and produce action potentials. The neuron's membrane potential, which combines inputs from several synapses, is a crucial variable in these models. These inputs are commonly modeled as currents or changes in membrane potential. They can be excitatory or inhibitory. The membrane potential of the neuron rises in response to synaptic inputs until it exceeds a certain threshold. A spike is produced by the neuron, which is said to "fire" after this threshold is crossed. After this spike, the neuron goes into a refractory period, during which it is momentarily unable to fire again, and the membrane potential is restored to a baseline level [32]. Figure 9.3 shows the details of the leaky integrate-and-fire model.

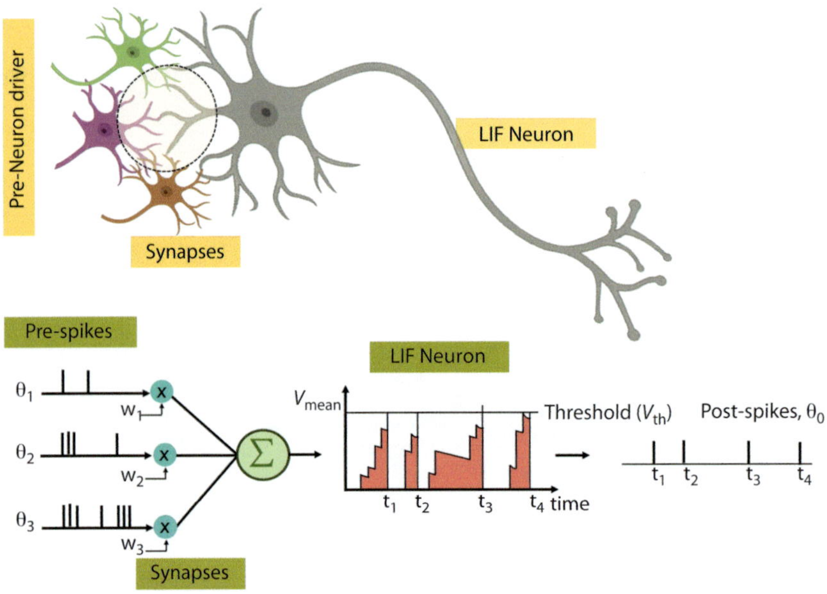

FIGURE 9.3 Schematic diagram of a leaky integrate-and-fire (LIF) neuron model. In the upper part, it shows the connections from other input neurons to an LIF neuron. The lower part shows the integration of the spikes from input neurons. The LIF neuron fires an action potential (spike) when the integration of these pre-spikes reaches a threshold. After firing, the membrane potential undergoes a reset to a resting potential.

Integrate-and-fire models are particularly good at including synaptic plasticity rules like spike-timing-dependent plasticity (STDP) or Hebbian learning. These criteria enable the model to simulate key components of the brain's learning and memory processes by modifying the strength of synapses in response to the timing and patterns of neuronal spikes. Because of their simplicity, these models are perfect for modeling massive networks of neurons, allowing scientists to investigate emergent characteristics of neural circuits and collective behaviors that would be computationally impractical with more sophisticated models.

These models, which show how complex patterns of brain activity and information processing can emerge from the integration of synaptic inputs and the creation of spikes, have proven essential in breaking down the fundamentals of neural computation. They act as a link between the macroscopic events of neural rhythms and oscillations seen in the brain and the microscopic world of molecular and synaptic alterations [33].

9.3.2.3 Spiking Neural Network Models

In computational neuroscience, spiking neural network (SNN) models offer a comprehensive method for comprehending synaptic plasticity. These models are unique in that they capture the temporal dynamics of actual neural networks, which makes them very useful for researching the intricate processes of learning and synaptic alterations in the brain. The theory that neurons communicate by releasing spikes, or action potentials, which are distinct occurrences in time, lies at the heart of SNN models. SNNs embrace the biological reality of neurons firing in an all-or-nothing fashion, in contrast to standard artificial neural networks that depict neuronal activation with continuous values. SNNs may now process and transfer information more like real neurons thanks to this technique. The capacity of SNNs to model spike-timing-dependent plasticity (STDP), a type of synaptic plasticity in which the strength of synaptic connections is influenced by the timing of spikes between presynaptic and postsynaptic neurons, is one of its important characteristics. The foundation of STDP is the idea that a synapse is strengthened (LTP) if a presynaptic neuron spikes before a postsynaptic neuron spikes, and weakened (LTD) if the order is reversed. SNN models neatly reflect this timing-dependent mechanism, which is essential for learning and memory formation in biological systems [1].

SNNs have been used in numerous branches of artificial intelligence and neuroscience study. They shed light on how learning and memory develop at the network level, how complicated computations may be carried out by brain circuits, and how abnormalities in synaptic plasticity might result in neurological conditions. In the field of artificial intelligence, SNNs provide inspiration for new machine learning methods, especially in domains like time series prediction and speech recognition that demand the processing of temporal data.

9.3.2.4 Deep Learning Models for Synaptic Plasticity

Synaptic plasticity is being studied more and more with deep learning models, a subset of machine learning, especially when combined with fMRI (functional magnetic resonance imaging) and EEG (electroencephalography) data. These models possess the capability to decipher intricate neurological patterns and derive significant insights concerning synaptic alterations in the brain [7].

Deep learning models are able to examine brain electrical activity in the context of EEG and detect patterns linked to synaptic plasticity. The electrical impulses derived from neural activity are reflected in EEG data, which can be complex and multidimensional. EEG recording parsing is a good fit for deep learning algorithms because of its capacity to handle big datasets and spot hidden patterns in the information. They can assist in identifying patterns that may indicate a strengthening or weakening of the synapses [13].

Similar to this, deep learning models are employed in fMRI research to examine blood oxygen level-dependent (BOLD) data, offering insights into the functional connectivity of the brain and the areas responsible for synaptic plasticity. These models are able to clarify the intricate connections between various brain regions as well as the ways in which variations in synaptic strength affect brain activity as a whole.

Deep learning in conjunction with fMRI and EEG data provides a promising way to comprehend synaptic plasticity's dynamic nature. This method improves our understanding of fundamental brain processes and may have therapeutic uses in the diagnosis and treatment of disorders involving dysregulated synaptic plasticity, such as neurodegenerative illnesses.

9.3.3 Advantages and Limitations of Computational Models

Synaptic plasticity may be effectively studied using computational models, which yield important new information on how the brain develops and changes over time. Like every scientific instrument, they do have some benefits and drawbacks, though.

9.3.3.1 Advantages

The advantages of computational models for studying synaptic plasticity are mentioned below:

Complex System Analysis: Computational models facilitate the analysis of intricate systems that are challenging to study experimentally. They can integrate extensive data and simulate interactions across different scales, from molecules to neural networks.

Hypothesis Testing and Exploration: These models allow researchers to test hypotheses regarding synaptic plasticity mechanisms in a controlled, virtual environment. The ability to explore potential outcomes before conducting experiments in the lab can save time and resources.

Bridging Different Levels of Understanding: Computational models can establish connections between molecular and cellular-level changes in synaptic plasticity and higher-level phenomena such as learning and memory. This aids in comprehending how synaptic changes impact overall brain function.

Predictive Power: These models can predict the behavior of neural systems under various conditions. This predictive capability is particularly valuable in neuroscience, where direct experimentation may pose challenges.

Educational Utility: Computational models serve as effective educational tools, offering students and newcomers to the field an intuitive understanding of synaptic plasticity and neural processes.

9.3.3.2 Limitations

The limitations of computational models in the study of synaptic plasticity are mentioned below:

Oversimplification: To make the models computationally tractable, certain simplifications and assumptions are often necessary. This can lead to models that may not fully capture the complexity of biological systems.

Dependency on Quality of Data: The accuracy of computational models is heavily dependent on the quality and completeness of the input data. Incomplete or inaccurate data can lead to misleading conclusions.

Computational Resource Intensity: Detailed models, especially those simulating large networks or complex dynamics, can require significant computational resources, limiting their accessibility and scalability.

Gap between Model and Reality: There's often a gap between the model's predictions and real-world biological phenomena [34]. This gap can be due to the inherent differences between a simplified, virtual environment and the complex, noisy biological reality.

Interpretation Challenges: The results generated by computational models can sometimes be difficult to interpret, requiring a deep understanding of both the model and the biological system.

Dynamic Biological Processes: Biological systems are dynamic and constantly changing, which can be challenging to accurately represent in a static model [35,36].

9.4 TECHNIQUES FOR STUDYING SYNAPTIC PLASTICITY

Studying synaptic plasticity encompasses a range of methods from behavioral neuroscience to molecular biology. To measure synaptic responses in neurons and track changes in synaptic strength in response to stimuli, electrophysiological techniques like patch-clamp recording are essential [37]. This method can be used in in vivo and in vitro environments (brain slices, for example). Another effective technology is calcium imaging, which makes it possible to see the calcium fluxes inside neurons, which are a sign of synaptic activity. Molecular biology methods such as immunohistochemistry and Western blotting facilitate the detection and measurement of proteins, such as receptors and signaling molecules, that are involved in synaptic plasticity. Furthermore, the study of synaptic plasticity has been completely transformed by optogenetics, a technique that employs light to regulate neurons genetically altered to express light-sensitive ion channels. This approach allows for precise manipulation of specific neural circuits learning, memory, and behavior, providing a holistic view of synaptic plasticity [38].

9.4.1 In-Silico Approaches

The use of in-silico methods has become essential in the research of neuroplasticity because they provide a window into intricate brain functions that are difficult to investigate through experimentation. These computational approaches cover a wide spectrum of methodologies, ranging from large-scale neural network modeling to intricate simulations of molecular dynamics in synaptic contacts. They make it possible to investigate and forecast how small-scale alterations in synapses affect larger-scale brain processes like memory and learning. Through the utilization of machine learning and data mining techniques, in-silico models examine extensive and complex datasets, revealing the mechanisms and patterns that underlie neuroplasticity. These virtual techniques are becoming more and more important as computational resources and algorithms advance, enhancing traditional lab-based research and offering fresh perspectives on how the brain functions [39].

The two most common approaches, the molecular dynamic simulation and the Monte Carlo simulation, are discussed in the text below.

9.4.1.1 The Molecular Dynamic Simulation

Molecular dynamics (MD) simulation is a potent computational method that has gained importance in the molecular research of synaptic plasticity. With this method, scientists can watch the intricate dynamics of chemical interactions inside and outside synapses by emulating the physical motions of atoms and molecules over time.

MD simulations can be used to investigate the interactions and temporal changes of different proteins, neurotransmitters, and receptors involved in synaptic transmission in the setting of synaptic plasticity. Simulations, for example, can shed light on the interactions between neurotransmitter molecules, such as glutamate, and receptors, such as NMDA or AMPA receptors, and how these interactions cause the receptors' conformations to alter, affecting ion channel activity.

Furthermore, understanding the function of intracellular signaling molecules and their cascades—which are essential for processes like long-term depression (LTD) and long-term potentiation (LTP)—is made possible by MD simulations. These simulations can demonstrate how certain kinases and phosphatases are activated by calcium ions that enter through NMDA receptors, changing the phosphorylation states of proteins and ultimately the synaptic strength.

The capacity of MD simulations to capture the dynamic and frequently fleeting interactions at the atomic level—interactions that may be difficult to perceive through experimental methods—is one of its main advantages in the study of synaptic plasticity. This degree of specificity can be useful in developing precise models of synapse activity and in forecasting how medicines, mutations, and other changes would affect synaptic behavior.

It's crucial to remember, though, that the quality of the original model and the available computer power have a significant impact on how accurate MD simulations are. Although these simulations provide deep insights, it is better to use them in conjunction with experimental evidence to confirm and deepen our understanding of synaptic plasticity.

9.4.1.2 Monte Carlo Simulations

An invaluable resource for comprehending synaptic plasticity is the class of computational algorithms known as Monte Carlo simulations, which use repeated random sampling to get numerical results. This method works especially well in situations involving complicated systems and uncertainty—qualities that are inherent in biological brain networks. Monte Carlo simulations can shed light on the stochastic aspects of synapse modification and the stochastic nature of synaptic transmission in the study of synaptic plasticity.

Investigating the probability release of neurotransmitters at synaptic terminals is a crucial use of Monte Carlo simulations in synaptic plasticity. The release of neurotransmitters is an essentially random process that is impacted by a number of variables, such as the concentration of calcium ions and the availability of vesicles. Numerous scenarios, each representing a potential state of a synapse during transmission, can be simulated using Monte Carlo methods. Researchers can obtain a statistical understanding of neurotransmitter release patterns and how they affect synapse strength by combining different instances.

Furthermore, the kinetics of receptor activation and ion channel behavior in response to neurotransmitters are studied using Monte Carlo simulations. The simulations can assist in forecasting the possibility of various outcomes under variable settings, like changes in neurotransmitter concentration or receptor sensitivity, because these molecular interactions are equally probabilistic. Understanding the mechanisms underpinning synaptic plasticity phenomena, such as long-term depression (LTD) and long-term potentiation (LTP), where the strength of synaptic connections is altered, is very important.

The capacity of Monte Carlo simulations to simulate the cumulative impact of synaptic alterations over time is another useful feature. This ability is essential for comprehending the mechanisms involved in learning and memory. They can model the way that discrete, probabilistic alterations in synapses result in substantial behavioral shifts in brain networks. In large-scale neural network models, where synaptic interactions are extremely complicated and variable, this is especially crucial.

Monte Carlo simulations are not without restrictions, despite their benefits. The quality of the underlying model and the quantity of simulations conducted have a significant impact on the accuracy of the findings. Furthermore, these simulations can be computationally demanding, needing a significant amount of computing power for intricate synapse models or complicated networks.

9.4.1.3 Agent-Based Modeling

An innovative and effective method for researching synaptic plasticity is agent-based modeling (ABM), which enables the simulation of intricate interactions at the level of individual neurons and synapses. In ABM, every neuron or synapse is represented as a separate "agent" with its own set of rules dictating interactions and behavior. This approach is especially helpful for investigating how emergent behaviors at the network level can originate from local modifications at the synapse level. By modifying the strength of synaptic connections in response to particular activity patterns, ABMs can mimic the processes of long-term depression (LTD) and long-term potentiation (LTP), which are important features of synaptic plasticity. They may also take into account elements like receptor dynamics, synaptic tagging, and neuromodulators. Through the use of ABMs, scientists can learn more about the collective dynamics of brain networks and how a complex interaction of multiple synaptic events leads to the processes of learning and memory. This method is particularly useful for determining how changes in synaptic plasticity affect neurological conditions and for creating focused treatments.

9.4.2 In Vitro Techniques

In vitro techniques enable researchers to examine synaptic function in a controlled environment and offer vital insights into the mechanisms underlying synaptic plasticity. These investigations mostly use electrophysiological techniques, like patch-clamp recording, which allow the assessment of synaptic potentials and currents in individual neurons or brain slices. Long-term depression (LTD) induction and long-term potentiation (LTP) in brain slices are two methods that are useful in figuring out the cellular and molecular underpinnings of learning and memory. Additionally, to examine alterations in protein expression, receptor trafficking, and intracellular signaling pathways connected to synaptic modifications, biochemical tests, immunocytochemistry, and fluorescence imaging, including calcium imaging, are utilized. These in vitro methods provide an in-depth understanding of the dynamic alterations that take place at synapses, which is essential for understanding the intricate mechanisms that underlie synaptic plasticity.

9.4.2.1 Electro Psychological Recordings

When examining synaptic plasticity, electrophysiological recordings are essential because they provide a clear understanding of how synaptic neurons interact and change over time. Postsynaptic responses to synaptic activity can be measured by researchers using methods like as patch-clamp and field potential recordings in brain slices or cultured neurons. These techniques are crucial for studying basic mechanisms that underlie learning and memory, including long-term depression (LTD) and long-term potentiation (LTP). Electrophysiological approaches enable the precise investigation of synaptic efficacy, the function of various neurotransmitter receptors, and the involvement of certain ion channels in synaptic plasticity by applying controlled electrical stimuli and monitoring the resulting synaptic responses.

9.4.2.2 Calcium Imaging

Since calcium ions play a crucial role in neural signaling, calcium imaging is a powerful tool for researching synaptic plasticity. This technique visualizes the dynamics of calcium within neurons by utilizing fluorescent dyes or calcium markers that have been genetically encoded. Calcium imaging provides real-time imaging of the fluctuations in calcium levels that occur within neurons as a result of synaptic alterations or input. Researchers can use this method to see how different neural networks or individual neurons exhibit different patterns of brain activity. Scientists can determine the strength and patterns of synaptic connections by examining these calcium signals, which provides important information about the molecular mechanisms behind learning and memory [18].

9.4.2.3 Optogenetics for Precise Modulation

With optogenetics, neuronal activity may be modulated with an unprecedented level of accuracy, revolutionizing the study of synaptic plasticity. This novel method entails genetically engineering neurons to develop light-sensitive ion channels or pumps, which can be stimulated or inhibited in response to particular light wavelengths. Because of this degree of control, scientists are able to directly investigate the causal connections between neural activity and synaptic alterations by modifying neuronal activity in real time.

Optogenetics can be used to silence or activate certain neurons or brain circuits in the context of synaptic plasticity, allowing researchers to watch the changes in synaptic strength that occur. Deciphering the mechanisms behind long-term depression (LTD) and long-term potentiation (LTP), the neuronal correlates of memory and learning will benefit greatly from this. Scientists can alter synaptic connections and see how these changes affect information processing, brain network behavior, and synaptic connectivity by activating specific neurons.

Optogenetics has applications in synaptic plasticity studies that go beyond fundamental science. It may have therapeutic uses, providing methods for fixing damaged neuronal circuits in mental illnesses or regaining synaptic function in neurodegenerative diseases. With continued advancements, optogenetic technology holds great potential for expanding our comprehension of the intricate processes of synaptic plasticity and providing novel therapeutic approaches for brain illnesses.

9.4.3 In Vivo Imaging

By allowing researchers to examine synaptic plasticity directly in living brains, in vivo imaging methods have opened up new avenues of inquiry and provided a window into the dynamic mechanisms underlying memory and learning. Two-photon microscopy is one technique that makes it possible to view synaptic architecture and functions in real time in an intact brain [40]. Researchers can watch changes in synaptic strength and the creation, removal, and remodeling of synapses by using fluorescent markers. This method is especially useful for comprehending how synaptic plasticity affects the entire brain circuit and behavior, providing information on how experiences and external circumstances affect the neural connections in a living thing.

9.4.3.1 Functional Magnetic Resonance Imaging (fMRI)

Particularly in the context of the human brain, functional magnetic resonance imaging (fMRI) has emerged as a crucial tool for comprehending synaptic plasticity at the macroscopic level. Despite not being able to measure synaptic activity directly, fMRI can identify alterations in brain blood flow and oxygenation levels, which are intimately related to neural activity. The hemodynamic response is the term used to describe the increase in blood flow that occurs when a particular region of the brain shows enhanced synaptic activity. fMRI maps these alterations to reveal which brain regions are active during particular cognitive functions, learning procedures, or memory retrieval. With the use of this technology, neuronal networks and pathways linked to different types of learning and plasticity have been identified. It also provides a non-invasive tool to investigate the ways in which experiences, training, and rehabilitation affect brain function [3]. Our knowledge of the neurological correlates of synaptic plasticity has greatly benefited from fMRI research, which shows how the brain reorganizes itself in response to learning new abilities, changing environments, and new information.

9.4.3.2 Two-Photon Microscopy

Because two-photon microscopy can scan deep within the living brain with little injury to the tissue, it has become an invaluable tool for understanding synaptic plasticity. By using two lower-energy photons rather than one high-energy photon to ignite fluorescent molecules, this method allows for the detailed visualization of synapse and neural structures. The primary benefit of two-photon microscopy over traditional fluorescence microscopy is its ability to reach deeper into brain tissue while minimizing photobleaching and phototoxicity.

Two-photon microscopy provides a dynamic perspective of synaptic plasticity by allowing researchers to watch the creation, removal, and alteration of synapses in real time. Dendritic spines, which are tiny projections on neurons where synaptic connections are created and altered, may be tracked thanks to it. This method is particularly useful for researching how disease processes, learning, and memory impact synaptic connections and architecture in living animals. Because of its depth penetration and great resolution, two-photon microscopy is a vital technique for furthering our understanding of the mechanisms underlying synaptic plasticity in the intact brain.

9.4.3.3 Real-Time Monitoring of Synaptic Dynamics

Monitoring synaptic dynamics in real time is essential to comprehending the basic mechanisms behind memory, learning, and brain plasticity. Technological developments in imaging and recording have greatly improved our ability to watch synaptic alterations in real time. By visualizing synaptic features like dendritic spines in living neurons, techniques such as two-photon fluorescence microscopy make it possible to monitor the development, removal, and structural remodeling of synapses in real time. This technique offers a window into the dynamic nature of synaptic connections within intact brain circuits in conjunction with the use of fluorescent synaptic markers.

Concurrently, high temporal resolution synaptic response measurements have been made possible by the development of electrophysiological methods like patch-clamp recording. Rapid variations in synapse strength, such as those that take place during long-term potentiation (LTP) or depression (LTD), can be detected using these techniques.

In addition, novel approaches such as optogenetics have created new opportunities for the real-time control and observation of synaptic activity. Researchers can cause synaptic alterations and observe the results instantly by manipulating neuronal activity with light, establishing causal relationships between neuronal activity and synaptic plasticity [13].

9.5 SOFTWARE TOOLS FOR ANALYZING BRAIN DATA

9.5.1 Electrophysiological Signal Processing Tools

9.5.1.1 EEGLAB

Description: EEGLAB is an open-source environment for electrophysiological signal processing. It provides an interactive MATLAB toolbox for visualization, artifact removal, and analysis of EEG, MEG, and ECoG (ElectroCorticoGraphy) data.
Supported Data Types: EEG, MEG, ECoG
License and Availability: Freely available, Open source
Website: https://sccn.ucsd.edu/eeglab/index.php
Unique Features: Interactive MATLAB toolbox for signal processing, artifact removal tools, analysis of EEG, MEG, and ECoG data.
User Type: Neuroscientists, EEG researchers

9.5.1.2 Chronux

Description: Chronux is a MATLAB library that preprocesses, explores, and analyzes neural data, including point process and continuous data.
Supported Data Types: Point process, continuous data
License and Availability: Freely available, Open source
Website: http://chronux.org/
Unique Features: Time-frequency analysis, neural data exploration
User Type: Neuroscientists, Data Analysts

9.5.2 Neuroimaging Analysis Tools

9.5.2.1 SPM (Statistical Parametric Mapping)

Description: SPM deals with EEG, fMRI, MEG, PET, and SPECT data sequences. It is widely used for statistical parametric mapping and supports multiple neuroimaging data types.
Supported Data Types: EEG, fMRI, MEG, PET, SPECT
License and Availability: Freely available, Open source
Website: https://www.fil.ion.ucl.ac.uk/spm/
Unique Features: Statistical parametric mapping, PET and SPECT support
User Type: Researchers, Clinicians, Neuroscientists

9.5.2.2 FSL (FMRIB Software Library)

Description: FSL is a library of tools for analyzing EEG, MRI, and fMRI data. It provides advanced analysis tools and supports various neuroimaging modalities.
Supported Data Types: EEG, MRI, fMRI

License and Availability: Freely available, Open source
Website: https://fsl.fmrib.ox.ac.uk/fsl
Unique Features: Advanced analysis tools, Connectivity analysis
User Type: Researchers, Neuroscientists

9.5.2.3 BrainVoyager QX

Description: BrainVoyager QX is a tool that handles MRI, EEG, and MEG data. It incorporates statistical, numerical, and image processing tools, requiring a HASP license for access.
Supported Data Types: MRI, EEG, MEG
License and Availability: Requires HASP license, Open source
Website: https://www.brainvoyager.com/
Unique Features: Real-time analysis, Advanced visualization
User Type: Neuroimaging Researchers, Clinicians

9.5.2.4 Turbo BrainVoyager

Description: Turbo BrainVoyager is real-time processing software designed for fMRI data. It operates online and necessitates a HASP license for usage.
Supported Data Types: fMRI
License and Availability: Requires HASP license, Open source
Website: https://www.brainvoyager.com/
Unique Features: Online real-time processing
User Type: Neuroimaging researchers

9.5.2.5 MRIcron

Description: MRIcron serves as an image viewing toolbox specifically for MRI data. It supports multi-planar views and comes with built-in statistical functions.
Supported Data Types: MRI
License and Availability: Freely available, Open source
Website: https://people.cas.sc.edu/rorden/mricron/index.html
Unique Features: Multi-planar views, Statistical functions
User Type: Researchers, Clinicians

9.5.2.6 FreeSurfer

Description: FreeSurfer is dedicated to reconstructing the brain surface from MRI data. It additionally overlays fMRI data on the reconstructed surface.
Supported Data Types: MRI, fMRI
License and Availability: Freely available, Open source
Website: https://surfer.nmr.mgh.harvard.edu/
Unique Features: Brain surface reconstruction, fMRI overlay
User Type: Neuroimaging researchers, Clinicians

9.5.3 Medical Image Segmentation Tools

9.5.3.1 itk-SNAP

Description: itk-SNAP is a 3D medical image segmentation software. It handles MRI and CT scan data.
Supported Data Types: MRI, CT scan
License and Availability: Freely available, Open source
Website: http://www.itksnap.org/

Unique Features: Intuitive segmentation, CT scan support
User Type: Researchers, Medical imaging professionals

9.5.4 Statistical and Graphical Visualization Tools

9.5.4.1 Bioelectromagnetism MATLAB Toolbox

Description: The Bioelectromagnetism MATLAB Toolbox aids in the visualization and measurement of event related potential (ERP). It accepts EEG and MRI data.
Supported Data Types: EEG, MRI
License and Availability: GNU General Public License
Website: https://eeg.sourceforge.net/bioelectromagnetism.html
Unique Features: Event-related potential measurement, EEG and MRI visualization
User Type: Neuroscientists, EEG researchers

9.5.4.2 MNE (Magnetoencephalography) Python

Description: MNE provides pre-processing tools and data conditioning utilities specifically for MEG and EEG data.
Supported Data Types: MEG, EEG
License and Availability: Freely available, Open source
Website: https://mne.tools/stable/index.html
Unique Features: Advanced pre-processing, Data conditioning
User Type: Neuroscientists, EEG and MEG researchers

9.5.5 Functional Connectivity Analysis Tools

9.5.5.1 eConnectome

Description: eConnectome is a MATLAB package designed for imaging brain functional connectivity. It handles EEG, MEG, and ECoG data.
Supported Data Types: EEG, MEG, ECoG
License and Availability: GNU General Public License, Open source
Website: https://www.nitrc.org/projects/econnectome
Unique Features: Functional connectivity analysis, Multi-modal data support
User Type: Neuroimaging researchers, Connectivity analysts

9.5.5.2 BrainNet Viewer

Description: BrainNet Viewer is a visualization tool focused on constructing both structural and functional networks from filtered or processed data.
Supported Data Types: Filtered or processed data
License and Availability: GNU General Public License
Website: https://www.nitrc.org/projects/bnv/
Unique Features: Network construction, Structural and functional views
User Type: Neuroscientists, Connectivity analysts

9.5.5.3 CONN

Description: CONN is a MATLAB toolbox designed for computation, display, and fMRI data analysis, with a focus on functional connectivity analysis and network visualization.
Supported Data Types: fMRI
License and Availability: MIT/X Consortium License
Website: https://web.conn-toolbox.org/
Unique Features: Functional connectivity analysis, Network visualization
User Type: Neuroscientists, Connectivity analysts

9.5.6 Simulation and Connectivity Manipulation Tools

9.5.6.1 The Virtual Brain

Description: The Virtual Brain simulates brain behavior by manipulating brain connectivity and associated network parameters.

Supported Data Types: N/A

License and Availability: Freely available, Open source

Website: https://www.thevirtualbrain.org/tvb/zwei

Unique Features: Brain behavior simulation, Network manipulation

User Type: Neuroscientists, Computational modelers

9.6 CHALLENGES IN STUDYING AND MODULATING SYNAPTIC PLASTICITY

Studying and modulating synaptic plasticity presents a multitude of challenges, stemming from both the inherent complexity of neural mechanisms and the limitations of current technologies. Some key challenges are discussed in the text below.

9.6.1 Variability in Synaptic Plasticity Across Individuals

Due to the intricate interactions between genetic, environmental, and developmental factors, studying and modulating synaptic plasticity in different individuals can be quite challenging. The molecular mechanisms underlying synaptic plasticity can be greatly influenced by an individual's unique genetic makeup, resulting in variations in the way synapses form, grow, or decrease. Another level of intricacy is added by epigenetic variables, which are impacted by experiences and the environment and alter gene expression and synapse function. Age-related variables further exacerbate this variability; it is well known that synaptic plasticity is increased in the early stages of development and decreases with age, impacting learning and memory functions throughout the lifetime. Environmental influences, including learning experiences, stress, and trauma, also play a critical role in shaping synaptic plasticity. Such diversity presents significant challenges in research, making it difficult to generalize findings across different populations. Moreover, in therapeutic contexts, this variability necessitates personalized approaches to modulate synaptic plasticity effectively. Treatments that prove beneficial in one individual might not yield the same results in another, or could even be counterproductive. Thus, understanding and accounting for individual variability is crucial in developing effective strategies for studying and harnessing synaptic plasticity in diverse human populations [39].

9.6.2 Ethical Considerations in Computational Neuroscience Research

Although using animal models to study a subject might yield insightful results, there are practical and ethical issues when applying these findings to humans. Humans make it very difficult to conduct intrusive investigations on synaptic plasticity, while non-invasive methods such as fMRI provide only oblique measurements of synaptic activity [41].

9.6.3 Integrating Computational Models with Experimental Data

Integrating computational models with experimental data in studying synaptic plasticity presents a significant challenge due to the complexity of accurately simulating biological processes. Computational models often simplify these processes, which can lead to discrepancies when compared with the nuanced, real-world data obtained from experiments. Bridging this gap requires continuous refinement of models and validation against empirical findings.

9.6.4 Challenges in Translating Findings from Computational Models to Clinical Applications

There are various obstacles involved in transferring research from computer models to clinical settings in the field of neuroscience. Despite the fact that computer models provide insightful information about complicated neurological processes, such as synaptic plasticity, the complexity and unpredictability of the human brain make it difficult to apply these models directly to clinical settings. First of all, there is often a discrepancy between model predictions and actual human brain activity since computational models sometimes rely on assumptions and simplifications that may not accurately reflect biological reality. Individual differences in brain shape and function, which are impacted by elements including genetics, environment, and life experiences and which computer models may not always fully account for, amplify this mismatch.

Moreover, there may be difficulties related to the size at which models function. Many models concentrate on particular facets of neuroactivity (e.g., synaptic transmission) at the microscopic level. However, understanding how these processes affect the brain's overall functioning is necessary for clinical applications. Completing this comprehension gap between the micro and macro levels is not easy [42].

The practical and ethical issues involved in adapting results from computational models to human subjects present another major obstacle. Although models can identify possible interventions, lengthy and expensive clinical trials are necessary to turn these suggestions into safe and efficient treatments.

Finally, there are logistical and communication issues arising from the translation's interdisciplinary nature, which calls for cooperation among neuroscientists, physicians, computer scientists, and others. A coordinated and cooperative effort involving a variety of specialized domains is required to ensure that computational results are appropriately evaluated and implemented in clinical research [43].

9.7 FUTURE DIRECTIONS AND EMERGING TECHNOLOGIES

9.7.1 Advances in Computational Models for Synaptic Plasticity

The development of computational models for synaptic plasticity has greatly improved our knowledge of the brain processes underlying memory and learning [43]. These more sophisticated and complicated models provide insights into the intricate processes occurring at the synapse level and their implications for the general functioning of the brain.

9.7.1.1 High-Resolution Models

Modern computational models now incorporate detailed representations of synaptic components, including specific neurotransmitter receptors, ion channels, and intracellular signaling pathways. This high-resolution approach allows for a more accurate simulation of synaptic dynamics.

9.7.1.2 Integration of Molecular and Electrophysiological Data

Recent models successfully integrate molecular biology data with electrophysiological findings, providing a more comprehensive view of synaptic plasticity. This integration bridges the gap between biochemistry and electrical activity in neurons [44].

9.7.1.3 Spike-Timing-Dependent Plasticity (STDP) Models

Advances in modeling STDP have shed light on how the precise timing of neuronal spikes influences synaptic strength.

These models have been crucial in understanding the temporal aspect of synaptic changes and their role in encoding information.

9.7.1.4 Network-Level Simulations

There's a growing focus on large-scale simulations that represent entire neural circuits or brain regions. These models are crucial for understanding how local synaptic changes translate into global patterns of neural activity and behavior.

9.7.2 Integration of Artificial Intelligence and Machine Learning in Neuroscience Research

Understanding intricate neurological processes and solving difficult neuroscience problems has advanced significantly as a result of the incorporation of artificial intelligence (AI) and machine learning into computational neuroscience. The study of big datasets is one of the most direct uses of AI and machine learning in computational neuroscience. The neuroscience field produces large volumes of data, and machine learning algorithms are excellent at finding patterns and connections in this data that a human analyst could miss.

9.7.3 Reinforcement Learning in Understanding Synaptic Dynamics

Reinforcement learning, an important idea in artificial intelligence and machine learning, is being used more and more to comprehend synaptic dynamics, providing fresh insights into how neural networks learn and adapt. Learning and memory are believed to be supported by synaptic plasticity, which strengthens or weakens synapses in the brain. This process is similar to the concepts of reinforcement learning. Here, synapses and neurons modify their responses in response to environmental input, much as algorithms modify actions to optimize rewards. Researchers can investigate how neural circuits might optimize behavior in response to inputs, resulting in the creation of efficient neural pathways, by modeling synaptic alterations as a process of reinforcement learning [44]. This approach helps in understanding the biological basis of learning, particularly the role of dopamine as a neuromodulator that signals reward prediction errors, a concept central to reinforcement learning. Incorporating these principles into computational models of synaptic plasticity provides a powerful framework for unraveling the complex mechanisms of learning and decision-making in the brain, bridging the gap between artificial intelligence and neurobiology.

9.7.4 Predictive Modeling for Disease Progression

This strategy is especially important for diseases like multiple sclerosis, Parkinson's, and Alzheimer's, where management and therapy can be greatly impacted by early detection. Predictive models are capable of identifying patterns and risk factors specific to neurological illnesses by integrating a variety of datasets, including genetic, neuroimaging, clinical, and environmental aspects. These models are skilled at deciphering intricate relationships found in the data, which may reveal biomarkers and early warning indicators of illness. Predictive algorithms, for instance, may examine genetic data, alterations in brain structure, and outcomes of cognitive tests to identify those at high risk of Alzheimer's disease before clinical symptoms appear. Applying machine learning methods, such as deep learning, improves the capacity to handle big datasets and yields predictions that are more precise and complex. The application of predictive modeling in neurology offers a viable route towards personalized medicine, enabling customized therapies and improving patient outcomes in neurological illnesses despite obstacles linked to data variability and model generalizability [43].

9.7.5 Application of Quantum Computing in Simulating Complex Synaptic Interactions

A new frontier in computational neuroscience is the use of quantum computing to simulate intricate synaptic interactions, which has the potential to solve some of the trickiest problems relating to brain function. When modeling the complex dynamics of synaptic processes, quantum computing offers substantial advantages over traditional computing due to its exceptional speed and ability to handle large volumes of data.

9.7.5.1 Molecular Scale Modeling

The molecular and quantum mechanisms that underlie synaptic connections, such as neurotransmitter release, receptor binding, and ion channel dynamics, may be precisely simulated thanks to quantum computing. Because of the enormous processing load, ordinary computers are unable to handle this degree of detail.

9.7.5.2 Managing Complexity

Each of the billions of neurons and trillions of synapses that make up the human brain has its own set of characteristics. Due to their higher processing capability than classical computers, quantum computers are better able to handle the intricacy of these neural networks.

9.7.5.3 Neuroplasticity Simulation

A more accurate simulation of synaptic plasticity—the gradual strengthening or weakening of synapses in response to activity—can be achieved by quantum computing. This involves simulating the influence of several neuromodulators as well as the stochastic nature of synaptic transmission.

9.7.5.4 Drug Discovery

By simulating drug interactions with synaptic proteins at the quantum level, pharmacology can greatly benefit from the use of quantum computing in the field of neurology. This may result in the creation of more focused and efficient therapies for neurological conditions.

9.7.6 Potential Breakthroughs in Enhancing Synaptic Plasticity Using Computational Tools

The incorporation of computational methods foreshadows potential advances in synaptic plasticity research, especially in personalized neurotherapeutics. These techniques have the potential to offer personalized treatments for conditions such as depression, schizophrenia, and Alzheimer's disease by mimicking unique brain profiles. Additionally, they promise improvements in brain-computer interfaces (BCIs), making them more flexible and useful, particularly for people with neurological problems. Computational modeling of synaptic plasticity in neurorehabilitation may result in novel recovery strategies for stroke and trauma patients. These tools also have the potential to inspire cutting-edge machine learning algorithms, provide insights into the course of disease, and speed up medication discovery. The potential of computational modeling to improve synapse-on-a-chip technology could transform drug screening and synaptic function research.

9.8 CONCLUSION

We talked about the important role computational neuroscience plays in our knowledge of synapse dynamics and neural network behavior, especially with models like integrate-and-fire and Hodgkin-Huxley. The development of brain-computer interfaces, neural simulation, and data analysis have all benefited from the incorporation of AI and machine learning in computational neuroscience. We also discussed some of the difficulties in examining synaptic plasticity, such as individual differences, technological constraints, and the complexity of brain networks [12].

Methods such as calcium imaging, two-photon microscopy, and electrophysiological recordings, both in vitro and in vivo, have been recognized for their critical roles in monitoring cellular and molecular changes in synapses. It also highlighted how valuable computational tools are for finding drug targets and supporting personalized medicine and how they may be used in drug discovery and development.

The study of synaptic plasticity has significant significance for the field of neurodegenerative diseases, especially when conducted using sophisticated computational and experimental techniques. Synapse dysfunction is a common feature of neurodegenerative diseases such as Alzheimer's, Parkinson's, and Huntington's. A better comprehension of synaptic plasticity mechanisms can shed light on the pathophysiological causes of various illnesses and provide an important understanding of how disruptions in neural transmission result in deficits in cognition and motor function. Having this knowledge is crucial for creating focused pharmacological therapies [4]. The potential of computational models to expedite drug development through the simulation of pharmacological effects on synaptic functioning and disease progression is particularly noteworthy. Furthermore, by identifying synaptic alterations as early biomarkers of neurodegeneration, these researches can aid in the development of early diagnosis and intervention techniques. Personalized medicine also has significant benefits since specific synaptic profiles can allow for tailored therapeutic methods informed by computational insights. In summary, the progress made in comprehending synaptic plasticity is not only expanding our understanding of brain function but also playing a crucial role in determining future approaches to treating neurodegenerative illnesses.

In the above discussion, a number of unmet research needs were brought up regarding synaptic plasticity and its potential implications in computational neuroscience and neurodegenerative illnesses. One major gap is applying theoretical and computational models to clinical practice, particularly in the context of neurodegenerative illnesses. Although computational models provide insightful information, the complexity of human brain function and individual diversity makes it difficult to apply these findings to real-world situations. The integration of information from diverse sources, including genetic, neuroimaging, and behavioral investigations, to develop thorough models of synapse function is another area that requires more research. Furthermore, although emerging technologies such as optogenetics and quantum computing offer fresh perspectives on synaptic dynamics, methodological and technological obstacles still restrict their practical use. For example, problems with error rates and the need for very low temperatures are encountered in the field of quantum computing. Furthermore, there is much to be discovered about the use of AI and machine learning to comprehend intricate brain functions like synaptic plasticity. It is still difficult to create algorithms that accurately capture the stochastic aspects of synaptic transmission and plasticity. Finally, more thought and investigation should be given to the ethical ramifications of using AI and sophisticated computational methods in neuroscience, particularly in fields like brain-computer interfaces and cognitive improvement. Closing these gaps will be essential to improving our knowledge of the brain and creating useful remedies for neurological conditions.

REFERENCES

1. Richa Shukla, Sonu Pahal, Ayushi Gupta, Princy Choudhary, Krishna Misra and Sangeeta Singh. Modulation of GPCR receptors common to gut inflammatory diseases and neuronal disorders, Alzheimer's and Parkinson's diseases as druggable targets through withania somnifera bioactives: an in silico study. *Journal of Biomolecular Structure and Dynamics*, 41(10):4485–4503, 2023.
2. Princy Choudhary, Rameshwar Sankhala and Sangeeta Singh. Integration of significant signaling pathways involved in neural-fate differentiation of stem cells. In *2023 10th International Conference on Computing for Sustainable Global Development (INDIACom)*, pp. 694–701. IEEE, 2023.
3. Wickliffe C Abraham and Mark F Bear. Metaplasticity: the plasticity of synaptic plasticity. *Trends in Neurosciences*, 19(4):126–130, 1996.
4. LF Abbott and Wade G Regehr. Synaptic computation. *Nature*, 431(7010):796–803, 2004.
5. Hojjat Adeli, Samanwoy Ghosh-Dastidar and Nahid Dadmehr. Alzheimer's disease: models of computation and analysis of EEGs. *Clinical EEG and Neuroscience*, 36(3):131–140, 2005.
6. Yves Agid, György Buzsáki, David M Diamond, Richard Frackowiak, Jay Giedd, Jean-Antoine Girault, Anthony Grace, Jeremy J Lambert, Husseini Manji, Helen Mayberg, et al. How can drug discovery for psychiatric disorders be improved? *Nature Reviews Drug Discovery*, 6(3):189–201, 2007.
7. Mohammad Asif, Sudhakar Mishra, Majithia Tejas Vinodbhai, and Uma Shanker Tiwary. Emotion recognition using temporally localized emotional events in eeg with naturalistic context: Dens# dataset. IEEE Access, 11:39913–39925, 2023.

8. Venkata Subba Rao Atluri, Melissa Hidalgo, Thangavel Samikkannu, Kesava Rao Venkata Kurapati, Madhavan Nair. Synaptic plasticity and neurological disorders in neurotropic viral infections. *Neural Plasticity*, 2015:138979, 2015.
9. Avik Atta, Ayushi Gupta, Princy Choudhary, Shrey Dwivedi and Sangeeta Singh. Inhibition of lingo1 as a therapeutic target to promote axonal regeneration and repair for neurological disorders. *3 Biotech*, 13(11):372, 2023.
10. Thomas M Bartol Jr, Cailey Bromer, Justin Kinney, Michael A Chirillo, Jennifer N Bourne, Kristen M Harris and Terrence J Sejnowski. Nanoconnectomic upper bound on the variability of synaptic plasticity. *Elife*, 4:e10778, 2015.
11. Mark F Bear and Robert C Malenka. Synaptic plasticity: LTP and LTD. *Current Opinion in Neurobiology*, 4(3):389–399, 1994.
12. Mohammad Asif, Diya Srivastava, Aditya Gupta and Uma Shanker Tiwary. Inter subject emotion recognition using spatio-temporal features from eeg signal. In *2023 27th International Computer Science and Engineering Conference (ICSEC)*, 1–4. IEEE, 2023.
13. Asif M., Vinodbhai, M.T., Mishra, S., Gupta, A., and Tiwary, U.S.. Emotion recognition in VAD space during emotional events using CNN-GRU hybrid model on EEG signals. In Zaynidinov, H., Singh, M., Tiwary, U. S., Singh, D. (Eds.), Intelligent Human Computer Interaction. IHCI 2022. Lecture Notes in Computer Science, vol 13741. Springer, Cham, 2023. https://doi.org/10.1007/978-3-031-27199-1_8
14. Hagai Bergman and Günther Deuschl. Pathophysiology of Parkinson's disease: from clinical neurology to basic neuroscience and back. *movement Disorders: Official Journal of the Movement Disorder Society*, 17(S3):S28–S40, 2002.
15. Dean V Buonomano. Decoding temporal information: a model based on short-term synaptic plasticity. *Journal of Neuroscience*, 20(3):1129–1141, 2000.
16. Chandra V, Pandav R, Laxminarayan R, et al. Neurological Disorders. In Jamison DT, Breman JG, Measham AR, et al., editors. Disease Control Priorities in Developing Countries. 2nd edition., The International Bank for Reconstruction and Development / The World Bank, Washington, DC, 2006. Chapter 32. Available from: https://www.ncbi.nlm.nih.gov/books/NBK11793/ Co-published by Oxford University Press, New York.
17. Abhijit Chaudhuri and Peter O Behan. Fatigue in neurological disorders. *The Lancet*, 363(9413): 978–988, 2004.
18. Princy Choudhary, Shivani Gupta, Richa Shukla, Ayushi Gupta, Sonu Pahal and Sangeeta Singh. Regulation of neuronal repair and regeneration through inhibition of oligodendrocyte myelin glycoprotein (OMGP). *Journal of Biomolecular Structure and Dynamics*, 40(24):13936–13952, 2022.
19. Samuel Frazer Cooke and Timothy VP Bliss. Plasticity in the human central nervous system. *Brain*, 129(7):1659–1673, 2006.
20. Ana Salomé Correia, Armando Cardoso and Nuno Vale. Oxidative stress in depression: the link with the stress response, neuroinflammation, serotonin, neurogenesis and synaptic plasticity. *Antioxidants*, 12(2):470, 2023.
21. Princy Choudhary, Amit Chaudhary, Sonu Pahal, Ayushi Gupta and Sangeeta Singh. Network pharmacological evaluation of withania somnifera bioactive phytochemicals for identifying novel potential inhibitors against neurodegenerative disorder. *Journal of Biomolecular Structure and Dynamics*, 40(21):10887–10898, 2022.
22. Jianfeng Feng. *Computational Neuroscience: A Comprehensive Approach*. CRC Press, 2003.
23. Andrea Gómez-Palacio-Schjetnan and Martha L Escobar. Neurotrophins and synaptic plasticity. *Neurogenesis and Neural Plasticity*, 15:117–136, 2013.
24. Sheena A Josselyn and Peter V Nguyen. CREB, synapses and memory disorders: past progress and future challenges. *Current Drug Targets-CNS & Neurological Disorders* 4(5):481–497, 2005.
25. A Kimberley McAllister, Lawrence C Katz and Donald C Lo. Neurotrophins and synaptic plasticity. *Annual Review of Neuroscience*, 22(1):295–318, 1999.
26. Pradeep J Nathan, Stuart R Cobb, Bai Lu, Edward T Bullmore and Ceri H Davies. Studying synaptic plasticity in the human brain and opportunities for drug discovery. *Current Opinion in Pharmacology*, 11(5):540–548, 2011.
27. Charlotte A Oomen, Heleen Soeters, Nathalie Audureau, Lisa Vermunt, Felisa N Van Hasselt, Erik MM Manders, Marian Jöels, Paul J Lucassen and Harm Krugers. Severe early life stress hampers spatial learning and neurogenesis, but improves hippocampal synaptic plasticity and emotional learning under high-stress conditions in adulthood. *Journal of Neuroscience*, 30(19):6635–6645, 2010.
28. Alex Pavlides, S John Hogan and Rafal Bogacz. Computational models describing possible mechanisms for generation of excessive beta oscillations in Parkinson's disease. *PLoS Computational Biology*, 11(12):e1004609, 2015.

29. Sudhakar Mishra et al.. Dataset on emotion with naturalistic stimuli (DENS) on Indian Samples. bioRxiv (2021): 2021-8.
30. Donald G Rainnie, Richard Bergeron, Tammy J Sajdyk, Madhvi Patil, Donald R Gehlert and Anantha Shekhar. Corticotrophin releasing factor-induced synaptic plasticity in the amygdala translates stress into emotional disorders. *Journal of Neuroscience*, 24(14):3471–3479, 2004.
31. Michael T Rogan and Joseph E LeDoux. Emotion: systems, cells, synaptic plasticity. *Cell*, 85(4): 469–475, 1996.
32. Eric L Schwartz. *Computational Neuroscience.* Mit Press, 1993.
33. Terrence J Sejnowski, Christof Koch and Patricia S Churchland. Computational neuroscience. *Science*, 241(4871):1299–1306, 1988.
34. Sen Song, Kenneth D Miller and Larry F Abbott. Competitive hebbian learning through spike-timing-dependent synaptic plasticity. *Nature Neuroscience*, 3(9):919–926, 2000.
35. Cornelis J Stam. Modern network science of neurological disorders. *Nature Reviews Neuroscience*, 15(10):683–695, 2014.
36. Klaas E Stephan, Torsten Baldeweg and Karl J Friston. Synaptic plasticity and dysconnection in schizophrenia. *Biological Psychiatry*, 59(10):929–939, 2006.
37. David Sterratt, Bruce Graham, Andrew Gillies, Gaute Einevoll and David Willshaw. *Principles of Computational Modeling in Neuroscience.* Cambridge University Press, 2023.
38. Thomas C Südhof. Towards an understanding of synapse formation. *Neuron*, 100(2):276–293, 2018.
39. Keith Tully and Vadim Y Bolshakov. Emotional enhancement of memory: how norepinephrine enables synaptic plasticity. *Molecular Brain*, 3:1–9, 2010.
40. Robert S Zucker and Wade G Regehr. Short-term synaptic plasticity. *Annual Review of Physiology*, 64(1):355–405, 2002.
41. Thomas V Wiecki and Michael J Frank. Neurocomputational models of motor and cognitive deficits in parkinson's disease. *Progress in Brain Research*, 183:275–297, 2010.
42. Nikolaus Kriegeskorte and Pamela K Douglas. Cognitive computational neuroscience. *Nature Neuroscience*, 21(9):1148–1160, 2018.
43. Martine T Usdin, Peggy F Shelbourne, Richard M Myers and Daniel V Madison. Impaired synaptic plasticity in mice carrying the huntington's disease mutation. *Human Molecular Genetics*, 8(5):839–846, 1999.
44. Felice Iasevoli, Carmine Tomasetti and Andrea de Bartolomeis. Scaffolding proteins of the post-synaptic density contribute to synaptic plasticity by regulating receptor localization and distribution: relevance for neuropsychiatric diseases. *Neurochemical Research*, 38:1–22, 2013.

10 Tools of Artificial Intelligence in the Improvement of Synaptic Impairment in Different Neurodegenerative Diseases

Sonu Pahal
University of Arkansas for Medical Sciences, University of Arkansas at Little Rock
Little Rock, AR

Vishvender Pahal
All India Institute of Medical Sciences
Raebareli, Uttar Pradesh, India

*Meenakshisundaram Balasubramaniam**
University of Arkansas for Medical Sciences, Little Rock, AR

10.1 INTRODUCTION

Neurodegenerative disorders (NDDs) are a very convoluted and confined group of disorders like Alzheimer's disease (AD), Parkinson's disease (PD), Huntington's disease (HD), and amyotrophic lateral sclerosis (ALS). Synaptic impairment is the common ground in all these ailments, and substantial research has been done using a variety of model organisms. For instance, AD seed proteins Aβ and tau lead to synaptic dysfunction and cognitive loss in animal models, altered synaptic vesicle release in the HD mice model, age-dependent synapse loss in rats, and sex-dependent synaptic density and transmission in mice. These studies shed light on the vitality of synapses in triggering and accelerating neuronal damage and as a critical target to mitigate the onset of NDDs. Yet the mechanisms underlying synaptic impairment and timely detection in NDDs remain poorly understood [1–3].

Traditional approaches for the treatment of synaptic impairment in neurodegenerative disorders have various limitations. These approaches, such as manually analyzing scientific data, tend to be tedious and prone to errors. Furthermore, these methods might face problems in handling massive amounts of scientific data.

These approaches include histological analysis of synaptic region tissue samples, electrophysiological recording of electrical signals in neurons to analyze their synaptic functions, behavioral analysis, and biochemical assays to quantify the level of neurotransmitters in synapses [4–6].

AI offers solutions to these problems by handling diverse and complex scientific data, thus offering a more comprehensive understanding of synaptic impairment. With predictive modeling, AI makes identifying potential therapeutic targets and small drug molecules quick and precise. These tools are trained to effectively analyze enormous amounts of data, revealing hidden trends and

DOI: 10.1201/9781003464648-12

TABLE 10.1
List of AI Tools Used in Synaptic Impairment Research

S. No.	AI Tool	Current Use/Might be Used as
1.	Machine learning models	Predicting patterns of synaptic impairment patterns.
2.	Deep learning models	Analyzing the structure of synapses.
3.	Neural networks	Modeling of synaptic interactions and changes in their connectivity.
4.	Reinforcement learning	Increase effectiveness of neurostimulation procedures and improve synaptic dysfunction.
5.	Natural language processing (NLP)	Literature mining for analyzing synaptic plasticity.
6.	Brain-computer interfaces (BCIs)	Real-time monitoring of synaptic activity.
7.	Data mining and integration	Combining diverse datasets for in-depth analysis.
8.	Quantum computing	For simulating synaptic interactions at quantum level.

connections that might be missed by conventional methods. AI tools that might be useful in combating synaptic impairment include machine learning (ML) models, deep learning models, natural language processing (NLP), and robotics [7,8]. A list of AI tools that might be useful in synaptic impairment research is provided in Table 10.1.

10.2 ARTIFICIAL INTELLIGENCE TOOLS

Artificial neural networks (ANNs), designed based on the structure and functions of the human brain, have artificial neurons networked together with tunable connectivity. These connections vary as an outcome of learning and resemble the synaptic changes in the human brain [9]. The advanced capabilities of AI in processing intricate neurological data and generating models, as evidenced by IBM's Blue Gene Processor simulating 8×10^6 mouse cortical neurons and 8000 synapses per neuron, might help researchers to access their synaptic dysfunction hypothesis prior to conducting extensive studies on animals [10].

A recent study used a neural circuit model with 2400 neurons to mimic synaptic processing incorporating both short- and long-term plasticity mechanisms. Using MATLAB for statistical analysis, they showed that synaptic impairment impedes acquiring new memories and modulates brain dynamics hindering memory recall [11].

Another AI tool, deep continuous local learning (DECOLLE), spans the gap between biological and neural networks by rejecting the need for memory overhead and employing local error functions for learning. Its novel synaptic plasticity rule offers linear scalability and compatibility with ML libraries addressing the difficulties in spiking neuron models [12]. Recent research applied a spiking neural network model showing that synaptic loss reduces the sensitivity of the neural network preceding cognitive decline. When this state was avoided by a compensatory mechanism, it led to firing rate homeostasis and restored altered network features [13].

10.2.1 Neural Network Simulator Tools

Modeling and simulating the intricate synaptic connections and understanding the influence of synaptic impairment on the entire network require neural network simulator tools (Figure 10.1). These tools might help uncover key factors in synaptic impairment. Brian2, Neural Simulation Tool (NEST), Nengo, BindsNET, NEURON, and Brian2Genn make up the tools on the list [14–20].

10.2.2 Neuronal Reconstruction Software

AI-driven neuronal reconstruction aids in precise morphological insights, comprehending synaptic impairment and timely understanding of structural changes in neurons. Such software,

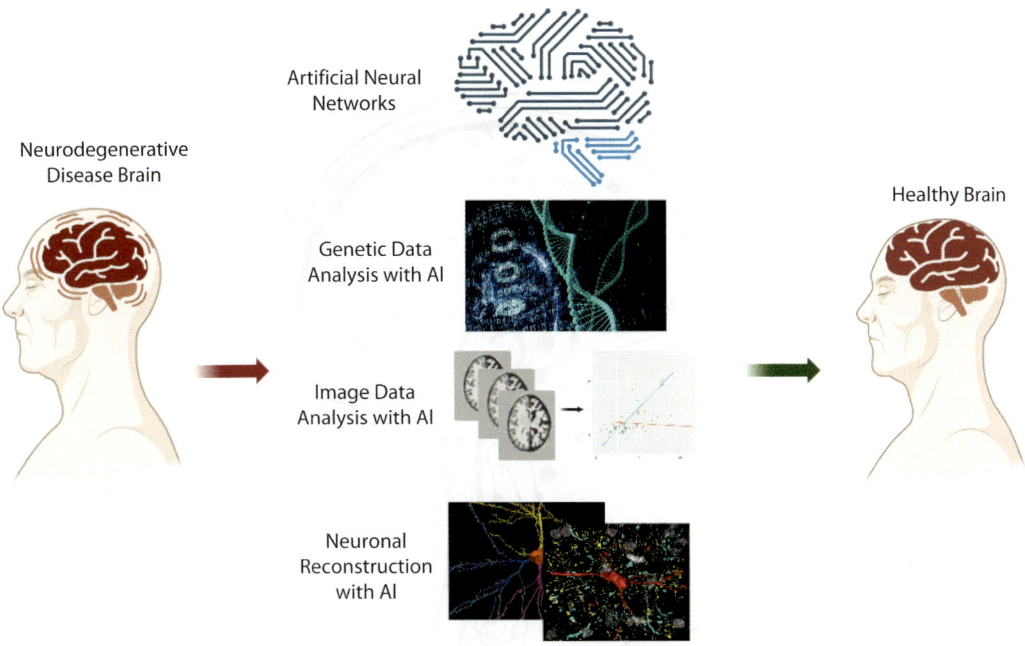

FIGURE 10.1 AI tools in alleviating synaptic dysfunction in various neurodegenerative disorders.

IMARIS [http://www.bitplane.com/imaris/imaris] and neuTube 1.0 [14], help reconstruct three-dimensional neuron structure, making quantification and synaptic distribution feasible. IMARIS has been used by researchers in tissue imaging and analysis (Figure 10.1) for understanding neurodegeneration events like microglia engulfment of synapses in tau pathology [21], synaptic proteome alteration analysis in tau pathology [22], and synaptic loss due to microglial phagocitation [23].

10.2.3 Neuroimaging Analysis with ML and AI Tools

Traditional image analysis methods can be time-consuming and subjective. ML algorithms can automate the analysis of brain images from techniques like MRI, CT scans, and microscopy. This aids in identifying patterns, anomalies, and structural changes that might be indicative of neurological conditions. AI models enable the quantification of structural changes in the brain, including the measurement of volumes and thickness of specific regions. This quantitative analysis aids in tracking disease progression and assessing the impact of treatments (Figure 10.1). For example, ML algorithms can be trained on labeled datasets to differentiate between healthy and diseased brains. This classification capability is valuable for early and accurate diagnosis of neurodegenerative diseases, such as Alzheimer's or Parkinson's. ML and AI have emerged as a powerful tool in the field of biomarker discovery for neurodegenerative diseases. ML algorithms are used to integrate and analyze diverse datasets, including genomic, proteomic, imaging, and clinical data. This holistic approach allows for the identification of complex relationships and patterns that may be indicative of early disease stages or progression. Similarly, ML and AI are used to analyze neuroimaging data, such as MRI and PET scans, to identify subtle patterns associated with neurodegenerative changes. This can lead to the discovery of imaging biomarkers that reflect structural, functional, or metabolic alterations in the brain [24–26].

Another current use of ML and AI is to understand disease progression and treatment response from MRI brain images from patients diagnosed with glioblastoma. ML algorithms are currently being used for tumor segmentation, radiomic analysis, treatment response evaluations, and survival predictions in glioblastoma patients. Tumor segmentation involves differentiating or identifying tumor regions from surrounding healthy tissues in the brain [27,28]. This is critical for identifying treatment plans.

Over and above, ML and AI play a transformative role in drug discovery, offering innovative approaches to accelerate the identification of potential drug candidates, optimize drug development processes, and enhance therapeutic outcomes. ML models analyze biological and genetic data to identify potential drug targets, such as proteins or genes associated with specific diseases. AI assists in validating these targets by predicting their relevance and potential impact on disease pathways [29].

A wide range of ML and AI tools are available, catering to various aspects of data analysis, model development, and deployment. Some of the commonly used ML and AI tools include Pandas, a Python library for data manipulation and analysis [30], and NumPy, which provides support for large, multi-dimensional arrays and matrices, along with mathematical functions for data processing and analysis [31]. ML libraries such as Scikit-learn, for classical ML algorithms such as classification, regression, clustering, and dimensionality reduction [32]; TensorFlow, an open-source ML framework developed by Google, widely used for deep learning applications [33]; and PyTorch, a deep learning framework, developed by Facebook's AI Research lab, known for its dynamic computation graph [34]. Furthermore, Keras, an open-source deep learning library that acts as an interface for TensorFlow, Theano, and Microsoft Cognitive Toolkit [35], and MXNet, is a flexible deep learning framework that supports both symbolic and imperative programming [36]. These tools serve various purposes within the ML and AI landscape, and the choice of tools often depends on the specific requirements of a project, the type of task, and the preferences of the developers and data scientists involved.

Recently an ML algorithm was created using retinal images, which could decipher normal cognition and mild cognitive impairment with high specificity and sensitivity. Thus, such tools and algorithms might also have a high capability of replacing conventional techniques in the area of synaptic impairment [37,38].

10.2.4 Quantitative Proteomics Tools

To better understand the alterations in protein expression profiles linked to synaptic impairment, different quantitative proteomics tools such as MaxQuant, Skyline, Scaffold, and PDQuest come into play. MaxQuant is being used in research for analyzing synaptic protein expression alterations in the human ALS cortex [39], cerebrospinal fluid (CSF) depressive disorders [40], and synaptic degeneration in AD patients [41]. Another tool, Skyline is also being used in the comparative analysis of histopathology of cerebral cortex in ND motor disease [42] and CSF protein level in PD and dementia with Lewy body (DLB) [43].

10.3 AI TOOLS IN DRUG DISCOVERY FOR SYNAPTIC IMPAIRMENT

Drug discovery for neurodegenerative disorders is by far the most difficult area of research. Tremendous expansion of biological data generated with modern experimental technologies has made the contribution of AI tools vital to derive significant inferences from this data and enhance drug discovery in this area. A list of AI tools that are being used in research and might help in combating synaptic impairment are provided in Table 10.2.

AI offers promising opportunities for synaptic impairment drug discovery research. Our understanding of this specific research area can by enhanced further by incorporating proteomics, transcriptomics, and genomics data by using different AI tools. New AI models and tools must be developed for a better understanding of experimental data, and these data insights should be shared in the research community to fast-track synaptic impairment research. Furthermore, tailored methods led by AI-powered precision medicine, which

TABLE 10.2
Table Illustrates AI Tools That Aid in Synaptic Impairment Drug Discovery

S. NO.	AI Tool	In Use/Might be Used as	Reference
1.	Chemo-informatics and quantitative structure-activity relationship (QSAR) models	Help in analyzing chemical structures better, and improve efficacy of small molecule drug discovery.	[44,45]
2.	IBM Watson	May help in analyzing big datasets and fast-track drug discovery by finding new biomarkers and drug targets.	[46]
3.	DeepChem	A deep learning library that might help in predicting therapeutic targets for synaptic impairment drug-discovery.	[47]
4.	Benevolent AI	May help in formulating new hypotheses and predict potential new therapeutic targets from available biomedical data.	[48]
5.	DeepMind	Helps in predicting protein structures and supports new drug development based on reinforcement and deep learning.	[49]

considers individual differences, will probably reshape treatment plans for conditions associated with synapse dysfunction.

10.4 LIMITATIONS OF USING AI TOOLS IN SYNAPTIC IMPAIRMENT RESEARCH

10.4.1 Interpretability

AI models are frequently referred to as "black boxes" and "opaque" models as the rationale behind processing performed between data input and output isn't well understood [50–52]. This causes a lack of trust in AI predictions, particularly in clinical settings, which rely upon responsibility, reliability, and transparency of the physician [53].

10.4.2 Generalization and Dataset Imbalances

Most AI algorithms are trained using restricted datasets and fail when exposed to new unseen data due to overfitting [50]. The toughest challenge lies in data composition and standardization. One possible strategy is to use large-scale, diverse training datasets representative of the heterogeneity of disease processes and population dynamics [51,53,54]. However, large-scale datasets can lead to false discoveries due to the amplification of small-sized effects and further decrease interpretability [55].

10.5 ETHICAL CONSIDERATIONS AND PRIVACY

Using AI in patient care risks physician bias, accountability, patient privacy, and safety [50,53]. Physician bias arises when AI-suggested diagnoses are favored over clinical acumen while ignoring inconsistencies of AI models. This leads to incorrect diagnoses, risking patient safety and losing accountability [50]. Although most patient information datasets today are anonymized, algorithms exploit data patterns, even if the patterns are irrelevant to the output [50,53]. Future developments must focus on ensuring privacy while enabling meaningful analysis.

As AI models are meant to learn continuously, closer to their task capacity the accuracy of subsequent tasks may be jeopardized compared to previous tasks. Keeping in mind the brain's natural forgetting behavior and synapse complexity, AI models may slightly differ in the implementation issues, indicating the necessity of a balance between computational efficiency and depth [56].

10.6 FUTURE DIRECTIONS

10.6.1 Using AI-based Real-Time Sensors

To better tackle the problem of synaptic impairment, combining wearable sensors and AI algorithms might fetch us a cutting-edge strategy. Analyzing data in real time may offer more insightful information about continuous fluctuations in synapse functioning, thus allowing for early detection of changes and development of personalized treatment plans [57]. Devices that can measure electrophysiological signals in the brain [58,59] and wearable electroencephalogram (EEG) devices that can sense brain activity [60] might be useful in research. In a study, researchers used AI to assess the short-term memory data generated from an EEG sensor, and they found that the AI system achieved a high level of accuracy in identifying images presented to the participants. Specifically, 90.12% of the displayed images were correctly identified by the AI. Additionally, when evaluating the responses of emotionally responsive individuals, the AI system achieved an accuracy of 90.51% in accurately answering questions related to time and image type. These findings indicate that the AI system, when processing EEG data, demonstrated a notable ability to recognize and interpret information related to short-term memory tasks, offering promising results in the field of cognitive assessment and brain-computer interface applications [61].

10.6.2 AI-based Personalized Medicine

Personalized treatments enabled by AI can tailor therapies to specific patient characteristics. AI approaches can aid in the discovery of patient-specific responses to synaptic impairment drugs, hence improving medical outcomes by analyzing hereditary, medical, and lifestyle data. Although there aren't any particular AI tools for synaptic impairment in personalized medicine, AI is used by companies like IBM Watson for the purposes of drug discovery to search medical data for possible medications aimed at synaptic dysfunction. By analyzing brain alterations, neuroimaging AI tools—which are frequently used in conditions like Alzheimer's disease—indirectly support tailored medication. Datasets on gene expression associated with synapse function are available from the Allen Brain Atlas. Personalized medicine research in neurology is supported by neuroinformatic tools and libraries that integrate different neuroscience data with artificial intelligence analysis. It's important to stay informed of new developments, and scientists can find updated data by visiting the websites of particular tools.

10.6.3 Developing Drug Repurposing and Virtual Screening AI Tools

The discovery of promising treatments for synaptic dysfunction can be sped up with the use of AI-based virtual screening and drug repurposing initiatives. AI speeds up the drug discovery process and saves time and money by utilizing pre-existing drug inventories and forecasting novel applications for approved molecules.

AI tools that come under drug repurposing include BenevolentAI [62] and Atomwise [63]. AI tools such as DeepChem [47], IBM Watson for drug discovery [64], VLife MDS [65], BioSolveIT SeeSAR [www.biosolveit.de/SeeSAR], AutoDock Vina [66], and SurflexDock [67] will be useful in virtual screening of new druggable small molecules for synaptic impairment.

10.7 CONCLUSION

AI is an important technique for treating synaptic dysfunction in neurodegenerative illnesses like Parkinson's and Alzheimer's. It is anticipated to support human judgment as quantitative approaches become essential to neuroimaging biomarkers testing, particularly in complicated pattern identification. AI is excellent at deciphering intricate brain networks and provides early detection capabilities that are vital for treatment. The promise is twofold: it can help with accurate diagnosis and tailored treatment. The complexities of brain anatomy, the diversity of illnesses, and the lack of standardized protocols are some of the challenges faced when using AI tools in a neurodegeneration context. These obstacles must be overcome for smooth integration into clinical practices. Deep learning algorithms for early detection of NDDs using neuroimaging, AI for tailored treatment plans, and extensive data analysis for drug response prediction are the future areas of drug discovery. To ensure successful integration into clinical practices, ethical issues, patient privacy, and legal frameworks must be considered. To maintain quality across a demographic range, collaborative attempts are crucial for standardized data gathering. To summarize, AI has great potential for use in neurodegenerative disorders. AI tools give leverage to medical practitioners to draw as much details from the available diagnostic findings, thus reducing variations in understanding and improving accuracy. To fully realize this potential, continued breakthroughs in AI, moral frameworks, and cooperative activities are required to uncover efficient detection and treatment methods for synaptic dysfunction.

REFERENCES

1. Subramanian, J., and M.-È. Tremblay, Synaptic Loss and Neurodegeneration. 2021, Frontiers Media SA. p. 681029.
2. Wishart, T.M., S.H. Parson, and T.H. Gillingwater, *Synaptic vulnerability in neurodegenerative disease*. Journal of Neuropathology & Experimental Neurology, 2006. **65**(8): p. 733–739.
3. Tzioras, M., et al., *Synaptic degeneration in Alzheimer disease*. Nature Reviews Neurology, 2023. **19**(1): p. 19–38.
4. Verstraelen, P., et al., *Image-based profiling of synaptic connectivity in primary neuronal cell culture*. Frontiers in Neuroscience, 2018. **12**: p. 389.
5. Rohrbough, J., and K. Broadie, *Electrophysiological analysis of synaptic transmission in central neurons of Drosophila larvae*. Journal of Neurophysiology, 2002. **88**(2): p. 847–860.
6. Glasgow, S.D., et al., *Approaches and limitations in the investigation of synaptic transmission and plasticity*. Frontiers in Synaptic Neuroscience, 2019. 11: p. 20.
7. Davenport, T., and R. Kalakota, *The potential for artificial intelligence in healthcare*. Future Healthcare Journal, 2019. **6**(2): p. 94.
8. Xu, Y., et al., *Artificial intelligence: A powerful paradigm for scientific research*. The Innovation, 2021. **2**(4): p. 100179.
9. Surianarayanan, C., et al., *Convergence of artificial intelligence and neuroscience towards the diagnosis of neurological disorders—A scoping review*. Sensors, 2023. **23**(6): p. 3062.
10. Frye, J., R. Ananthanarayanan, and D.S. Modha, *Towards real-time, mouse-scale cortical simulations*. CoSyNe: Computational and Systems Neuroscience, Salt Lake City, Utah, 2007.
11. Li, K.T., D. Ji, and C. Zhou, *Memory rescue and learning in synaptic impaired neuronal circuits*. iScience, 2023. **26**(7): p. 106931.
12. Kaiser, J., H. Mostafa, and E. Neftci, *Synaptic plasticity dynamics for deep continuous local learning (DECOLLE)*. Frontiers in Neuroscience, 2020. **14**: p. 424.
13. Bachmann, C., et al., *Firing rate homeostasis counteracts changes in stability of recurrent neural networks caused by synapse loss in Alzheimer's disease*. PLoS Computational Biology, 2020. **16**(8): p. e1007790.
14. Feng, L., T. Zhao, and J. Kim, *neuTube 1.0: A new design for efficient neuron reconstruction software based on the SWC format*. eNeuro, 2015. **2**(1): p. ENEURO.0049-14.2014.
15. Stimberg, M., R. Brette, and D.F.M. Goodman, *Brian 2, an intuitive and efficient neural simulator*. eLife, 2019. **8**: p. e47314.
16. Gewaltig, M.-O., and M. Diesmann, *Nest (neural simulation tool)*. Scholarpedia, 2007. **2**(4): p. 1430.
17. Bekolay, T., et al., *Nengo: A python tool for building large-scale functional brain models*. Frontiers in Neuroinformatics, 2014. **7**: p. 48.
18. Hazan, H., et al., *Bindsnet: A machine learning-oriented spiking neural networks library in python*. Frontiers in Neuroinformatics, 2018. **12**: p. 89.
19. Hines, M.L., and N.T. Carnevale, *NEURON: A tool for neuroscientists*. Neuroscientist, 2001. **7**(2): p. 123–135.
20. Stimberg, M., D.F.M. Goodman, and T. Nowotny, *Brian2GeNN: Accelerating spiking neural network simulations with graphics hardware*. Scientific Reports, 2020. **10**(1): p. 410.

21. Benetatos, J., et al., *PTEN activation contributes to neuronal and synaptic engulfment by microglia in tauopathy*. Acta Neuropathologica, 2020. **140**: p. 7–24.
22. Dejanovic, B., et al., *Changes in the synaptic proteome in tauopathy and rescue of tau-induced synapse loss by C1q antibodies*. Neuron, 2018. **100**(6): p. 1322–1336.e7.
23. Dundee, J.M., et al., *P2Y6 receptor-dependent microglial phagocytosis of synapses mediates synaptic and memory loss in aging*. Aging Cell, 2023. **22**(2): p. e13761.
24. Stefano, G.B., Artificial Intelligence as a Tool for the Diagnosis and Treatment of Neurodegenerative Diseases. 2023, MDPI. p. 938.
25. Tăuţan, A.-M., B. Ionescu, and E. Santarnecchi, *Artificial intelligence in neurodegenerative diseases: A review of available tools with a focus on machine learning techniques*. Artificial Intelligence in Medicine, 2021. **117**: p. 102081.
26. Myszczynska, M.A., et al., *Applications of machine learning to diagnosis and treatment of neurodegenerative diseases*. Nature Reviews Neurology, 2020. **16**(8): p. 440–456.
27. Cè, M., et al., *Artificial intelligence in brain tumor imaging: A step toward personalized medicine*. Current Oncology, 2023. **30**(3): p. 2673–2701.
28. Xu, J., et al., *Applications of artificial intelligence based on medical imaging in glioma: Current state and future challenges*. Frontiers in Oncology, 2022. **12**: p. 892056.
29. Vora, L.K., et al., *Artificial intelligence in pharmaceutical technology and drug delivery design*. Pharmaceutics, 2023. **15**(7): p. 1916.
30. McKinney, W., Pandas, python data analysis library. http://pandas.pydata.org, 2015: p. 3–15.
31. Oliphant, T.E., Guide to Numpy. Vol. 1. 2006, Trelgol Publishing USA.
32. Kramer, O., and O. Kramer, Scikit-learn. Machine learning for evolution strategies, 2016: p. 45–53.
33. Pang, B., E. Nijkamp, and Y.N. Wu, *Deep learning with tensorflow: A review*. Journal of Educational and Behavioral Statistics, 2020. **45**(2): p. 227–248.
34. Imambi, S., K.B. Prakash, and G. Kanagachidambaresan, PyTorch. Programming with TensorFlow: Solution for Edge Computing Applications, 2021. p. 87–104.
35. Gulli, A., and S. Pal, Deep Learning with Keras. 2017, Packt Publishing Ltd.
36. Chen, T., et al., *Mxnet: A flexible and efficient machine learning library for heterogeneous distributed systems*. arXiv preprint arXiv:1512.01274, 2015.
37. Jack, C.R. Jr, et al., *Introduction to revised criteria for the diagnosis of Alzheimer's disease: National institute on aging and the Alzheimer association workgroups*. Alzheimer's & Dementia: The Journal of the Alzheimer's Association, 2011. **7**(3): p. 257.
38. Wisely, C.E., et al., *A convolutional neural network using multimodal retinal imaging for differentiation of mild cognitive impairment from normal cognition*. Ophthalmology Science, 2024. **4**(1): p. 100355.
39. Laszlo, Z.I., et al., *Synaptic proteomics reveal distinct molecular signatures of cognitive change and C9ORF72 repeat expansion in the human ALS cortex*. Acta Neuropathologica Communications, 2022. **10**(1): p. 156.
40. Al Shweiki, M.R., et al., *Proteomic analysis reveals a biosignature of decreased synaptic protein in cerebrospinal fluid of major depressive disorder*. Translational Psychiatry, 2020. **10**(1): p. 144.
41. Hesse, R., et al., *Comparative profiling of the synaptic proteome from Alzheimer's disease patients with focus on the APOE genotype*. Acta Neuropathologica Communications, 2019. **7**(1): p. 214.
42. Louis, E.D., et al., *Histopathology of the cerebellar cortex in essential tremor and other neurodegenerative motor disorders: Comparative analysis of 320 brains*. Acta Neuropathol, 2023. **145**(3): p. 265–283.
43. Lerche, S., et al., *CSF protein level of neurotransmitter secretion, synaptic plasticity, and autophagy in PD and DLB*. Movement Disorders, 2021. **36**(11): p. 2595–2604.
44. Xu, J., and A. Hagler, *Chemoinformatics and drug discovery*. Molecules, 2002. **7**(8): p. 566–600.
45. Muratov, E.N., et al., *QSAR without borders*. Chemical Society Reviews, 2020. **49**(11): p. 3525–3564.
46. Lee, K.Y., and J. Kim, *Artificial intelligence technology trends and IBM Watson references in the medical field*. Korean Medical Education Review, 2016. **18**(2): p. 51–57.
47. Ramsundar, B., Molecular Machine Learning with DeepChem. 2018, Stanford University.
48. Dube, S., *Benevolent AI for All*, in An Intuitive Exploration of Artificial Intelligence: Theory and Applications of Deep Learning. 2021, Springer. p. 285–291.
49. Beattie, C., et al., Deepmind lab. arXiv preprint arXiv:1612.03801, 2016.
50. Mirkin, S., and B.C. Albensi, *Should artificial intelligence be used in conjunction with neuroimaging in the diagnosis of Alzheimer's disease?* Frontiers in Aging Neuroscience, 2023. **15**: p. 1094233.
51. Ganesh, S., et al., *Exploring Huntington's disease diagnosis via artificial intelligence models: A comprehensive review*. Diagnostics, 2023. **13**(23): p. 3592.

52. Patwekar, M., et al., *Harnessing artificial intelligence for enhanced Parkinson's disease management: Pathways, treatment, and prospects.* Trends in Immunotherapy, 2023. **7**(2): p. 2395.
53. Marzi, S.J., et al., *Artificial intelligence for neurodegenerative experimental models.* Alzheimers Dement, 2023. **19**(12): p. 5970–5987.
54. Naji, Y., et al., *Artificial intelligence and multiple sclerosis: Up-to-Date review.* Cureus, 2023. **15**(9): p. e45412.
55. Cope, T.E., et al., *Advances in neuroimaging to support translational medicine in dementia.* J Neurol Neurosurg Psychiatry, 2021. **92**(3): p. 263–270.
56. Laborieux, A., et al., *Synaptic metaplasticity in binarized neural networks.* Nature Communications, 2021. **12**(1): p. 2549.
57. Shajari, S., et al., *The emergence of AI-based wearable sensors for digital health technology: A review.* Sensors, 2023. **23**(23): p. 9498.
58. Wang, X., Z. Liu, and T. Zhang, *Flexible sensing electronics for wearable/attachable health monitoring.* Small, 2017. **13**(25): p. 1602790.
59. Huang, S., et al., *Flexible electronics: Stretchable electrodes and their future.* Advanced Functional Materials, 2019. **29**(6): p. 1805924.
60. Ha, M., S. Lim, and H. Ko, *Wearable and flexible sensors for user-interactive health-monitoring devices.* Journal of Materials Chemistry B, 2018. **6**(24): p. 4043–4064.
61. Swapna, M., et al., *Bio-signals in medical applications and challenges using artificial intelligence.* Journal of Sensor and Actuator Networks, 2022. **11**(1): p. 17.
62. Savage, N., *Tapping into the drug discovery potential of AI.* Nature.com, 2021.
63. Thomas, U., *AI-based drug discovery company atomwise sets its sights on inflammatory disease market.* GEN Edge, 2023. **5**(1): p. 683–687.
64. Strickland, E., *IBM Watson, heal thyself: How IBM overpromised and underdelivered on AI health care.* IEEE Spectrum, 2019. **56**(4): p. 24–31.
65. VLife Sciences Technologies: MDS 3.5 (2008) Molecular Design Suite, Pvt. Ltd., Pune, 2010.
66. Eberhardt, J., et al., *AutoDock Vina 1.2. 0: New docking methods, expanded force field, and python bindings.* Journal of Chemical Information and Modeling, 2021. **61**(8): p. 3891–3898.
67. de Almeida Filho, J.L., and J.H. Fernandez. *MDR surflexdock: a semi-automatic webserver for discrete receptor-ensemble docking.* in *Brazilian Symposium on Bioinformatics.* 2019. Springer.

11 Computational Neuroscience

Simulating the Systems Biology of Synaptic Plasticity

Princy Choudhary
National Centre for Cell Science
Pune, Maharashtra, India
and
National Institute of Mental Health and Neurosciences
Bengaluru, Karnataka, India

Mohammad Asif and Diya Srivastava*
Indian Institute of Information Technology Allahabad
Prayagraj, Uttar Pradesh, India

11.1 INTRODUCTION

Neurons possess the remarkable ability to respond differently to specific patterns of inputs, including temporal and spatial. This reciprocation, explicitly termed as information or memory storage, is not inherent in neurons but readily develops as the organism interacts with its environment. The plasticity of synapses and ion channels is believed to be responsible for this learning and memory process. Therefore, to comprehend how neurons store information, it is crucial to understand how the spatial and temporal patterns of synaptic input, combined with the activity of the receiving neuron, lead to changes in neuronal plasticity, which in turn alter the patterns of neuronal activity. Synaptic plasticity is a fundamental process that underlies learning and memory formation in the brain. It plays a crucial role in shaping the connectivity and functionality of neural circuits, allowing them to adapt and respond to environmental stimuli. Understanding the mechanisms of synaptic plasticity is essential for unraveling the mysteries of brain function and developing new therapeutic strategies for neurological disorders.

Recent research has uncovered the involvement of numerous molecules and intricate interactions among them in the process of plasticity. Consequently, simple conceptual models are no longer sufficient to explain the complex molecular mechanisms underlying neuronal plasticity. Instead, a deeper analysis requires quantitative computational models that account for the dynamics and nonlinear interactions of molecules. These computational models are essential for gaining a comprehensive mastery of the underlying processes at a molecular level [1,2]. Two crucial components of neuronal plasticity play a significant role in information processing: the plasticity of intrinsic excitability and synaptic plasticity. The plasticity of intrinsic excitability refers to changes in the properties of ion channels within a neuron, while synaptic plasticity involves alterations in the strength of connections between two neurons. While our knowledge of the signaling pathways involved in synaptic plasticity is more extensive, many of these pathways also contribute to the plasticity of intrinsic excitability. Modern technical advances have hastened the creation of signaling pathway models. The first advancement is live cell imaging, which offers critical limits on kinetics (rate constants) and spatial organization of molecules. This information is crucial for investigating various aspects of information processing, including

 DOI: 10.1201/9781003464648-13

synaptic specificity and the discrimination of spatio-temporal patterns. The second breakthrough is the rapid growth of computer technology and design, which is critical given synaptic plasticity models' temporal and spatial scale. Simulating large-scale spatial structures with high resolution over extended periods necessitates performing trillions of calculations [3,4]. To comprehend the functioning, controlling, adaptability, and robustness of a biological system, it is necessary to examine and scrutinize both the individual components and the dynamic interactions among them across various temporal and spatial scales. This comprehensive characterization is essential for interpreting how the system operates and maintains its functionality in different contexts. This degree of insight is achieved in systems biology by integrating computer modeling and theoretical analysis with extensive experimental inquiry. By integrating these approaches on a large scale, important questions about the functional organization of complex cellular signaling networks have been effectively addressed. Since the signaling cascades driving synaptic plasticity are a subset of those investigated by systems biologists, the following methodologies may be used to understand synaptic plasticity computationally [4,5].

11.2 ROLE OF SYSTEMS BIOLOGY IN UNDERSTANDING SYNAPTIC PLASTICITY IN NEURODEGENERATIVE DISEASES

Neurodegenerative diseases, such as Alzheimer's, Parkinson's, and Huntington's, are a major public health concern worldwide. These conditions are associated with the progressive loss of neurons and synaptic dysfunction, leading to cognitive impairment, motor dysfunction, and other debilitating symptoms. Despite significant advances in analyzing the molecular and cellular mechanisms underlying these diseases, effective therapies remain elusive [6,7].

The study of synaptic plasticity, the ability of neurons to change their strength and connectivity in response to experience, holds significant relevance in the context of neurodegenerative diseases. Synaptic plasticity is essential for learning, memory, and other cognitive functions, and its dysfunction is implicated in the pathogenesis of neurodegenerative diseases [8].

The traditional approach to synaptic plasticity has focused on individual components, such as receptors, signaling pathways, and genes. However, this reductionist approach fails to capture the complexity of synaptic plasticity and its interactions with other cellular processes.

Systems biology provides an alternative approach to contemplate synaptic plasticity by integrating molecular, cellular, and systems-level data into mathematical and computational models. This approach allows innovators to analyze the intricate interactions and feedback loops involved in synaptic plasticity and identify key components and pathways contributing to neurodegenerative pathologies.

The aim of this chapter is to investigate the role of systems biology in studying synaptic plasticity and its implications for neurodegenerative diseases. Specifically, this chapter will address the following topics:

- The use of systems biology simulations in drug discovery and target identification for neurodegenerative diseases.
- The predictive modeling of synaptic plasticity outcomes under pathological conditionsinherent to neurodegenerative disorders [9].
- The personalized medicine approach using systems biology simulations to tailor therapeutic interventions based on individual patient data.
- The learning of synaptic plasticity through systems biology to unravel the mechanisms underpinning synaptic dysfunction in neurodegenerative diseases.

Overall, this chapter aims to provide a comprehensive formulation of the role of systems biology in examining synaptic plasticity and its potential for developing effective therapies for neurodegenerative diseases [10].

11.3 SYSTEM BIOLOGY FOUNDATION

Systems biology represents a holistic approach to complex biological phenomena, emphasizing the interdependence and interactions of various components within a system. The multidisciplinary discipline of system biology examines the intricate relationships within biological systems to figure out how these relationships affect how the system behaves and functions [11]. It aims to predict and uncover emergent characteristics of cells, animals, and ecosystems by combining data from computational modeling, bioinformatics, and molecular biology.

11.3.1 Basic Principles and Components of Systems Biology

A key idea in systems biology is the holistic view, which highlights the importance of comprehending biological systems as a whole as opposed to concentrating only on their separate parts. This method is essential because biological systems are complex, and the linkages and interactions between many components can result in behaviors and functions that are not evident when these components are investigated separately [11–13].

11.3.1.1 Holism vs. Reductionism

Systems biology stresses the study of biological systems as a whole, in contrast to classical biology, which frequently concentrates on individual components (like genes or proteins).

It acknowledges that examining a system's constituent elements separately will not provide a complete understanding of its behavior and features.

11.3.1.2 Interdisciplinary Approach

To investigate biological systems thoroughly, systems biology incorporates ideas and techniques from a variety of academic fields, including biology, chemistry, physics, mathematics, computer science, and engineering.

11.3.1.3 Emergent Properties

The idea that the total is greater than the sum of its parts is expressed in this principle. Emergent characteristics, which give rise to new functions or behaviors that individual system components lack, are the result of interactions among biological systems' constituent parts [14].

11.3.1.4 Networks and Interactions

Systems biology studies how genes, proteins, and other molecules interact with one another to understand how these networks work and control biological processes. It adopts a network-centric perspective, treating biological entities as interconnected nodes and their interactions as edges. This network thinking is particularly relevant in the analysis of neural systems, where the connectivity between neurons and synapses forms intricate networks. By mapping and analyzing these networks, systems biology provides insights into the organizational principles that govern synaptic plasticity across the entire neural architecture [15].

11.3.1.5 Dynamic Systems

Biological systems are dynamic, exhibit changes over time, and undergo continuous changes over time. The goal of systems biology is to comprehend these temporal shifts and the underlying mechanisms [14].

11.3.1.6 Quantitative Analysis and Modeling

Systems biology mostly depends on quantitative data and computer models to simulate and forecast system behaviors under various circumstances.

11.3.1.7 Iterative and Integrative

Systems biology is characterized by iterative cycles of modeling and experimentation, in which models are used to forecast future experimental results, and experimental data is used to improve models.

11.3.1.8 Feedback and Control

Systems biology explores the role of feedback loops and regulatory mechanisms in maintaining homeostasis and orchestrating responses to internal and external stimuli. Comprehending these feedback systems is crucial for deciphering the resilience and adaptability of biological systems.

11.3.1.9 Integration of Multiple Components

A holistic perspective considers all biological system components—genes, proteins, cells, tissues, organs, and organisms—to be interrelated and dependent on one another. This implies that modifications made to one component of the system may have a significant impact on other components [16].

11.3.1.10 Emphasis on Interactions

The key is the interactions among various biological components. These connections may be environmental, cellular, molecular, or hereditary. The goal of systems biology is to map out these relationships and comprehend how they affect the system's overall behavior and function.

11.3.1.11 Systems-Level Analysis

Systems-level exploration, including whole-genome sequencing, proteomics, and metabolomics, which offer extensive data sets that capture the state of every component inside the system, is frequently incorporated into holistic analysis [17].

11.3.1.12 Computational Modeling

Computational models are frequently used to simulate and forecast system behavior because of the complexity and volume of data they include. By combining different kinds of data, these models can shed light on how biological systems work as a whole.

11.3.1.13 Transdisciplinary Collaboration

One may need to draw on insights from computer science, physics, chemistry, biology, and other fields to comprehend complex systems. As such, a holistic perspective frequently calls for collaboration across disciplines. Systems biology encourages collaboration across diverse scientific disciplines, fostering an interdisciplinary approach to biological research. By bringing together experts from molecular biology, genetics, physics, mathematics, and computer science, systems biology leverages a collective grasp to address the complexities of synaptic plasticity. This collaborative ethos enhances the capacity to develop comprehensive computational models that bridge the gap between molecular events and network-level phenomena in neural systems [18,19].

11.3.1.14 Application in Biotechnology and Medicine

In fields such as personalized medicine, where knowledge of the intricate relationships between genes, environment, and lifestyle can result in more individualized, more successful therapies, the use of this holistic approach is becoming more and more significant.

11.3.2 Integration of Molecular-, Cellular-, and Network-Level Interactions in Neural Systems

The integration of molecular-, cellular-, and network-level interactions lies at the core of systems biology, particularly in the context of understanding synaptic plasticity within neural systems.

This intricate integration offers a comprehensive view of the mechanisms governing the dynamic changes in synaptic strength and plasticity.

A good illustration of the systems biology method taking a holistic perspective is the integration of interactions at the molecular, cellular, and network levels in neural systems. Molecular elements such as receptors and neurotransmitters greatly influence cell signaling in the nervous system. Fundamental to neuron function, these chemical interactions result in cellular behaviors like excitability and synaptic plasticity. The focus goes beyond individual neurons and instead looks at how these cells combine to form complex networks, which are necessary for complex behavior generation and information processing. Learning, memory, and general brain function are all affected by these dynamically modulating networks, which are influenced by a variety of factors, including genetic and environmental ones [19]. Field experts hope to create comprehensive models that can explain brain processes and dysfunctions, providing insights into neurological illnesses and directing the development of targeted therapeutics by combining data across multiple levels, from molecules to networks. This integrative approach highlights the importance of systems biology in deciphering the complexity of the brain [20,21].

11.3.2.1 Multi-scale Modeling

In systems biology, a cardinal approach involves modeling phenomena across multiple hierarchical scales, spanning the molecular intricacies of individual synapses to the broader interactions occurring between neurons in intricate neural circuits. This strategy recognizes synaptic plasticity as a phenomenon influenced not merely by discrete molecular events but also by the collective dynamics emerging from the intricate interplay of interconnected neurons. Integrating diverse scales enhances the fidelity of computational models, encapsulating the richness and complexity of synaptic plasticity [21].

11.3.2.2 Data Integration

An indispensable facet of systems biology is the synthesis of diverse datasets arising from various experimental modalities. Genomic, proteomic, and imaging data are amalgamated to cultivate a comprehensive formulation of the molecular and cellular determinants influencing synaptic plasticity. This integrative methodology empowers scientists to formulate computational models that leverage a spectrum of experimental data, thereby providing a more precise representation of the intricate processes governing synaptic function.

11.3.2.3 Cross-Talk Between Components

Within systems biology, the acknowledgment prevails that the molecular, cellular, and network strata are not discrete entities but interconnected through intricate cross-talk. Signaling cascades initiated at the molecular level traverse cellular domains, exerting a discernible impact on the behavior of entire neural networks. Scrutinizing this cross-talk is imperative for comprehending how alterations in individual synapses can precipitate cascading effects, thereby influencing the overarching functionality of neural systems.

11.3.2.4 Emergent Properties

Systems biology focuses on how molecules, cells, and networks interact. It looks at how complex behaviors and properties emerge that can't be predicted just by studying the individual parts on their own. Synaptic plasticity, when examined through the systems biology lens, exhibits emergent properties resulting from dynamic interactions across various scales. A nuanced knowledge of these emergent properties is pivotal for capturing the quintessence of synaptic plasticity within computational models.

In summation, the confluence of molecular-, cellular-, and network-level interactions constitutes the sine qua non of systems biology's approach to synaptic plasticity. By embracing the intricacies and interconnected nature of neural systems, examiners can engender computational models that faithfully replicate the dynamic and emergent nature characterizing synaptic plasticity, thereby fostering a profound comprehension of the operational dynamics and adaptive proclivities inherent to these intricate biological systems [22].

11.4 MOLECULAR MECHANISMS OF SYNAPTIC PLASTICITY

The capacity of synapses, the connections between neurons, to become stronger or weaker with time is known as synaptic plasticity. This alteration is dependent on the activity of the neurons involved and is essential for learning and memory. It symbolizes the ability of the brain to change and rearrange neural connections in reaction to experiences.

11.4.1 Overview of Key Molecular Players in Synaptic Plasticity

The molecular components involved in synaptic plasticity are crucial for signal transmission and structural modifications at synapses. Neurotransmitters like glutamate and GABA play essential roles in excitatory and inhibitory synapses, respectively, influencing synaptic plasticity. The glutamate-binding NMDA and AMPA receptors are vital for synaptic plasticity, with AMPA receptors quickly modulated during plasticity. The significant role of NMDA receptors in synaptic strengthening makes them particularly important. Calcium ions (Ca^{2+}), particularly through NMDA receptors, trigger various neuronal signaling pathways critical for synaptic plasticity. Enzymes such as calcineurin, PKA, and CaMKII control the phosphorylation statuses of proteins, impacting synaptic strength. Adhesion molecules like neuroligins and neurexins help in organizing and stabilizing synaptic connections. Cytoskeletal proteins, including actin and tubulin, facilitate structural alterations at synapses during plasticity. Neurotrophic factors such as BDNF control synaptic plasticity by influencing neuron survival, development, and differentiation. Immediate early genes (IEGs) like c-Fos and Arc quickly regulate synaptic strength in response to synaptic activity. Long-term variations in synaptic strength rely on mRNA and local protein synthesis at the synapse, while epigenetic factors such as histone acetylation and DNA methylation can impact gene expression patterns crucial for sustained synaptic plasticity. Figure 11.1 depicts how different kinds of synaptic plasticity regulate neuronal activity involving different receptors.

11.4.2 Signaling Pathways Involved in Synaptic Plasticity

There are various significant signaling mechanisms that regulate the potency and strength of synaptic connections associated with synaptic plasticity. They have an extensive impact on learning, memory, and neural transmission.

- In the **NMDA Receptor Pathway**, the binding of glutamate to the NMDA receptor, Ca^{2+} ions are allowed to drift into neurons. This influx of calcium is significant in activating numerous downstream signaling pathways. Depending on the intensity and duration of the signal, it could lead to long-term potentiation (LTP) or long-term depression (LTD).
- During the **Calcium/Calmodulin-Dependent Protein Kinase II (CaMKII) Pathway**, Ca^{2+} influx via NMDA (N-methyl-D-aspartate) receptors causes activation of CaMKII. It is essential for the phosphorylation of AMPA receptors and other proteins, causing alterations in synaptic strength.
- cAMP Response Element-Binding Protein (CREB) is a significant transcription factor in the **CREB Pathway**, which gets activated by different kinases like PKA, CaMKIV, and MAPK. It affects the regulation of genes associated with synaptic plasticity, imparting to the long-term modifications in synaptic strength [6].
- The **Protein Kinase A (PKA) Pathway** is also involved in the modulation of synaptic plasticity by regulating the activity of several proteins comprising ion channels and receptors. This signaling gets activated by cyclic AMP (cAMP) and PKA, major players known for neural development [6].

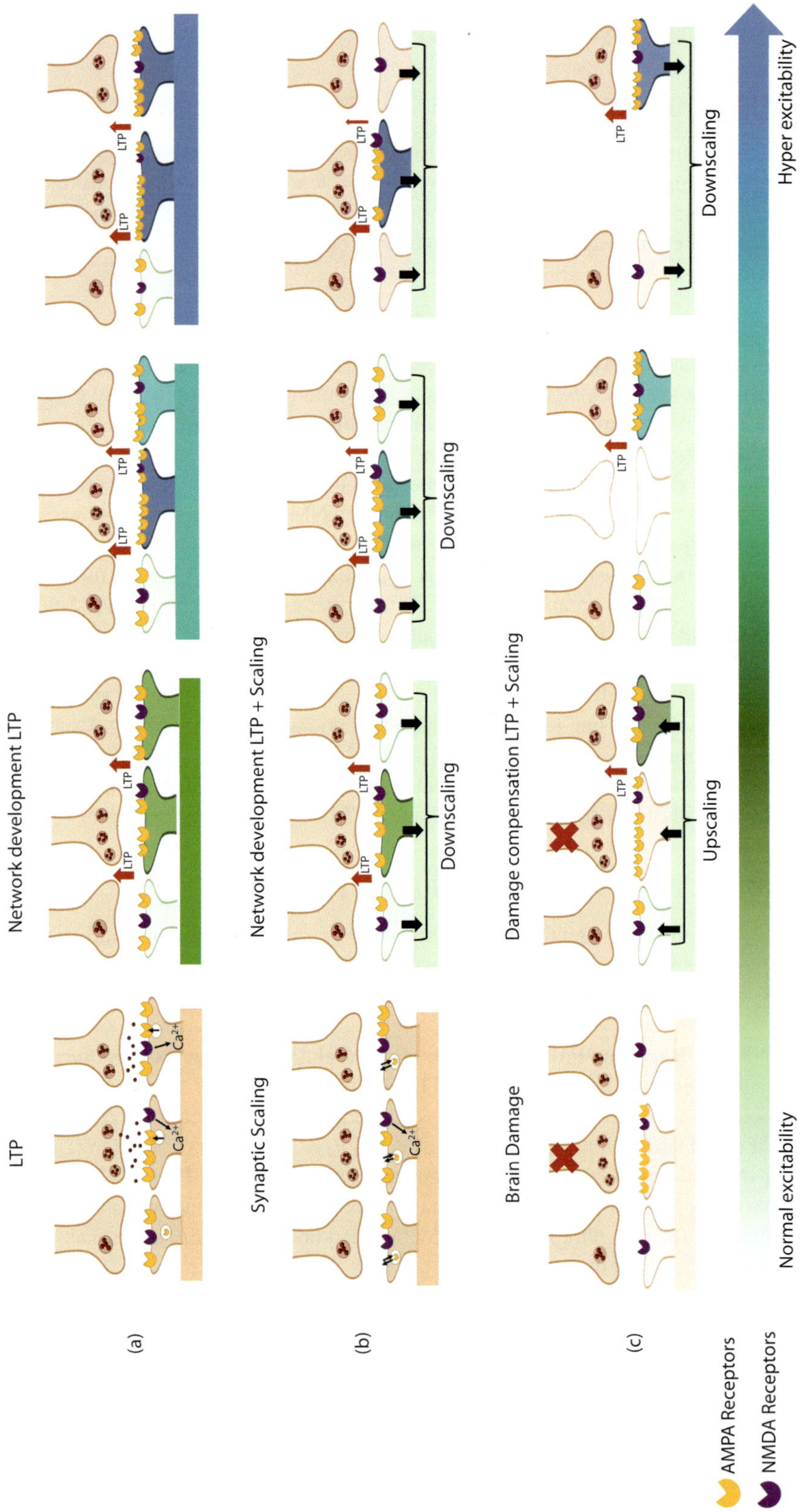

(a)
LTP
Ca2+
Network development LTP
LTP
(b)
Synaptic Scaling
Network development LTP + Scaling
Downscaling
(c)
Brain Damage
Damage compensation LTP + Scaling
Upscaling
Downscaling
Normal excitability
Hyper excitability
AMPA Receptors
NMDA Receptors

FIGURE 11.1 Regulation of neuronal activity by collaboration of different kinds of synaptic plasticity.

(a) LTP is specific to the input, as it only involves active synapses. This type of NMDA-mediated plasticity is caused by the entry of calcium into the post-synaptic terminal, which in turn leads to an increase in the expression of AMPA receptors and synaptic excitability, thus promoting further LTP expression. However, it is important to note that LTP is an anti-homeostatic form of plasticity, which means that it can promote uncontrolled enhancement of synaptic activity, leading to neuronal hyperexcitability and network instability during brain network development.
(b) Synaptic scaling is a type of plasticity that helps maintain a stable and balanced network of neurons independently of NMDA receptor activation. This process involves all the synapses in the network and is regulated by the expression of AMPA receptors, which can be either increased (upscaling) or decreased (downscaling). The balance between anti-homeostatic and homeostatic plasticity is crucial for optimal network organization and efficient information processing. The presence of both potentiated and silent synapses (represented by grey spines) allows for specific and segregated information processing, while also preventing an excessive increase in overall excitability.
(c) Acute disconnection due to brain damage deprives neurons of synaptic inputs. Synaptic scaling and LTP cooperate to restore neuronal excitability. This promotes initial hyperexcitability, which is induced by compensatory upscaling. Chronic reorganization is favored, which properly balances homeostatic and anti-homeostatic plasticity. Abbreviations: long-term potentiation (LTP); N-methyl-D-aspartate (NMDA); α-amino-3-hydroxy-5-methyl-4-isoxazolepropionic acid (AMPA).

- The **Mitogen-Activated Protein Kinase (MAPK/ERK) Pathway** is activated by many extracellular signals that affect the gene expression and protein synthesis for long-term synaptic modifications [6].
- The **Phospholipase C (PLC) Pathway** gets activated by GPCR (G-protein coupled receptors), PLC leads to the IP3 and DAG formation, which are secondary messengers causing the discharge of Ca^{2+} from intracellular stores and trigger protein kinase C (PKC), respectively, determining synaptic plasticity [6].
- In the **Brain-Derived Neurotrophic Factor (BDNF) Pathway**, BDNF binds to TrkB receptors, causing the activation of different downstream signaling pathways like PI3K/Akt and PLCγ, which has a significant role in neuronal survival, growth, and synaptic plasticity [6].
- Rho family GTPases, such as Rac, Rho, and Cdc42, in the **Rho GTPase Pathway** take part in cytoskeletal rearrangements that are critical for synaptic structural modifications during plasticity [23].

By exploring the neurotransmitter systems, signaling cascades, and recently discovered molecular players, we have gained a comprehensive understanding of how synaptic strength can be dynamically modulated in response to neural activity. It is important to continue examining these processes to further our familiarity of the brain and how it functions.

11.5 CELLULAR PROCESSES AND SYNAPTIC PLASTICITY

Cellular mechanisms including neurotransmitter release, receptor trafficking, structural modifications to the dendritic spine, and changes in synaptic strength are all involved in synaptic plasticity. Essential for learning and memory formation, neurons adapt by changing the size and shape of synapses, changing the density of receptors, and regulating signal transmission efficiency [1,3,24].

11.5.1 Neuronal Morphology and Its Impact on Synaptic Plasticity

11.5.1.1 Structural Plasticity

Structural plasticity encompasses the dynamic changes in dendritic spines and axonal processes, playing a pivotal role in synaptic plasticity. Dendritic spine dynamics involve the formation, enlargement, and retraction of spines, with a focus on comprehending the structural adaptations that contribute to long-term changes in synaptic efficacy. Additionally, axonal sprouting is explored as a mechanism influencing synaptic plasticity, shedding light on how alterations in axonal morphology contribute to network-level adaptations. This comprehensive examination provides insights into the intricate processes that underlie structural plasticity in the context of synaptic dynamics.

11.5.1.2 Synaptic Pruning

Developmental pruning involves a detailed observation of cellular processes contributing to synaptic pruning during developmental stages. The discussion encompasses the molecular players and activity-dependent mechanisms that play a crucial role in shaping the refinement of synaptic connections. Shifting the focus to the adult brain, the examination continues with an analysis of synaptic elimination, addressing the delicate balance between synaptogenesis and synaptic elimination. Insights are provided into the influence of experience and learning in sculpting neural circuits, offering a comprehensive formulation of the dynamic processes involved in both developmental and adult synaptic re-modeling.

11.5.2 Intracellular Signaling Cascades in Response to Synaptic Activity

11.5.2.1 Calcium Signaling

Delving into dynamic calcium dynamics, this analysis provides an in-depth examination of the intricate calcium fluctuations within neurons during synaptic activity. It inspects the sources of calcium influx, encompassing voltage-gated channels and ligand-gated channels. Shifting focus to calcium sensors and effectors, the examination further discusses the role of calcium-binding proteins as sensors. Additionally, it looks for downstream effectors like calmodulin and their pivotal involvement in transducing calcium signals to various intracellular pathways. This comprehensive research sheds light on the multifaceted aspects of calcium regulation within neurons and its crucial role in orchestrating cellular responses during synaptic activity.

11.5.2.2 Intracellular Messengers

This detailed cAMP-PKA pathway (Figure 11.2) provides insights into the cyclic adenosine monophosphate (cAMP)–protein kinase A (PKA) signaling cascade. It explores the activation of this pathway in response to synaptic activity and its modulation of synaptic strength. Shifting focus, the examination extends to the inositol trisphosphate (IP3) system, serving as a mediator of intracellular signaling. Insights are offered into its role in calcium release from internal stores and its significant contribution to synaptic plasticity. This comprehensive analysis provides a nuanced understanding of two key intracellular pathways, highlighting their roles in regulating synaptic activity and plasticity.

11.5.3 Role of Glial Cells in Modulating Synaptic Plasticity

11.5.3.1 Astrocytic Involvement

Astrocytes traverses into bidirectional communication with neurons, offering an in-depth review of the reciprocal interactions. It examines astrocytic processes, including gliotransmission, and their profound impact on synaptic efficacy. Shifting the focus to astrocytic calcium signaling, the examination contemplates its role in regulating neurotransmitter release and synaptic plasticity. Insights are provided into the dynamic interplay between astrocytes and synaptic function, emphasizing the intricate mechanisms through which astrocytes contribute to the modulation of neuronal activity and synaptic plasticity.

11.5.3.2 Microglial Dynamics

Microglial Dynamics focuses on the emerging role of microglia in synaptic pruning and maintenance, shedding light on the molecular mechanisms through which microglia actively shape

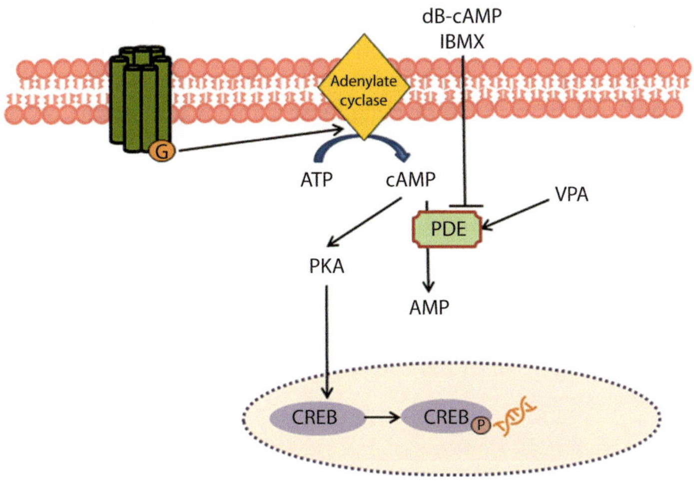

FIGURE 11.2 CAMP-PKA pathway.

Abbreviations: Protein kinase A (PKA); CREB protein (cAMP responsive element binding protein); Phosphodiesterases (PDE); 3-isobutyl-1-methylxanthine (IBMX); dibutyryl-cAMP (db-cAMP); Valproic acid (VPA).

Cyclic-adenosine-monophosphate (cAMP) is synthesized from ATP via adenylate cyclase, which active GPCR induces. cAMP activates PKA in the cytoplasm, which assists in the phosphorylation of CREB protein on reaching the nucleus. Post phosphorylation, CREB binds with CREB binding protein, and with the help of different co-factors, it binds to certain DNA sequences, thereby regulating different gene expressions. PDE is involved in the demolition of cAMP by converting it into AMP and then regulating the cytoplasmic concentration of cAMP.

synaptic connectivity. Shifting the focus to immune responses and neuroinflammation, the exploration encompasses the impact on synaptic plasticity. The discussion includes how microglial activation and the release of cytokines influence synaptic function, providing a comprehensive notion of the intricate interplay between microglia, immune responses, and synaptic plasticity in the context of neural dynamics.

11.5.4 Metabolic Regulation and Energy Homeostasis

11.5.4.1 Mitochondrial Dynamics

Mitochondrial dynamics within neurons provides an in-depth exploration of mitochondrial morphology and distribution. It looks towards the impact of changes in these factors on synaptic plasticity. Shifting the focus to mitochondrial bioenergetics, the examination encompasses the role of mitochondria in supplying energy for synaptic transmission and plasticity. Insights are offered into how mitochondrial bioenergetics influence cellular responses to synaptic activity, contributing to a comprehensive understanding of the pivotal role played by mitochondria in shaping synaptic dynamics.

11.5.4.2 Metabolic Pathways

The discussion encompasses the roles of these pathways in meeting the energy demands of synaptic plasticity. Shifting the focus to metabolic interactions with synaptic activity, the examination searches the intricate interplay between cellular metabolism and synaptic function. Insights are provided into how metabolic pathways influence cellular responses to changes in synaptic strength, contributing to a nuanced modeling of the dynamic relationship between cellular metabolism and synaptic activity.

This section provided a detailed analysis of cellular processes that contribute to the modulation of synaptic strength. The aim is to understand the structural, intracellular, and metabolic intricacies that collectively contribute to the adaptability and information-encoding capabilities of neural circuits through synaptic plasticity. By unraveling these complex phenomena, we aim to lay a comprehensive foundation for better comprehension.

11.6 NETWORK DYNAMICS AND SYNAPTIC PLASTICITY

Synaptic plasticity and network dynamics in the brain are intimately related. The connection patterns of the neural network are reshaped by variations in synaptic strength, which are impacted by neuronal activity. The process of adaptation facilitates memory development and learning. Thus, synaptic plasticity plays a vital role in the evolution of the network by reflecting experience and promoting cognitive processes.

11.6.1 Interactions Between Neurons Within Neural Circuits

The brain's learning and memory processes are shaped by synaptic plasticity, which is mostly dependent on interactions between neurons within neural circuits. At synapses, where neurotransmitters and receptors control synaptic strength, neurons communicate with one another. Neural activity patterns impact this plasticity, with mechanisms such as depression and long-term potentiation modifying synaptic efficacy. Accurate neuronal responses are ensured by balancing and refining circuit function through lateral inhibition and other feedback processes. Through these interactions, the brain can adjust to new experiences by changing synapses to reflect changes in cognition. In addition to aiding in the storage of information, this dynamic interaction is what drives neural networks' ongoing adaptation, which is essential for cognitive processes.

11.6.1.1 Local Network Interactions

This type of analysis focuses on exploring the local interactions between neurons within neural circuits and how synaptic plasticity at the microscale contributes to the overall dynamics of neural networks. Shifting the focus, the examination aims to understand the intricacies of short-range connectivity and its implications for information processing. This dual investigation provides insights into both the microscale dynamics of synaptic interactions and their broader impact on the information-processing capabilities of neural networks.

11.6.1.2 Long-Range Connectivity

This module deals with the investigation of long-range connectivity patterns and their influence on information transmission within neural networks. It aims to provide insights into the role of synaptic plasticity in shaping communication between different brain regions. Shifting the focus to dynamic interplay, the examination seeks to understand how synaptic changes dynamically contribute to the establishment of functional connectivity across distributed brain areas. This dual exploration provides a comprehensive understanding of both the structural aspects of connectivity patterns and the dynamic processes influencing the functional relationships within neural networks.

11.6.2 Network-Level Effects of Synaptic Plasticity

Synaptic plasticity has a major effect on the operation and adaptation of brain circuits at the network level. It's a crucial system that enables the brain to store and encode data. Neural circuits can be reshaped via synaptic plasticity, which alters the strength or weakness of connections between neurons through processes like long-term potentiation and long-term depression. Learning and memory are impacted by this reorganization, which changes how information moves throughout the network. Furthermore, the network's structure may be further altered via synaptic plasticity, which can result in the creation of new synaptic connections or the removal of old ones. The brain needs these dynamic changes in synaptic connections to learn new abilities, establish memories, and adjust to new experiences. Complex behaviors and cognitive processes stem from the net effect of these synaptic modifications [25].

11.6.2.1 Emergence of Network Patterns

This topic focuses on getting to know how synaptic plasticity contributes to the emergence of specific network patterns. It delves into the role of recurrent connections and feedback loops in shaping the dynamic behavior of neural circuits. Shifting attention, the examination seeks to understand the mechanisms by which synaptic plasticity influences the organization and structure of neural networks. By addressing both the dynamic aspects of network behavior and the structural organization influenced by synaptic plasticity, this analysis provides a comprehensive view of the multifaceted contributions of synaptic plasticity to neural network dynamics and organization.

11.6.2.2 Plasticity-Driven Circuit Adaptations

Plasticity-driven circuit adaptations focus on understanding how changes driven by synaptic plasticity at the synaptic level translate into functional adaptations at the circuit level. It explores insights into the self-organization of neural circuits based on activity-dependent plasticity.

The focus then shifts to feedback loops, aiming to understand the dynamic relationship between synaptic changes and circuit-level adaptations. This reciprocal relationship leads to robust and flexible neural network dynamics, emphasizing the intricate mechanisms through which synaptic plasticity influences the broader functional organization of neural circuits.

11.6.3 Emergent Properties in Neural Networks Influenced by Synaptic Changes

Neural plasticity and brain function are fundamentally dependent on emergent features in neural networks, which are shaped by changes in synapses. This idea finds out how small-scale changes in brain networks are influenced by individual synaptic modifications, resulting in intricate behaviors and cognitive processes that are not visible at the level of individual neurons.

The network's ability to adapt to new experiences and environmental changes is known as plasticity. When a new skill is learned, for instance, some synaptic connections are strengthened while others are weakened, resulting in the creation of new brain pathways. In a similar vein, the network's ability to keep information in memory is dependent on these changes stabilizing over time [26,27].

These synaptic alterations give rise to a wide range of emergent characteristics. They include decision-making, problem-solving, processing and interpreting complicated sensory inputs, and displaying adaptive behaviors. These characteristics result from the network's overall dynamics, which are shaped by the complex interactions between excitatory and inhibitory impulses, the right balance of which is essential for brain health.

Moreover, synaptic plasticity plays a role in the brain's ability to recover from damage (Figure 11.3). After an injury, the brain can self-reorganize, with neurons forging new connections to replace lost functionality. This neuroplasticity is important when it comes to healing from conditions like stroke or serious brain injury [28].

11.6.3.1 Synchronization and Oscillations

This property centers on how synaptic plasticity influences network synchronization and oscillatory behavior. It investigates the specific synaptic connections that contribute to the generation and modulation of rhythmic activity within neural networks. Shifting the focus to temporal dynamics, the changes seek to understand the role of synaptic plasticity in shaping the temporal aspects of neural networks. By addressing both the synchronization-related aspects and the temporal dynamics influenced by synaptic plasticity, the overall analysis provides a comprehensive view of how synaptic changes impact the coordinated activity and temporal patterns within neural networks.

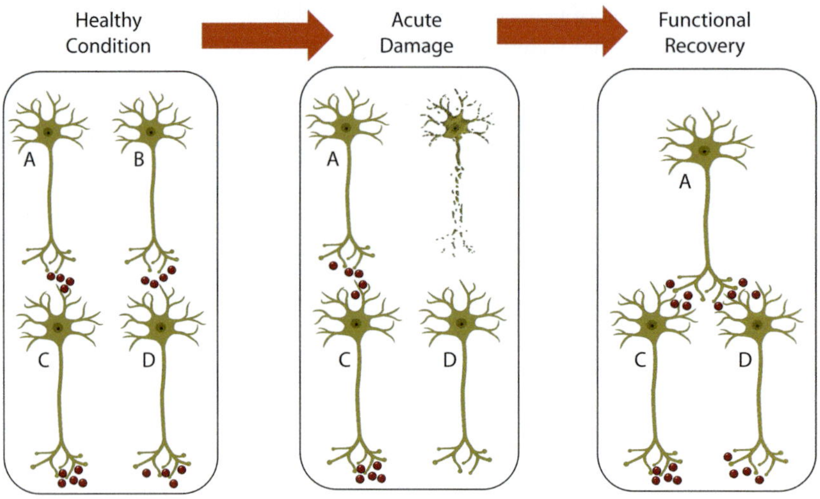

FIGURE 11.3 Synaptic plasticity is a process that helps in the recovery after neural damage. Under healthy conditions, neuronal excitatory connections are represented by a schematic model. Neurons C and D receive synaptic excitatory inputs from neurons A and B, respectively. However, if neuron B gets damaged, it deprives neuron D of excitatory synaptic input leading to disconnection and the appearance of symptoms. To recover from such a condition, increased excitability of the surviving A neuron is necessary. This process unmasks latent synaptic connections through LTP and restores the synaptic activity of neuron D. Therefore, clinical recovery is associated with the restoration of synaptic activity.

11.6.3.2 Information Coding and Decoding

This examination centers on analyzing how changes in synaptic strength contribute to the coding and decoding of information in neural networks. It explores insights into how plasticity mechanisms support learning and memory at the network level. Shifting the focus to representation and processing, the overall analysis aims to unveil the ways in which synaptic plasticity participates in the intricate processes of representing and processing information within complex neural networks. This provides a comprehensive understanding of the multifaceted contributions of synaptic plasticity to information processing and cognitive functions at the network level.

This section focused on the complex relationship between synaptic plasticity and network dynamics. It sheds light on how changes at the synaptic level affect neural circuits, shaping emergent properties and adaptive behaviors. Moreover, the network-level effects of synaptic plasticity may have implications for neurological and psychiatric disorders. Anomalous synaptic plasticity and function have been associated with neurological diseases such as Alzheimer's disease, schizophrenia, and autism spectrum disorders. Grasping these changes can help in the development of therapeutic strategies for these disorders and provide valuable insights into their pathophysiology [29].

11.7 COMPUTATIONAL MODELS OF SYNAPTIC PLASTICITY SYSTEMS

Computable representations of synaptic plasticity networks are essential for comprehending the brain's mechanisms behind memory and learning. Based on brain activity, these models mimic the processes of synapse strengthening and weakening. For the purpose of simulating how the time of neuronal firing affects synaptic alterations, they frequently include rules such as spike-timing-dependent plasticity (STDP). Through the investigation of how discrete synaptic modifications can result in intricate network behaviors, these computational frameworks contribute to our understanding of both normal brain function and illnesses involving synaptic malfunction. By using these models, scientists can predict the results of synaptic alterations and test theories, advancing the field of neuroscience.

11.7.1 System-Level Computational Models

Synaptic plasticity, a fundamental aspect of neural function, is explored through various computational models. Mastering the intricacies of these models is crucial for unraveling the dynamics of synaptic alterations.

11.7.1.1 Rate-Based Models

Rate-based models focus on the main variable of neuronal firing rates. These models are particularly suited for analyzing large networks due to their simplicity and reduced computing power requirements. They provide a clear illustration of how synapse strength changes impact brain networks' overall activity.

11.7.1.2 Spiking Neuron Models

Spiking neuron models go beyond rate-based models by re-creating individual action potentials, or spikes, of neurons. More detailed than their rate-based counterparts, these models are crucial for studying the timing-dependent aspects of synaptic plasticity and provide a more accurate depiction of brain dynamics [30].

11.7.1.3 Mean-Field Models

Mean-field models depict the typical neuronal activity in sizable populations. These models are useful for modeling how neural circuits manifest in behavior without the need to simulate every single neuron.

11.7.1.4 Network Models

Network models capture network dynamics by simulating interactions between neurons and synapses. They are essential for formulating how synaptic alterations affect memory and learning and can range from small circuits to models of entire brain regions.

11.7.2 Agent-Based Models Simulating Individual Components

In computational neuroscience, agent-based models (ABMs) provide a potent tool for examining synaptic plasticity. ABMs simulate individual components such as neurons and synapses as independent agents, offering a deep and dynamic understanding of brain processes. Their adaptability and flexibility make them useful for mimicking different brain regions and synaptic mechanisms, providing a meticulous approach to studying the complex processes of synaptic plasticity [31].

11.7.3 Network Models Capturing Interactions and Dynamics

Network models are essential tools in synaptic plasticity, capturing the intricate dynamics of brain systems by simulating synaptic connections between neurons. These models dynamically depict synaptic alterations, including long-term depression (LTD) and long-term potentiation (LTP), offering insights into how synaptic modifications impact learning, memory formation, and the general functioning of brain circuits.

11.7.4 Hybrid Models Integrating Molecular-, Cellular-, and Network-Level Details

Hybrid models in computational neuroscience seamlessly integrate information at the molecular, cellular, and network levels. These models provide a comprehensive modeling of synaptic plasticity by combining cellular processes, molecular interactions, and neuronal network connections. Their multiscale integration allows for intricate simulations, offering insights into how cellular and molecular alterations influence brain circuits' emergent behaviors. Hybrid models are invaluable for elucidating complex phenomena, including learning, memory, neurological diseases, and drug development.

This comprehensive study of computational models highlights their diverse applications in modeling the dynamic and multifaceted nature of synaptic plasticity within the field of computational neuroscience [32].

11.8 APPLICATIONS OF SYSTEMS BIOLOGY IN UNDERSTANDING SYNAPTIC PLASTICITY

The integrative approach of systems biology is essential to comprehending synaptic plasticity. Systems biology provides a complete notion of the molecular networks and pathways involved in synaptic alterations by integrating data from genetics, proteomics, and bioinformatics. This method helps to clarify the intricacies of synaptic adaptation processes, including long-term depression (LTD) and long-term potentiation (LTP). It is especially helpful in locating putative biomarkers and treatment targets for neurological conditions involving disruptions in synaptic plasticity. The comprehensive approach to systems biology improves our knowledge of how genetic and environmental factors contribute to the dynamic nature of synapse functioning in learning and memory [33].

11.8.1 Drug Discovery and Target Identification

Complicated methods such as computational modeling and high-throughput screening are used to find molecules that efficiently affect these targets. Following identification, prospective medication candidates are put through extensive pre-clinical research to ensure their safety and efficacy. After that, the most promising applicants move on to clinical trials. Neurodegenerative diseases are complex and multifactorial conditions, making drug discovery and target identification challenging. Network pharmacology, virtual screening, and molecular dynamic simulation have contributed significantly to this area. The traditional drug discovery approach involves screening large compounds' libraries for potential therapeutic activity [14,24,26,34,35].

Systems biology simulations offer an alternative approach to drug discovery by modeling synaptic dysfunction associated with neurodegenerative diseases. These simulations use mathematical and computational models to inspect complex molecular and cellular interactions involved in synaptic plasticity. By simulating the effects of different compounds on these interactions, experts can identify potential drug targets and optimize therapeutic efficacy. Treatments for neurological conditions where synaptic disruption is a hallmark, like depression, schizophrenia, and Alzheimer's disease, depend on this process. Patients suffering from these disorders may find hope in emerging medications that aim to improve or restore synaptic function by addressing the molecular foundations of synaptic plasticity. A thorough grasp of the intricate biology of synapses and the creative use of drug discovery methods is essential for the success of this strategy [36].

For example, systems biology simulations have been used to identify potential drug targets for Alzheimer's disease, a neurodegenerative disease characterized by the accumulation of amyloid beta ($A\beta$) plaques and tau protein tangles in the brain. By modeling the molecular interactions involved in $A\beta$ production and clearance, researchers identified several potential drug targets, including enzymes involved in $A\beta$ production and transporters involved in $A\beta$ clearance [37]. Similarly, systems biology simulations have been used to identify potential drug targets for Parkinson's disease, a neurodegenerative disease characterized by the loss of dopaminergic neurons in the brain. By modeling the molecular interactions involved in dopaminergic neuron degeneration, scientist identified several potential drug targets, including enzymes involved in dopamine metabolism and proteins involved in oxidative stress [38].

In addition to drug discovery, systems biology simulations also play a crucial role in target identification. By modeling the effects of different compounds on synaptic plasticity, field experts can identify the most promising drug targets for further study. This approach reduces the time and cost involved in drug development and increases the likelihood of success. Overall, the use of systems biology simulations in drug discovery and target identification holds significant promise for the development of effective therapies for neurodegenerative diseases. By providing a more efficient and targeted approach to drug development, these simulations offer the potential to improve patient outcomes and quality of life.

11.8.2 Predictive Modeling for Synaptic Plasticity Outcomes

Predictive modeling focuses on the predictive modeling of synaptic plasticity outcomes and its implications for developing prognostic tools in the context of neurodegenerative diseases. It explores the use of systems biology simulations in forecasting synaptic plasticity outcomes and the potential application of predictive modeling in identifying key components and pathways that contribute to neurodegenerative pathologies.

Synaptic plasticity predictive modeling is a developing field at the interface of data science and computational neuroscience. It entails building complex models to predict synaptic alterations and how neural network activity will be affected by them. These models combine a variety of datasets, including molecular, imaging, and electrophysiological data, and use cutting-edge machine-learning techniques to interpret intricate patterns and correlations. These models are especially good at modeling possible effects of adjustments to synapse strength under different circumstances, like altered brain activity, genetic differences, or the introduction of neuromodulators. This method is very helpful in both forecasting the course of neurological illnesses resulting from synaptic malfunction and comprehending the mechanics behind learning and memory. Furthermore, pharmaceutical exploration is benefiting greatly from predictive modeling, which directs the development and testing of novel medications that target synaptic processes. These models are constantly improving our ability to comprehend, anticipate, and control the complex mechanisms underlying synaptic plasticity, closing the knowledge gap between theoretical neuroscience and real-world applications in AI and medicine [39,40].

By integrating molecular, cellular, and systems-level data into mathematical and computational models, researchers can analyze the intricate interactions and feedback loops involved in synaptic plasticity and identify the factors that contribute to the development of neurodegenerative diseases. This approach has the potential to provide new insights into the mechanisms underlying synaptic dysfunction and to identify novel therapeutic targets. It highlights the potential of predictive modeling in developing prognostic tools for neurodegenerative diseases. By analyzing the data from individual patients, researchers can develop personalized models that predict disease progression and help clinicians to tailor treatment plans accordingly.

11.8.3 Personalized Medicine Approaches Based on Systems-Level Simulations

This approach discusses the use of individual patient data in systems biology simulations, which could help in developing tailored therapeutic interventions based on the unique pathophysiological characteristics of each patient.

Utilizing systems-level simulations, personalized medicine represents a revolutionary approach in healthcare, particularly in treating neurological illnesses. These patient-specific simulations provide insights into individual differences in disease progression and therapy response by incorporating patient-specific data, including genetic profiles, biomarkers, and clinical history, into extensive computer models. This strategy is especially important for diseases where synaptic plasticity is critical, including depression, epilepsy, and Alzheimer's. Treatment tactics can be customized by using systems-level simulations to forecast how a patient's brain circuitry could react to different therapeutic interventions. This approach helps identify new therapeutic targets while also improving the efficacy of currently available medicines. A new era in treating neurological conditions is being heralded by personalized medicine through systems-level simulations, which capture the complexity of synaptic interactions and their impact on neural function. This represents a significant step towards more efficient, effective, and individualized healthcare solutions.

11.8.4 Understanding Neurological Disorders

Systems biology provides a comprehensive framework for modeling the molecular and cellular underpinnings of neurological disorders associated with synaptic plasticity dysregulation. By integrating data from diverse sources, including genomics, proteomics, and clinical data, systems biology contributes to unraveling the complex mechanisms involved in disorders such as Alzheimer's, Parkinson's, and schizophrenia. This deeper understanding fosters the development of targeted therapies and interventions.

11.8.5 Cognitive Enhancement and Brain-Machine Interfaces

Exploring the intricacies of synaptic plasticity through systems biology contributes to the development of cognitive enhancement strategies. By deciphering the molecular and network-level dynamics, experts can identify potential avenues for enhancing cognitive functions such as memory and learning. Additionally, insights gained from systems biology can inform the design and optimization of brain-machine interfaces, enhancing their effectiveness in translating neural signals into actionable outputs.

11.9 ADVANCES IN EXPERIMENTAL TECHNIQUES FACILITATING SYSTEMS BIOLOGY SIMULATIONS

Recent developments in experimental approaches have greatly expanded the capabilities of systems biology simulations. Driven by technologies such as single-cell RNA sequencing, single-cell examination reveals cellular heterogeneity and improves our comprehension of complex cellular interactions. With its enhanced sensitivity and throughput, mass spectrometry-based proteomics offers a more profound knowledge of protein dynamics, which helps create precise cellular signaling networks. Field experts can now see cellular structures at previously unheard-of resolutions thanks to super-resolution microscopy, which also makes it easier to create more accurate models of cellular organization. Analyzing the activities of genes and their roles in cellular networks is made easier by the targeted gene editing capabilities of CRISPR/Cas9 technology. Dynamic studies of cellular processes are aided by optogenetics and microfluidics, which provide precise control over cellular activity and, correspondingly, provide controlled settings [40].

11.9.1 High-Throughput Technologies for Data Generation

High-throughput technologies are essential in the computational research of synaptic plasticity because they provide large-scale information that offers important insights into the dynamics of neuronal connections and the molecular mechanisms underpinning synaptic plasticity. Electrophysiological high-throughput screening is a crucial technology that enables the simultaneous recording of electrical activity in many neurons or synapses using automated patch-clamp systems. This makes it possible to effectively characterize synaptic responses and plasticity under a range of experimental setups. Additionally, sophisticated imaging techniques like high-throughput fluorescence microscopy make real-time monitoring of structural alterations and synaptic events across a large population of synapses easier. Moreover, high-throughput data produced by methods such as RNA sequencing (RNA-seq) are useful for computational models to examine changes in gene expression linked to synaptic plasticity. Collectively, these technologies contribute to a systems-level understanding of synaptic plasticity, aiding in the development and refinement of computational models that simulate the intricate dynamics of neuronal networks and their adaptive responses.

11.9.2 Imaging Techniques Providing Insights into Dynamic Cellular Processes

Advanced imaging techniques have completely transformed our ability to watch and comprehend dynamic cellular processes with great spatial and temporal precision. One such method is confocal microscopy, which visualizes cellular structures in three dimensions by removing out-of-focus light by point-by-point scanning and a pinhole. The use of fluorescent probes to tag particular molecules allows for the real-time tracking of cellular activities in fluorescence microscopy, particularly in live-cell imaging. Beyond the diffraction limit, super-resolution microscopy—which uses methods like structured illumination microscopy (SIM) and stimulated emission depletion (STED)—provides previously unheard-of insights into cellular structures. Protein-protein interactions and dynamic intracellular signaling processes can be observed using time-lapse imaging and fluorescence resonance energy transfer (FRET) microscopy. Furthermore, light-sheet microscopy and spinning disc confocal microscopy provide quick and less phototoxic imaging for dynamic processes. When taken as a whole, these imaging technologies offer priceless insights into the subtleties of cellular dynamics, leading to advancements in cell biology, neuroscience, and other sciences [41].

11.9.3 Electrophysiological Methods Complementing Computational Simulations

Especially in neuroscience, electrophysiological techniques are invaluable adjuncts to computational simulations, creating a synergistic approach to deciphering cellular processes' complexities. Real-time measurements of neuronal electrical activity are made possible by electrophysiology using methods such as patch-clamp recording and extracellular multi-electrode arrays. These measurements provide important information about action potentials, synaptic transmission, and network dynamics. The iterative process of model validation and refinement, which compares simulated predictions with experimentally collected data to improve models' accuracy and biological relevance, demonstrates the synergy between computational simulations and modeling. In addition to validating computer models, electrophysiological recordings aid in parameter estimation, guaranteeing that simulated behaviors correspond with actual physiological responses. This dynamic interaction extends to the testing of hypotheses, allowing scientists to confirm computer model predictions through experimentation. Moreover, by evaluating the effect of pharmacological interventions on cellular excitability, electrophysiological techniques aid in the testing and development of drugs [42]. By directing the creation of more realistic and predictive computer models and offering experimental validation based on the physiological reality of living cells, this integrative method promotes a deeper knowledge of neuronal function. Combined, these approaches increase our understanding of neuronal and cellular processes, opening new avenues for experimentation in neuroscience and other fields.

11.9.3.1 Optogenetics

The advent of optogenetics represents a groundbreaking development in experimental techniques. Optogenetic tools allow precise control of neuronal activity with spatial and temporal precision, enabling scientist to modulate synaptic plasticity in a targeted manner. The integration of optogenetic data into systems biology simulations enhances the realism and applicability of computational models.

11.9.3.2 Multi-Omics Approaches

The integration of multi-omics approaches, combining genomics, transcriptomics, proteomics, and metabolomics, provides a holistic view of synaptic plasticity. These techniques offer comprehensive insights into the molecular players and signaling pathways involved. Integrating multi-omics data into systems biology simulations enhances the granularity and complexity of models, reflecting the intricate nature of synaptic plasticity mechanisms.

11.9.3.3 Emerging Technologies

Continued advancements in emerging technologies, such as CRISPR gene editing and single-cell omics, contribute to the refinement of experimental techniques. These technologies offer unprecedented precision in manipulating genetic and molecular components, providing more targeted and specific data for systems biology simulations.

11.9.3.4 Collaborative Platforms and Data-Sharing Initiatives

The establishment of collaborative platforms and data-sharing initiatives fosters a culture of collaboration and knowledge exchange. Open access to experimental data enhances the reproducibility and transparency of studies, facilitating the integration of diverse data sets into systems biology simulations.

11.9.3.5 Integration of Experimental and Computational Research

The convergence of experimental and computational research methodologies is a key trend. Experimental data generated through advanced techniques directly inform and validate computational models, creating a symbiotic relationship that advances our knowledge regarding synaptic plasticity.

11.9.3.6 Future Prospects and Technological Integration

The ongoing evolution of experimental techniques holds promise for further advancements in systems biology simulations. The integration of emerging technologies, coupled with enhanced collaborative efforts, will continue to shape the landscape of synaptic plasticity modeling, providing novel opportunities for comprehensive and integrative studies.

11.10 CHALLENGES AND LIMITATIONS

11.10.1 Complexities in Integrating Multi-Scale Data

Within computational neuroscience, the integration of multi-scale data presents significant challenges, particularly in the context of synaptic plasticity. The challenges in integrating data across molecular, cellular, and network levels are multifaceted. Data heterogeneity poses a significant obstacle due to diverse forms, resolutions, and contexts, making harmonization for insightful analysis a complex task. Successful integration often relies on sophisticated computational tools and procedures, demanding high computing resources. Interpreting integrated data further presents challenges, requiring deep subject expertise and advanced analytical methods to ensure correctness and biological relevance. Scalability issues arise as models effective at one scale may not seamlessly apply to another, exacerbated by the rapid evolution of data-generating technologies leading to increased volume and complexity. There is a risk of oversimplification in multi-scale data integration, potentially compromising biological accuracy for computational efficiency. Additionally, practical challenges in collaboration, including communication difficulties, conflicting goals, and data protection concerns, can hinder effective multidisciplinary collaboration essential for successful integration efforts.

11.10.2 Validation and Refinement of Computational Models

The validation and refinement of computational models, particularly those examining synaptic plasticity, face numerous obstacles and limitations. There are many challenges associated with computational models in biological modeling. Obtaining complete and high-quality experimental data for model validation poses difficulties, which may introduce inaccuracies in predictions. The inherent complexity of biological systems often leads to the oversimplification of models, ignoring critical subtleties and behaviors. Scalability issues arise as models validated under specific scales or conditions may lack accuracy when applied to other scales or conditions. The dynamic nature of computational methods and biological knowledge requires frequent model updates, making maintenance resource-intensive. Additionally, interdisciplinary collaboration faces challenges due to communication barriers and disparate approaches, hindering the improvement and validation of models. Addressing these challenges requires collaborative efforts, incorporating diverse expertise and fostering effective communication across disciplines.

11.10.3 Ethical Considerations in Systems Biology Simulations of Synaptic Plasticity

Ethical considerations become paramount in systems biology simulations of synaptic plasticity, given their potential impact on our understanding of human cognition and behavior. The application of computational models in formulating synaptic plasticity raises various ethical considerations. Privacy and consent issues emerge due to the utilization of private and sensitive data, including genetic information, necessitating stringent measures for data confidentiality. Risks of misuse and misinterpretation of simulation results, particularly in anticipating neurological or mental problems, pose concerns related to potential discrimination or stigmatization. In therapeutic contexts, ethical considerations revolve around patient safety and the responsible application of treatments based on simulation outcomes. Broader societal ramifications, including implications for ideas of free choice and personal accountability, must be carefully considered. The rapid development in this field highlights the need for proactive governance and ethical foresight to ensure that the benefits of systems biology in familiarizing with synaptic plasticity are realized responsibly and ethically. Establishing frameworks and guidelines becomes crucial to navigating these ethical challenges and promoting responsible conduct in computational neuroscience.

11.11 FUTURE DIRECTIONS IN SIMULATING SYNAPTIC PLASTICITY SYSTEMS

11.11.1 Integration of Emerging Technologies

The integration of cutting-edge technologies, particularly optogenetics and CRISPR, holds immense promise for advancing the field of mimicking synaptic plasticity systems. Optogenetics, leveraging light-sensitive proteins, enables precise regulation of neuronal activity, facilitating the alteration of synaptic strength with high spatiotemporal accuracy. Incorporating optogenetic methods into synaptic plasticity simulations allows for a dynamic and context-dependent understanding of how specific synaptic connections contribute to network dynamics. Furthermore, the revolutionary genome editing technique CRISPR has the potential to enhance models by accurately modifying genes associated with synapse function. Utilizing CRISPR-edited cellular components in simulations allows for the investigation of the effects of genetic modifications on synaptic plasticity, offering valuable insights into the molecular mechanisms underlying memory and learning. The synergy between these technologies and advanced imaging techniques enables real-time monitoring of synaptic changes in response to optogenetic and CRISPR manipulations. This integration of cutting-edge experimental tools with computational simulations enhances the predictive power of models and fosters a more comprehensive modeling of the molecular and cellular mechanisms governing synaptic plasticity. Collaborative efforts between experimentalists and computational modelers leveraging these technologies are poised to open new avenues in deciphering the complexities of synaptic plasticity, fostering breakthroughs in neurobiology and cognitive science [20,43].

11.11.2 Enhanced Collaboration between Computational Modelers and Experimentalists

A new era in synaptic plasticity research has commenced with enhanced cooperation between experimentalists and computational modelers. Breaking down conventional silos, this interdisciplinary synergy promotes a thorough and holistic approach to comprehending the complexities of synaptic plasticity. Armed with advanced algorithms and simulations, computational modelers provide theoretical frameworks tested and refined by empirical data from experimentalists, ensuring biological applicability.

This collaboration is further strengthened by the utilization of cutting-edge technologies such as optogenetics and CRISPR. While CRISPR enables focused genetic alterations for exploring the nervous system, optogenetics offers precise control over neuronal activity, providing dynamic input for computer models. This cooperative strategy accelerates the development of innovative therapeutic strategies for neurological illnesses while concurrently enhancing our basic notion of synaptic plasticity. As computational models become more sophisticated and incorporate detailed biological data, and experimental techniques become more precise and high-throughput, this collaborative future holds the potential for groundbreaking discoveries in neuroscience and cognitive science.

11.11.3 The Role of Systems Biology in Unraveling Synaptic Plasticity in Health and Disease

Systems biology provides a comprehensive framework for understanding the dynamic interplay of molecular and cellular components underlying synaptic function, a crucial aspect in comprehending synaptic plasticity in health and disease. By integrating experimental data, computational models, and cutting-edge technology, systems biology offers a thorough analysis of the intricate networks controlling synaptic plasticity. It illuminates the coordinated molecular mechanisms supporting memory, learning, and cognitive functions in a healthy state. Importantly, systems biology reveals abnormalities in synaptic plasticity networks associated with diseases, particularly neurological disorders, shedding light on the molecular mechanisms behind conditions such as schizophrenia, Parkinson's disease, and Alzheimer's disease. Contributing to our knowledge regarding synaptic plasticity, systems biology clarifies the complex interactions among genes, proteins, and signaling pathways. It holds the potential to identify new therapeutic targets for neurological disorders and develop personalized medicine strategies based on individual molecular profiles [44].

11.12 CONCLUSION

By simulating the systems biology of synaptic plasticity, important insights have been obtained that have revolutionized our comprehension of the cellular and molecular principles underlying this basic process. Through sophisticated simulations and computational models, scientists have clarified the complex dynamics of synaptic plasticity networks. The significance of elements, including neurotransmitter release, receptor kinetics, and intracellular signaling pathways in determining synaptic strength, has been brought to light by these models. These models have been improved through the incorporation of experimental data from optogenetics, CRISPR-based methods, and high-throughput technologies. This has allowed for real-world validation and a more complex contemplation of synapse activity. Additionally, simulations have revealed the function of synaptic plasticity in memory, learning, and cognitive processes, offering a systems-level knowledge of how brain circuits are integrated. Significantly, these simulations have shown abnormalities in synaptic plasticity networks linked to a range of neurological illnesses, expanding our knowledge of diseases including schizophrenia and Alzheimer's. In summary, modeling the systems biology of synaptic plasticity has improved our notion of the normal neural function and may lead to personalized therapy based on the unique molecular profiles that underlie synaptic health and dysfunction.

Learning more about the systems biology of synaptic plasticity will have a significant impact on both healthy brain development and neurological illness dysfunction. Simulations provide a thorough analysis of the cellular and molecular mechanisms governing synaptic plasticity, which advances our knowledge of learning, memory, and cognitive functions. This information not only improves our basic comprehension of the healthy functioning of the brain, but it also offers important new perspectives on the abnormalities linked to neurological illnesses. Finding abnormal synaptic plasticity in diseases including schizophrenia, Parkinson's, and Alzheimer's provides insight into possible treatment targets. Contemplating the systems biology of synaptic plasticity will be important for formulating the development of the healthy brain as well as the malfunctioning of neurological illnesses. The cellular and molecular mechanisms behind synaptic plasticity are thoroughly understood thanks to simulations, which improve our knowledge regarding learning, memory, and cognitive processes. This knowledge not only advances our fundamental knowledge of how the brain functions normally but also provides significant new insights into the anomalies associated with neurological diseases. Discovering aberrant synaptic plasticity in conditions like Alzheimer's, Parkinson's, and schizophrenia offers information on potential therapeutic targets.

There is a strong call to action for more research, cooperation, and innovation as we stand at the nexus of computational neuroscience and systems biology. By stretching the boundaries of our knowledge, the dynamic interplay between these domains has already produced revolutionary insights into the intricacies of brain function and malfunction. Researchers, computational modelers, and experimentalists must work together to fully realize the potential of this interdisciplinary convergence by exchanging knowledge, techniques, and data. It is necessary to create and improve computational models that incorporate the most recent developments in optogenetics, CRISPR-based methods, and high-throughput technologies in addition to capturing the complexities of synaptic plasticity.

REFERENCES

1. Marcelo P Coba, Andrew J Pocklington, Mark O Collins, Maksym V Kopanitsa, Rachel T Uren, Sajani Swamy, Mike DR Croning, Jyoti S Choudhary, and Seth GN Grant. Neurotransmitters drive combinatorial multistate postsynaptic density networks. *Science Signaling*, 2(68):ra19–ra19, 2009.
2. Mark O Collins, Lu Yu, Marcelo P Coba, Holger Husi, Iain Campuzano, Walter P Blackstock, Jyoti S Choudhary, and Seth GN Grant. Proteomic analysis of in vivo phosphorylated synaptic proteins. *Journal of Biological Chemistry*, 280(7):5972–5982, 2005.
3. Gäel Daoudal and Dominique Debanne. Long-term plasticity of intrinsic excitability: Learning rules and mechanisms. *Learning & Memory*, 10(6):456–465, 2003.
4. Jeanette Hellgren Kotaleski and Kim T Blackwell. Modeling the molecular mechanisms of synaptic plasticity using systems biology approaches. *Nature Reviews Neuroscience*, 11(4):239–251, 2010.
5. Hiroaki Kitano. Systems biology: A brief overview. *Science*, 295(5560):1662–1664, 2002.
6. Princy Choudhary, Rameshwar Sankhala, and Sangeeta Singh. Integration of significant signaling pathways involved in neural-fate differentiation of stem cells. In *2023 10th International Conference on Computing for Sustainable Global Development (INDIACom)*, pp. 694–701. IEEE, 2023.
7. Mohammad Asif, Diya Srivastava, Aditya Gupta, and Uma Shanker Tiwary. Inter subject emotion recognition using spatio-temporal features from EEG signal. In *2023 27th International Computer Science and Engineering Conference (ICSEC)*, pp. 1–4. IEEE, 2023.
8. LF Abbott and Wade G Regehr. Synaptic computation. *Nature*, 431(7010):796–803, 2004.
9. Wickliffe C Abraham and Mark F Bear. Metaplasticity: The plasticity of synaptic plasticity. *Trends in Neurosciences*, 19(4):126–130, 1996.
10. Hojjat Adeli, Samanwoy Ghosh-Dastidar, and Nahid Dadmehr. Alzheimer's disease: Models of computation and analysis of EEGs. *Clinical EEG and Neuroscience*, 36(3):131–140, 2005.
11. Alan Aderem. Systems biology: Its practice and challenges. *Cell*, 121(4):511–513, 2005.
12. Yves Agid, György Buzsáki, David M Diamond, Richard Frackowiak, Jay Giedd, Jean-Antoine Girault, Anthony Grace, Jeremy J Lambert, Husseini Manji, Helen Mayberg, et al. How can drug discovery for psychiatric disorders be improved? *Nature Reviews Drug Discovery*, 6(3):189–201, 2007.
13. Mohammad Asif, Sudhakar Mishra, Majithia Tejas Vinodbhai, and Uma Shanker Tiwary. Emotion recognition using temporally localized emotional events in EEG with naturalistic context: Dens# dataset. IEEE Access, 11:39913–39925, 2023.
14. Avik Atta, Ayushi Gupta, Princy Choudhary, Shrey Dwivedi, and Sangeeta Singh. Inhibition of lingo1 as a therapeutic target to promote axonal regeneration and repair for neurological disorders. *3 Biotech*, 13(11):372, 2023.
15. Venkata Subba Rao Atluri, Melissa Hidalgo, Thangavel Samikkannu, Kesava Rao Venkata Kurapati, Madhavan Nair. Synaptic plasticity and neurological disorders in neurotropic viral infections. *Neural Plasticity*, 2015:138979, 2015.

16. Thomas M Bartol Jr, Cailey Bromer, Justin Kinney, Michael A Chirillo, Jennifer N Bourne, Kristen M Harris, and Terrence J Sejnowski. Nanoconnectomic upper bound on the variability of synaptic plasticity. *elife*, 4:e10778, 2015.
17. Mark F Bear and Robert C Malenka. Synaptic plasticity: LTP and LTD. *Current Opinion in Neurobiology*, 4(3):389–399, 1994.
18. Hagai Bergman and Günther Deuschl. Pathophysiology of Parkinson's disease: From clinical neurology to basic neuroscience and back. *Movement Disorders: Official Journal of the Movement Disorder Society*, 17(S3):S28–S40, 2002.
19. Kim T Blackwell and Joanna Jedrzejewska-Szmek. Molecular mechanisms underlying neuronal synaptic plasticity: Systems biology meets computational neuroscience in the wilds of synaptic plasticity. *Wiley Interdisciplinary Reviews: Systems Biology and Medicine*, 5(6):717–731, 2013.
20. Alexandra K Brooks and Thomas Gaj. Innovations in crispr technology. *Current Opinion in Biotechnology*, 52:95–101, 2018.
21. Dean V Buonomano. Decoding temporal information: A model based on short-term synaptic plasticity. *Journal of Neuroscience*, 20(3):1129–1141, 2000.
22. Vijay Chandra, Rajesh Pandav, Ramanan Laxminarayan, Caroline Tanner, Bala Manyam, Sadanand Rajkumar, Donald Silberberg, Carol Brayne, Jeffrey Chow, Susan Herman, et al. Neurological disorders. In Jamison DT, Breman JG, Measham AR, et al., editors. *Disease Control Priorities in Developing Countries.* 2nd edition. Washington (DC): The International Bank for Reconstruction and Development/ The World Bank; 2006. Chapter 32. Available from: https://www.ncbi.nlm.nih.gov/books/NBK11793/ Co-published by Oxford University Press, New York.
23. Kimberley F Tolias, Joseph G Duman, and Kyongmi Um. Control of synapse development and plasticity by rho gtpase regulatory proteins. *Progress in Neurobiology*, 94(2):133–148, 2011.
24. Princy Choudhary, Shivani Gupta, Richa Shukla, Ayushi Gupta, Sonu Pahal, and Sangeeta Singh. Regulation of neuronal repair and regeneration through inhibition of oligodendrocyte myelin glycoprotein (OMGP). *Journal of Biomolecular Structure and Dynamics*, 40(24):13936–13952, 2022.
25. Mario Stampanoni Bassi, Ennio Iezzi, Luana Gilio, Diego Centonze, and Fabio Buttari. Synaptic plasticity shapes brain connectivity: Implications for network topology. *International Journal of Molecular Sciences*, 20(24):6193, 2019.
26. Princy Choudhary, Amit Chaudhary, Sonu Pahal, Ayushi Gupta, and Sangeeta Singh. Network pharmacological evaluation of withania somnifera bioactive phytochemicals for identifying novel potential inhibitors against neurodegenerative disorder. *Journal of Biomolecular Structure and Dynamics*, 40(21):10887–10898, 2022.
27. Jianfeng Feng. *Computational Neuroscience: A Comprehensive Approach*. CRC press, 2003.
28. M. Asif, M.T. Vinodbhai, S. Mishra, A. Gupta, and U.S. Tiwary. Emotion recognition in VAD space during emotional events using cnn-gru hybrid model on eeg signals. In Zaynidinov, H., Singh, M., Tiwary, U.S., Singh, D. (eds) *Intelligent Human Computer Interaction*, IHCI 2022. Lecture Notes in Computer Science, vol 13741, 2023.
29. Jorrit J Hornberg, Frank J Bruggeman, Hans V Westerhoff, and Jan Lankelma. Cancer: A systems biology disease. *Biosystems*, 83(2–3):81–90, 2006.
30. Felice Iasevoli, Carmine Tomasetti, and Andrea de Bartolomeis. Scaffolding proteins of the post-synaptic density contribute to synaptic plasticity by regulating receptor localization and distribution: Relevance for neuropsychiatric diseases. *Neurochemical Research*, 38:1–22, 2013.
31. Robert S Zucker and Wade G Regehr. Short-term synaptic plasticity. *Annual Review of Physiology*, 64(1):355–405, 2002.
32. Xinyou Yin and Paul C Struik. Modeling the crop: From system dynamics to systems biology. *Journal of Experimental Botany*, 61(8):2171–2183, 2010.
33. Jin-Hui Wang, Gladys YP Ko, and Paul T Kelly. Cellular and molecular bases of memory: Synaptic and neuronal plasticity. *Journal of Clinical Neurophysiology*, 14(4):264–293, 1997.
34. Khalid Bashir Dar, Aashiq Hussain Bhat, Shajrul Amin, Bilal Ahmad Reshi, Mohammad Afzal Zargar, Akbar Masood, and Showkat Ahmad Ganie. Elucidating critical proteinopathic mechanisms and potential drug targets in neurodegeneration. *Cellular and Molecular Neurobiology*, 40:313–345, 2020.
35. Richa Shukla, Sonu Pahal, Ayushi Gupta, Princy Choudhary, Krishna Misra, and Sangeeta Singh. Modulation of gpcr receptors common to gut inflammatory diseases and neuronal disorders, alzheimer's and parkinson's diseases as druggable targets through withania somnifera bioactives: An in silico study. *Journal of Biomolecular Structure and Dynamics*, 41(10):4485–4503, 2023.
36. Levi B Wood, Ashley R Winslow, and Samantha Dale Strasser. Systems biology of neurodegenerative diseases. *Integrative Biology*, 7(7):758–775, 2015.

37. Christopher D Morrone, Mingzhe Liu, Sandra E Black, and JoAnne McLaurin. Interaction between therapeutic interventions for alzheimer's disease and physiological aβ clearance mechanisms. *Frontiers in Aging Neuroscience*, 7:64, 2015.
38. Panchanan Maiti, Jayeeta Manna, and Gary L Dunbar. Current understanding of the molecular mechanisms in parkinson's disease: Targets for potential treatments. *Translational Neurodegeneration*, 6:1–35, 2017.
39. Keith Tully and Vadim Y Bolshakov. Emotional enhancement of memory: How norepinephrine enables synaptic plasticity. *Molecular Brain*, 3:1–9, 2010.
40. Thomas C Südhof. Towards an understanding of synapse formation. *Neuron*, 100(2):276–293, 2018.
41. Armen R Kherlopian, Ting Song, Qi Duan, Mathew A Neimark, Ming J Po, John K Gohagan, and Andrew F Laine. A review of imaging techniques for systems biology. *BMC Systems Biology*, 2(1):1–18, 2008.
42. Cornelis J Stam. Modern network science of neurological disorders. *Nature Reviews Neuroscience*, 15(10):683–695, 2014.
43. Troy T Rohn, Nayoung Kim, Noail F Isho, and Jacob M Mack. The potential of crispr/cas9 gene editing as a treatment strategy for alzheimer's disease. *Journal of Alzheimer's Disease & Parkinsonism*, 8(3), 2018.
44. Ciro De Luca, Anna Maria Colangelo, Assunta Virtuoso, Lilia Alberghina, and Michele Papa. Neurons, glia, extracellular matrix and neurovascular unit: A systems biology approach to the complexity of synaptic plasticity in health and disease. *International Journal of Molecular Sciences*, 21(4):1539, 2020.

12 Machine Learning Approaches in Neurodegenerative Diseases

Anoop Kumar Yadav
National Institute of Technology
Patna, Bihar, India

Nilesh Kumar
University of Alabama at Birmingham
Birmingham, AL

*Anushree Tripathi**
National Institute of Technology
Patna, Bihar, India

12.1 INTRODUCTION

Neurodegenerative diseases (NDDs) encompass a harrowing spectrum of conditions that relentlessly erode the nervous system, with a particular onslaught on neurons within the central nervous system [1, 2]. This unforgiving degeneration paves the way for severe disabilities and, in the most extreme cases, an untimely demise. Amyotrophic lateral sclerosis, Huntington's disease (HD), Parkinson's disease (PD), multiple sclerosis (MS), and Alzheimer's disease (AD) are among the formidable adversaries that comprise this devastating array of disorders. Their insidious progression leaves an indelible mark, underscoring the urgent need for comprehensive solutions to combat these unrelenting foes. These conditions primarily affect older adults, but the incidence of these conditions is nonetheless not confined to the elderly and can manifest earlier. The prevalence of NDDs has notably increased in recent years, a trend expected to persist as the global population ages. The challenge in combating NDDs lies in the unknown origins of these diseases and the absence of definitive cures, rendering current treatments focused mainly on alleviating symptoms. The early identification and management of NDDs are crucial for mitigating health complications and reducing the risk of severe health outcomes.

NDDs are defined by the deterioration of neurons, leading to disorders that significantly impact the human body's functioning. Neurons, the fundamental units for transmitting signals in the body, facilitate communication within the nervous system through electrical and chemical means. The central nervous system (CNS) and peripheral nervous system (PNS), alongside the chemical control via the endocrine system, play pivotal roles in regulating bodily functions. Disruptions in neuronal signaling due to NDDs can impair specific body parts, leading to physical disorders. Early detection of these diseases is challenging, as symptoms often do not manifest until significant neuronal loss has occurred. The development of treatments has increasingly relied on imaging techniques like MRI and CT scans, utilizing artificial intelligence (AI) for diagnosis [3].

The detection of NDDs involves assessments by healthcare professionals or automated systems, each method having its own set of advantages and disadvantages. While manual assessments allow for expert discretion, early-stage ND symptoms can be subtle and difficult to distinguish [4]. Advances in deep learning (DL) offer promising avenues for the detection, classification, and management of NDDs, potentially easing the workload on healthcare professionals and minimizing human error [5, 6]. DL-based decision-making systems could significantly enhance disease detection and management efficiency. The vast amount of data generated in the ND field, from laboratory tests to patient reports, poses a challenge in data management and requires effective processing

DOI: 10.1201/9781003464648-14

methods. Modern clinics employ tools for data collection and analysis, integrating them with larger data systems for more reliable and efficient ND detection and diagnosis.

Despite the potential of DL in revolutionizing ND detection and classification, existing methods lack transparency, raising concerns about their interpretability and castability. Recent studies applying DL for ND detection and classification highlight the necessity of comprehensive review articles that summarize current research in the field [7]. This review article aims to evaluate DL-based methods for detecting and classifying various ND types, marking a significant contribution to the literature by encompassing ten ND types. This review is poised to assist research groups in ND detection and classification, outlining challenges and future research directions in this evolving field.

12.2 PARKINSON'S DISEASE

PD is a brain disorder that gets worse over time. It affects a part of the brain called the substantia nigra, which helps control movement. In Parkinson's, cells in this area die, leading to a shortage of a brain neurotransmitter called dopamine. Dopamine is like a messenger that helps our bodies move smoothly. Without enough dopamine, people with Parkinson's experience movement problems and other symptoms [8, 9].

Symptoms and Challenges in Diagnosis:

PD typically presents with a gradual onset of motor symptoms, including:

- **Rigidity**: Muscles feel stiff and inflexible, causing difficulty with movement.
- **Bradykinesia**: Slowness of movement, often accompanied by a reduced blinking rate and a shuffling gait.
- **Resting tremor**: A rhythmic trembling, usually affecting the hands or fingers, at rest.
- **Speech difficulties**: Slurred, hoarse, or weak speech due to impaired vocal cord control [10].
- **Facial masking**: Reduced facial expressions due to decreased muscle movement.
- **Postural instability**: Difficulty maintaining balance and a tendency to fall.

Diagnosing PD can be challenging as there is no single definitive test. Neurologists primarily rely on a detailed medical history and physical examination to assess symptoms and rule out other conditions. Imaging techniques like PET scans can sometimes help visualize the loss of dopamine neurons in the brain [4].

12.2.1 The Promise of Machine Learning in Parkinson's Disease

The limitations of traditional diagnosis methods highlight the potential of advanced technologies like machine learning (ML) and artificial intelligence (AI) in PD diagnosis and management. Recent research suggests that ML algorithms can analyze data such as speech patterns [10], handwriting, and movement patterns to detect subtle changes indicative of PD, even in the early stages [11]. This has the potential to improve diagnostic accuracy and facilitate earlier intervention.

12.3 ALZHEIMER'S DISEASE

AD is a brain disease that slowly worsens over time, robbing people of their memories and making it harder to think clearly. Unlike normal forgetfulness that comes with age, AD steadily gets worse, eventually making everyday tasks like getting dressed or eating a struggle [12].

12.3.1 Causes and Risk Factors

The exact cause of AD remains a mystery, but several factors are known to contribute to its development:

- **Genetics**: Certain genes, like APOE-ε4, increase the risk of AD. While not a guarantee, carrying these genes puts individuals at higher risk [12].
- **Lifestyle**: Several factors are known to increase the risk of AD, including lack of physical activity, high blood pressure, smoking, and unhealthy eating patterns [12].

Environmental factors: Exposure to toxins, head injuries, and air pollution may contribute to AD development [13].

12.3.2 Symptoms and Diagnosis

The hallmark symptoms of AD include:

Memory loss: Forgetting recent events, difficulty remembering names and faces, and problems with following instructions.
Language difficulties: Finding words, struggling to form complete sentences, and difficulty understanding spoken language.
Disorientation: Getting lost in familiar places, confusion about time and place.
Poor judgment: Making bad decisions about finances or safety.
Personality changes: Becoming withdrawn, irritable, or anxious.

According to the National Institute on Aging, there is currently no definitive test for AD. Diagnosis typically involves a combination of cognitive assessment, medical history, and sometimes brain imaging scans.

12.3.3 Treatment and Hope for the Future

Currently, AD has no cure, but certain medications can help alleviate symptoms and enhance the quality of life for many patients. Treatments may involve acetylcholinesterase inhibitors, which can help improve memory and cognitive functions, and memantine, which is used to manage behavioral symptoms.

Despite the challenges, research into AD is accelerating. Advancements in machine learning (ML) hold promise for earlier diagnosis based on language patterns, speech analysis, and gait assessments [14]. Additionally, researchers are actively investigating new medications and therapies aimed at slowing or halting disease progression.

AD presents a significant global health challenge, affecting millions worldwide. While the road ahead is long, ongoing research offers hope for improved diagnostic methods, more effective treatments, and ultimately, a cure for this devastating disease. Meanwhile significant advancement has been made regarding the detection of AD using ML techniques [15–18].

12.4 MULTIPLE SCLEROSIS

MS is a complex, chronic autoimmune disease that disrupts the CNS [19]. It's characterized by the deterioration of the myelin sheath, a fatty substance that insulates nerve fibers and facilitates rapid signal transmission. Damage to the myelin sheath disrupts communication between the brain, spinal cord, and other parts of the body, leading to a wide range of neurological symptoms [20].

12.4.1 Symptoms and Diagnosis

MS symptoms vary significantly depending on the affected nerve pathways. Common symptoms include:

Fatigue: An overwhelming tiredness that can significantly impact daily activities.
Muscle weakness and stiffness: Weakness in one or more limbs, often accompanied by muscle stiffness and spasticity.
Vision problems: Blurred vision, double vision, pain in the eye, and even vision loss.
Balance and coordination problems: Difficulty walking, maintaining balance, and coordinating movements.
Cognitive impairment: Problems with memory, concentration, and information processing.
Sensory disturbances: Numbness, tingling, or burning sensations in various parts of the body.
Speech problems: Slurred speech, difficulty finding words, or problems with voice control [20].

Diagnosing MS can be challenging due to the diverse nature of its symptoms. Neurologists typically rely on a combination of:

Detailed medical history: A thorough discussion of symptoms and any potential risk factors.
Neurological examination: Assessment of muscle strength, reflexes, coordination, and other neurological functions.
Imaging tests: MRI scans to visualize lesions in the brain and spinal cord that are indicative of MS damage.

12.4.2 Causes and Risk Factors

The exact cause of MS remains unknown. However, researchers believe it's a combination of genetic predisposition and environmental triggers:

Genetics: Certain genes increase the risk of MS, but a single gene mutation isn't enough to cause the disease.
Immune system dysfunction: In MS, the immune system erroneously targets and attacks the myelin sheath, leading to inflammation and subsequent damage.
Environmental factors: Vitamin D deficiency, smoking, and exposure to certain viruses may contribute to MS development [20].

12.4.3 Types of MS

There are five main types of MS, each with its own disease course:

Clinically Isolated Syndrome (CIS): An isolated neurological event that could potentially evolve into a complete case of MS.
Relapsing-Remitting MS (RRMS): This prevalent form is marked by recurrent episodes (relapses) succeeded by phases of partial or total recovery (remissions).
Primary Progressive MS (PPMS): A form of MS that is characterized by a continuous and gradual worsening of symptoms from the beginning, without distinct episodes of relapses and remissions.
Secondary Progressive MS (SPMS): Develops after an initial period of RRMS, when relapses become less frequent but disability gradually progresses [20].
Progressive-Relapsing MS (PRMS): A type of MS defined by an ongoing and steady decline in symptoms from the start, lacking clear periods of relapses and remissions.

While there is no cure for MS, treatment options can manage symptoms, prevent relapses, and slow disease progression. Research efforts are ongoing to unravel the mysteries of MS and develop new therapies that could potentially repair damaged myelin or halt the autoimmune attack. With continued research, there is hope for a brighter future for individuals living with MS [21, 22].

12.5 HUNTINGTON'S DISEASE

HD is a relentlessly progressive neurodegenerative disorder that casts a long shadow on families. It steals control, erodes memory, and disrupts emotions, leaving a devastating impact on individuals and their loved ones [5, 23]. HD primarily affects the CNS, particularly the basal ganglia and cerebral cortex, leading to a characteristic triad of symptoms: movement disorders, cognitive decline, and psychiatric disturbances [24].

Unlike most NDDs that strike later in life, HD often manifests in the prime of adulthood, typically between 30 and 50 years of age. The culprit behind this cruel onset is a faulty Huntington gene (HTT) inherited from a parent. This gene mutation results in the production of an abnormal huntingtin protein, which accumulates in brain cells, triggering a cascade of toxic effects [25].

HD's destructive dance unfolds through the relentless degeneration of vital brain regions:

Basal Ganglia: These structures are responsible for coordinating movement. Their deterioration leads to the hallmark symptom of HD—chorea, characterized by involuntary, jerky movements.

Cerebral Cortex: This region governs cognition, memory, and reasoning. HD-related damage disrupts these functions, leading to dementia and impaired judgment.

Cerebellum: This area controls balance and coordination. Cell loss in the cerebellum manifests as difficulties with walking, maintaining posture, and performing coordinated tasks.

Adding to the damage, HD disrupts the delicate balance of neurotransmitters in the brain.

Loss of GABA and acetylcholine: These neurotransmitters play a crucial role in muscle control. Their depletion contributes to movement problems.

Dopamine increase: While dopamine is essential for movement, excessive levels can paradoxically lead to abnormal movements and contribute to dementia.

Movement disorders: Chorea, the hallmark symptom, is accompanied by rigidity, tremor, and difficulty swallowing.

Cognitive decline: Memory loss, impaired reasoning, and difficulty concentrating become progressively worse.

Psychiatric disorders: Depression, anxiety, apathy, and even psychosis can significantly impact quality of life.

While there is currently no cure for HD, the discovery of the Huntington gene in 1983 marked a significant breakthrough. This discovery fueled research into the underlying mechanisms of the disease and opened doors for the development of potential therapies [26]. Ongoing research explores gene editing techniques, neuroprotective strategies, and gene silencing approaches, offering a glimmer of hope for individuals and families battling HD.

12.6 DOPAMINE CELL: STRUCTURE AND EFFECTS

Dopamine, a molecule with a seemingly simple structure, plays a surprisingly complex and crucial role in the human body. Found within the central nervous system, immune system, and renal system, dopamine acts as a chemical messenger, influencing a vast array of functions [27].

Dopamine's influence is most widely recognized within the central nervous system (CNS). Here, it acts as a neurotransmitter, influencing movement, perception, emotions, memory, and even prolactin secretion by the pituitary gland.

Dopamine transmission alterations link to neuropsychiatric conditions like ADHD, Tourette syndrome, schizophrenia, psychosis, depression. Moreover, NDDs, including Parkinson's, Huntington's, and MS, involve dopamine pathway dysfunction [27].

Dopamine's effects are mediated by specialized proteins called dopamine receptors. These receptors belong to the G protein-coupled receptor family and can be categorized into five subtypes (D1–D5) based on their structure and response to dopamine. Dopamine receptors can be further classified into two main groups: D1-like and D2-like families. D1-like receptors (D1 and D5) have a stimulating effect, increasing intracellular levels of a molecule called cyclic AMP (cAMP) by activating an enzyme called adenylate cyclase. Conversely, D2-like receptors (D2, D3, and D4) have an inhibitory effect, decreasing cAMP levels via the inhibition of adenylate cyclase [27].

12.6.1 Dopamine and Movement

In the CNS, dopamine plays a central role in regulating movement through its influence on the basal ganglia. The basal ganglia function like an orchestra conductor, coordinating movement signals.

Loss of dopamine disrupts this delicate balance, leading to motor dysfunction. The basal ganglia pathway can be simplified as two opposing pathways: direct and indirect. Dl receptor activation in the direct pathway increases GABA release, promoting movement. Conversely, D2 receptor activation in the indirect pathway inhibits GABA release, leading to decreased movement. A healthy balance between these pathways is crucial for normal movement. Disruptions in this balance can result in abnormal movements, including the characteristic bradykinesia (slowness of movement) observed in PD. While L-DOPA remains a cornerstone treatment for PD by replenishing dopamine levels, it can lead to side effects like Levodopa-induced dyskinesias (LID) characterized by involuntary movements. Recent research suggests that dopamine receptors may behave differently in disease states, highlighting the need for continued exploration of new treatment strategies.

Dopamine's reach extends beyond the CNS. In the immune system, dopamine interacts with immune cells in the spleen, bone marrow, and circulatory system, potentially modulating their activity. Within the kidneys, dopamine receptors on tubular cells influence blood flow, filtration rate, and sodium excretion, impacting blood pressure.

Overall dopamine's diverse roles throughout the body solidify its position as a master molecule. Understanding its intricate signaling pathways and interactions with different systems is critical for developing effective treatments for various disorders. Ongoing research continues to unravel the complexities of dopamine's actions, paving the way for a future where we can harness its power for improved health and well-being.

12.7 PLAQUE AND ACETYLCHOLINE IN ALZHEIMER'S DISEASE

Acetylcholine (ACh), a molecule present in the human brain, is a crucial neurotransmitter with a presence not just at the neuromuscular junction but also in the autonomic ganglia [28]. It acts beyond simple transmission, influencing how neurons respond to subsequent stimulation, a phenomenon known as neuromodulation. Within the central nervous system (CNS), ACh plays a diverse role, with models proposed to describe its complex actions [28].

Two main sources which contribute to brain Ach are projection neurons and local interneurons. Projection neurons, originating in nuclei scattered throughout the brain (laterodorsal tegmental areas, medial habenula, pedunculopontine tegmental nucleus, and basal forebrain complex), project long distances to influence faraway targets [29]. Local interneurons, on the other hand, operate within a smaller area, directly affecting nearby cellular targets [29]. ACh's critical function extends to activating motor neurons, essential for movement control in the nervous system [28].

The brain holds a network of cholinergic regions, each with distinct functionalities. These regions are vital for processes like arousal, concentration, memory, and learning. Disruptions within the brain's cholinergic system have been linked to memory-related disorders, with AD being a prime example [15]. AD is marked by the gradual decline of neuronal cells and the connections among them, defining it as a neurodegenerative condition. Abnormal protein deposits, known as plaques, are a hallmark feature of the disease. These plaques primarily consist of a protein fragment called amyloid beta (Aβ). The size and shape of these plaques can vary, and they originate from the breakdown of a larger protein called the Aβ-precursor protein. Another protein, tau, interacts with Aβ to form neurofibrillary tangles, which, along with plaques, are key diagnostic markers for AD [30]. Definitively diagnosing the disease often requires microscopic examination of brain tissue to confirm the presence of plaques and tangles.

12.8 MACHINE LEARNING METHOD OF NEURODEGENERATIVE DISEASES

12.8.1 Diagnosis of Parkinson's Disease

Diagnosing PD remains a complex challenge due to the absence of a single definitive test. Physicians must rely on a comprehensive evaluation to reach an accurate diagnosis. This evaluation involves assessing various aspects, including motor and non-motor functions, gait patterns, vocal quality, and brain imaging data. Specifically, neurologists may look for characteristic motor symptoms like bradykinesia (slowness of movement) and resting tremor. Examining gait patterns for abnormalities and analyzing vocal quality for changes like slurring can also provide valuable clues. In addition, advanced techniques like EEG for brainwave activity, high-order spectral analysis for specific frequency patterns, MRI scans for structural abnormalities, and SPECT imaging for dopamine function can offer further insights. By combining these assessments, doctors can increase the accuracy of diagnosing PD [31].

12.8.1.1 EEG Biomarker Based on Hjorth Feature

Researchers are actively exploring the use of biomarkers in diagnosing PD. One promising approach involves electroencephalography (EEG), a technique that measures electrical activity in the scalp. EEG offers a quantifiable way to assess brain function potentially relevant to Parkinson's. During an EEG evaluation, electrodes are placed on specific scalp locations to record voltage variations. Studies like the Pred+CT repository database have utilized 64 electrodes, including an additional AFz channel, with a sampling rate of 500 Hz. The raw EEG signals are then filtered using a band-pass filter (0.1 to 100 Hz) to focus on the relevant frequencies. Following this, preprocessing steps are employed to remove unreliable channels and eliminate unwanted artifacts from the data, often using a technique called independent component analysis (ICA). Once preprocessed, researchers can extract biomarkers from the EEG signal using Hjorth features. These features, categorized into activity, mobility, and complexity, provide insights into the amplitude and temporal characteristics of the EEG signal. By analyzing these features and comparing them between control groups and patients with and without medication (ON and OFF states), researchers can potentially identify patterns that differentiate healthy individuals from those with PD. This approach offers a non-invasive and potentially cost-effective way to aid in diagnosing PD. However, further research is needed to validate its effectiveness and establish it as a reliable diagnostic tool [7, 32].

The activity feature is given by following:

$$A_x = \sum_{n=1}^{N} [x[n] - \mu x]^2 / N$$

where

$x[n]$ is defined as discrete time signal sample
μ_x is statistical mean
N is length of signal

The mobility feature is given by the following:

$$M_x = sqrt\left(\left(var\ \dot{x}[n]\right) / \left(var\ x[n]\right)\right)$$

where,

$\dot{x}[n]$ is defined as a first derivative of the time signal
Var (.) is defined as a statistical variance operator

The complexity feature is defined as follows:

$$C_x = M\dot{x} / M_x$$

where M_x is defined as a signal mobility and $M\dot{x}$ defined as a mobility of its first derivative.

In the study by Coelho, Bruno Fonseca Oliveira, *et al.* EEG based biomarker has total 180 number of features obtained. The classification step is applied on obtained feature. The classification is done on MATLAB and learner app. There are three classifications namely SVM, KNN and Random Forest is performed. SVM gives the good result on CTRL*OFF group. KNN gives good result on CTRL*ON group. These result shows that Hjorth Feature better than the fast Fourier transform coefficient for identifying PD patient.

12.8.1.2 Metabolomics Data Biomarker Diagnosis

Metabolomics data markers diagnose Parkinson's better than physical symptoms. Metabolomics analyzes blood plasma and serum using cross-sectional studies. Techniques include GC-MS, CE-MS, and LC-MS with ionization modes. Machine learning approaches like random forests, boosting, discriminant analysis, regression, SVMs, and neural networks utilize metabolomics features. The neural network framework uses the CRANK-MS framework for accurate prediction of PD without a preselecting feature. CRANK-MS is good approach for predicting PD using blood plasma. The skin sebum metabolomics is also enhanced by the neural network approach. The metabolomics prediction model is also helpful for binary classification of diabetes, fatty liver disease, heart disease, and covid-19 disease [7].

12.8.1.3 Voice Analysis Diagnosis

Voice production is a multiplex motor activity carried out by humans in which more than hundred muscles are involved. Every vocal sound comes from many systems and subsystems like nasal activity, lungs, and the trachea. A voice disorder nowadays emerges as a most powerful tool for diagnosis of PD at an earlier stage. Hypokinetic dysarthria affects the breath-related problem and tremulousness, and an unstable voice gives the best result for diagnosis of the PD patient. For a diagnosis of PD, different voice-related features are measured. Table 12.1 shows the voice-related features with number of occurrences that is helpful for the diagnosis of PD. Based on these features, different machine learning classifiers are applied for the diagnosis of PD [11].

12.8.1.4 Multimodal Time Series Data for Parkinson's Disease Diagnosis

A combination of different types of data modalities shows their result to enhance the performance of a machine learning model. A multimodal time series combination helps for finding the disease. The PD diagnosis done via patient characteristics such as a bio sample, medical history, motor, and non-motor function data. Data is collected with the help of PPMI databases, which contain all the records of PD progression data. Multimodal time series covers five modalities: Subject Characteristics (SC), Bio-Samples (BS), Medication History (MH), Motor function (M), and Non-Motor (MN) function modalities. SC modalities belong to the non-time series data, and the rest of them belong to the time series data. The basic approach for better accuracy is to

TABLE 12.1
Frequency Feature Description with Number of Occurrences

Name	Feature Definition	Feature Description	Number of Occurrences
Jitter	—	F0 variation between cycle	62
Shimmer	—	Peak amplitude variation between cycle	60
HNR	Harmonic to noise ratio	Ratio between harmonic and non-harmonic content	59
F0	Fundamental frequency	Fundamental frequency estimated along with different approaches evaluated on frame	56
MFCC	Mel frequency cepstral coefficients	Obtained by inverse transformation of the Mel scaled spectrum of the signal	52
GNE	Glottal to noise excitation ratio	Extent of noise in speech with the help of linear and non-linear energy concepts	28
NHR	Noise to harmonic ratio	Ratio between noise	47

TABLE 12.2
Machine Learning Methods Used in Parkinson's Disease

Sl	Reference	Data Type	Method/Classifier	Accuracy
1	[32]	EEG	SVM, KNN, and RF	89.56 SVM(CTRL*OFF) 87.78 SVM(CTRL*ON) 89.56 SVM(OFF*ON)
2	[7]	Blood plasma and serum trace level	RF, extreme gradient boosting, linear discriminant analysis, logistics regression, SVM, neural network	AUC(ROC) – 0.843 ± 0.045 AUC(PR) – 0.896 ± 0.046 MCC – 0.530 ± 0.098
3	[11]	Voice data	Praat, DARTH-VAT	Most effective feature for PD speech impairment—MFCC, HNR, F0, Jitter, Shimmer, RPDE, DFP, DPI
4	[33]	PPMI	RF, LGBM, ET, SVC, SGD	94% – LGBM (for three class) 94% – RF (for four class)

determine the minimum set of features that are involved in classification process. The features with less involvement in PD diagnosis are identified and reduced using the XGBoost and SULOV (Synthetic Unlabeled Local Outlier Voting) approaches. This process helps in selecting the best set of features that provides more accurate results. Using this reduced set of features, various classifiers including SVM, RF, LR, DT, ET, GNB, LGBM, SGD, ADB, and KNN are applied for the purpose of classifying and predicting the progression of PD progression. These machine learning method performance is computed before and after optimization using Bayesian optimizer and compared between each other and extends the obtained framework by executing the three different XAI technique which include SHAP, LIME and SHAPASH. For the three-class case, LGBM obtained the maximum accuracy, and for the four-class case, RF gives the maximum accuracy [33].

Details of machine learning methods used in PD are shown in Table 12.2.

12.8.2 Diagnosis of Alzheimer's Disease

AD is very common NDD in older age. It is important to diagnosis AD before it goes into the advanced stage. There are various challenges to diagnosing AD. It is very helpful to find the following: the exact cause of memory loss, findings of thinking rate, good judgment accuracy, level of communication, frequency of mood change and behavior, self-care activity, biomarkers of brain tissue, ratio of brain volume, renal system status, and immune system status.

12.8.2.1 Cascade Machine Learning Pipeline

AD is a NDD in which cognitive decline of memory occurs. Early diagnosis is very important for preventing it from going to an advanced stage. There are various neuroimaging biomarker methodologies that can give reliable prediction information for the diagnosis of AD. MRI is one of them. In MRI, the brain's image biomarker is used for diagnosis of brain volume structure like hippocampus atrophy or ventricle enlargement. In cascade machine learning, pipelining hippocampal atrophy is used for the diagnosis of AD. In this approach, first, coronal T1 weighted image data is collected which contain AD affected and healthy people. Coronal T1 weighted image data is preprocessed by linear interpolation with respect to vocal size and resolution and then image gray scale normalization and crop process of the image is done. Secondly, segmentation is performed on the preprocessed image by the encoder and decoder [15]. The decoder used deep supervised methodology for improving the performance. Augmentation techniques such as translation, rotation, and flipping are applied to increase data sample. To update the gradient, the Adam optimizer is used. After segmentation of the image, the radiomics feature is extracted. The extracted features are three types: first order feature, morphological feature, and higher order texture feature. In all the three obtained feature types, we get many features. Only some of these features are responsible for the diagnosis of AD. So, we need to remove the rest of the features. For this purpose, we use the hierarchical clustering method and least absolute shrinkage and selection operator. Highly correlated features are removed by the hierarchical clustering method, and irrelevant and redundant features are removed by the least absolute shrinkage and selection operator. Then the different types of classifiers such as gaussian naïve bayes, random forest, support vector machine, and AdaBoost are applied on radiomic feature for diagnosis of AD in which the support vector machine gives the best performance.

12.8.2.2 Novel Approach Utilizing Machine Learning

The human brain comprises billions of neurons which are interconnected to each other and transmit information from one to another. When protein structures, which are also known as plaques and tangles, build between neurons, they break the connections and cause the death of neurons. These protein structures play an important role for the diagnosis of AD. There are various ways to measure disease; for example, covid-19 can be measured by X-ray and tumors can be measured by MRI. The study by *Uddin, Khandaker Mohammad Mohi, et al.* shows that a person with a larger ratio of brain volume has a smaller change of being diagnosed with AD and a person with a small ratio of brain volume has a greater chance of being diagnosed with AD [16]. The dataset utilized is the Open Access Series of Imaging Studies (OASIS). The description of dataset along with score function of each attribute is shown in Table 12.3.

The SelectKBest method is applied to identify the appropriate feature, and f class is used as a scoring function. Both classification and regression can be used by changing the score function. The machine learning approaches Gaussian NB, Decision Tree, Random Forest, Voting classifier, XGBoost and Gradient Boost are applied on these attributes. Gaussian NB gives the best accuracy among all the machine learning approaches.

12.8.2.3 Brain Tissue-Based Biomarker

Brain tissues are a collection of many types of cells including neurons. For diagnosis of AD using a brain tissue-based biomarker, firstly three AD GEO data set, GSE5281, GSE48350 and GSE1297 which contain AD and healthy control are used. Further, using six ranking algorithms such as maximum neighborhood component, degree, maximal clique centrality, closeness centrality, betweenness centrality, and stress centrality, the hub genes are identified from the PPI network. In all these

TABLE 12.3
Dataset along with Score Function of Each Attribute

Sn	Attribute	Description	Score Value
1	Visit	Number of visits during study	3.234565
2	MR delay	Delay	6.627021
3	M/F	Male/Female	13.413107
4	Age	Age in years	3.132662
5	EDUC	Education in years	13.485719
6	SES	Social and economic status	11.530848
7	MMSE	Mini-mental state examination	107.216426
8	CDR	Clinical dementia rating	436.908989
9	eTIV	Estimated total incremental volume	0.674019
10	nWBV	Normalize total brain volume	20.581034
11	ASF	Atlas scaling factor	0.381442

brain tissue parameters some features are important for the computation of AD and some features are not important for the computation of AD. So, we need to identify those features that are important for computation and remove the rest. This can be done with the help of LASSO and the ridge feature selection method. In the study by Alamro, Hind, *et al.* five overlapped genes are are found to be overlapping using LASSO and ridge feature selection method. [17].

12.8.2.4 Multimodal Neuroimaging Data

Cardinal clinical symptoms and protein biomarkers of disease are sometimes applied for classification purpose of NDD. The pathological accumulation of tau protein and amyloid-beta leads to tauopathies, with AD being classified under this category due to its protein-based pathology. Amyloid-beta, which can accumulate in the brain up to 20 years prior, is a significant factor in diagnosing AD, while tau protein buildup occurs around 15 years before diagnosis. Consequently, the detection of these proteins is crucial for identifying AD. Various studies have employed multimodal neuroimaging data for classification purposes, utilizing methods such as sparse multitask learning to differentiate between AD and healthy controls. Techniques like FDG-PET and CSF measurement have been integrated with the sparse hierarchical extreme learning machine fusion method to diagnose AD and distinguish it from healthy controls [18].

Details of machine learning methods used in AD are shown in Table 12.4.

12.8.3 Diagnosis of Multiple Sclerosis Disease

MS is a frequently seen NDD that is characterized by multiple plaques in the spinal cord [34]. There is no permanent treatment available, so there is a need for early diagnosis.

12.8.3.1 Transfer-Transfer Model with MSNet Using Deep Learning

This approach is used for the diagnosis of MS and normal myelitis. In myelitis, spinal cord lesions extending beyond three vertebral lengths were seen, and in MS, spinal cord lesions extending beyond two vertebral lengths were seen. A neurological image data may have of three diseases classes such as MS, myelitis, control patient. The MSNet model utilizes the new deep feature extraction methodology which amount to its main innovation. The feature extraction can be done using two pre-trained networks: Dense Net 201 and Resnet50. These networks are highly accurate, which gives good accuracy and performance. The obtained feature using densenet201 and ResNet50 was in large number in size in which only some of the feature involved in MS diagnosis. So, we can select only those features that are involved in MS diagnosis. For this purpose, a different feature section approach, such as ReliefF, Chi2, or neighborhood component analysis, is used. ReliefF assesses the contribution of features to classification performance by utilizing label information; the Chi2 feature selector measures the dependency between features; and neighborhood component analysis is applied for proximity information to optimize the classification

TABLE 12.4
Machine Learning Methods Used in Alzheimer's Disease

Sn	Reference	Data Type	Method/Classifier	Accuracy
1	[15]	Coronal T1 weighted image dataset	Gaussian naïve Bayes, RF, SVM, AdaBoost	95%
2	[16]	OASIS dataset	GNB, XGBoost, DT, RF, GB, Voting	96%
3	[17]	Three AD GEO dataset (brain tissue based)	RF, SVM, DNN, CNN	Prediction performance-AUC-0.979
4	[18]	Image	ML method, Fusion level of Abstraction, Explainability method	Better classify Alzheimer stages

performance of features and chooses the best feature out of them. Then the support vector machine and K-nearest neighbor classifier are applied to the selected feature. Post-processing is done on obtained results from the classifier and then an iterative majority voting scheme is applied to obtain the optimal performance [21].

12.8.3.2 Deep Learning Approach for Predicting Multiple Sclerosis Disease

Deep learning models, powered by artificial intelligence, play a pivotal role in the diagnosis of multiple sclerosis (MS). The hidden layer is instrumental in predicting overfitting and reducing model complexity. The deep learning approach for MS diagnosis unfolds as follows: Firstly, data can be imported from the public repository such as NCBI GEO, specifically the GSE17048 dataset [35]. In the study by Riveros, Carlos, *et al.* the mRNA expression profiling of 74 genes was acquired from 144 individuals, comprising 99 with MS and 45 healthy controls (GPL6947). The GEO2R tool facilitated the analysis of expression values, permitting gene expression visualization through the profile graph tab. The data preprocessing stage involved standardization, a feature normalization technique that centers the mean and scales the unit variance. Following standardization, a dimensionality reduction technique known as recursive feature elimination with cross-validation (REFCV) was applied to select the most important features for MS diagnosis. REFCV recursively selects features by considering smaller and smaller subsets, eliminating irrelevant features. Its objective is to identify the optimal number of features through a cross-validation loop [36]. The deep learning method is capable of presenting complex functions using a composition of simple functions. Neurons are the basic building blocks for complex function. The described architecture of the artificial neural network model is illustrated below [35], featuring 144 individuals with 35 neurons in the input layer, chosen through the RFECV method. The model includes a single dense hidden layer with 106 hidden neurons, utilizing "tanh" as the activation function, and is accompanied by a dropout layer with a dropout rate of 0.1. The output is processed through a second dense layer with a sigmoid activation function, which receives inputs from the dropout layer and categorizes them into either MS or healthy control groups. [22]. An artificial neural network with a single hidden layer gives the better accuracy as compared to the other four conventional learning techniques as shown in Table 12.5 (Table 12.6).

TABLE 12.5
Performance Result by Classifier

Sn	Classifier	Accuracy without Feature Selection	Accuracy with Feature Selection
1	k-neighbors	0.6896	0.8620
2	Gaussian naïve Bayes	0.7931	0.7931
3	C-support vector	0.7586	0.7931
4	Decision tree	0.6551	0.6896
5	ANN1	0.7931	0.8965

TABLE 12.6
Machine Learning Methods Used in Multiple Sclerosis Disease

Sn	Reference	Dataset	Method/Classifier	Accuracy
1	[21]	Magnetic resonance image dataset	Support vector machine, k-nearest neighbor, voting scheme	97.63%
2	[22]	GSE17048 microarray dataset	k-neighbors, Gaussian naïve Bayes, C-support vector, decision tree, ANN1	89.65% (ANN1)

12.9 CONCLUSION

The global incidence of neurodegenerative disorders is escalating, with PD, AD, MS, and HD among the most common. The hallmark of PD includes reduced dopamine levels coupled with elevated acetylcholine levels, in addition to the buildup of α-synuclein proteins playing a role in its progression. Neurotransmitter imbalances are a fundamental aspect in the development of these NDDs. AD is influenced by both genetic and non-genetic factors, with therapies aimed at improving cognitive function. In the case of MS, monitoring the myelin sheath's condition is crucial to prevent its progression. Once the myelin sheath is significantly damaged, leading to advanced stages of MS, recovery to its original state becomes unattainable, resulting in severe neurodegenerative effects. Symptoms play a vital role in identifying NDDs, manifesting as diverse changes in the body. Data collection involves various types of information related to disease regions and characteristics of healthy individuals versus those affected.

The application of machine learning and deep learning approaches has significantly enhanced the diagnosis of NDDs. While definitive treatments for these conditions are lacking, leveraging machine learning and deep learning for early-stage diagnosis proves invaluable in managing NDDs effectively. These advanced technologies offer promising avenues for early detection and intervention in neurodegenerative disorders, paving the way for improved patient outcomes and enhanced healthcare strategies.

REFERENCES

1. Chaki, J., and M. Woźniak, *Deep learning for neurodegenerative disorder (2016 to 2022): A systematic review*. Biomed Signal Process Control, 2023. **80**: p. 104223.
2. Przedborski, S., M. Vila, and V. Jackson-Lewis, *Neurodegeneration: What is it and where are we?* J Clin Invest, 2003. **111**(1): p. 3–10.
3. Schmidt-Erfurth, U., et al., *Artificial intelligence in retina*. Prog Retin Eye Res, 2018. **67**: p. 1–29.
4. Aybek, S., and D.L. Perez, *Diagnosis and management of functional neurological disorder*. BMJ, 2022. **376**: p. o64.
5. Tautan, A.M., B. Ionescu, and E. Santarnecchi, *Artificial intelligence in neurodegenerative diseases: A review of available tools with a focus on machine learning techniques*. Artif Intell Med, 2021. **117**: p. 102081.
6. Myszczynska, M.A., et al., *Applications of machine learning to diagnosis and treatment of neurodegenerative diseases*. Nat Rev Neurol, 2020. **16**(8): p. 440–56.
7. Zhang, J.D., et al., *Interpretable machine learning on metabolomics data reveals biomarkers for Parkinson's disease*. ACS Cent Sci, 2023. **9**(5): p. 1035–45.
8. Jankovic, J., *Parkinson's disease: Clinical features and diagnosis*. J Neurol Neurosurg Psychiatry, 2008. **79**(4): p. 368–76.
9. Poewe, W., et al., *Parkinson disease*. Nat Rev Dis Primers, 2017. **3**: p. 17013.
10. Sakar, B.E., et al., *Collection and analysis of a parkinson speech dataset with multiple types of sound recordings*. IEEE J Biomed Health Inform, 2013. **17**(4): p. 828–34.
11. Amato, F., et al., *Machine learning- and statistical-based voice analysis of Parkinson's disease patients: A survey*. Expert Syst Appl, 2023. **219**: p. 119651.
12. Alzheimer's Disease Facts and Figures. *Alzheimer's Disease facts and figures*. 2024 [cited 2024; Available from: https://www.alz.org/alzheimers-dementia/facts-figures.
13. Chin-Chan, M., J. Navarro-Yepes, and B. Quintanilla-Vega, *Environmental pollutants as risk factors for neurodegenerative disorders: Alzheimer and Parkinson diseases*. Front Cell Neurosci, 2015. **9**: p. 124.
14. Diogo, V.S., et al., *Early diagnosis of Alzheimer's disease using machine learning: A multi-diagnostic, generalizable approach*. Alzheimer's Res Ther, 2022. **14**(1): p. 107.
15. Zhou, K., et al., *A novel cascade machine learning pipeline for Alzheimer's disease identification and prediction*. Front Aging Neurosci, 2022. **14**: p. 1073909.
16. Uddin, K.M.M., et al., *A novel approach utilizing machine learning for the early diagnosis of Alzheimer's disease*. Biomed Mater Devices, 2023. **1**: p. 1–17.
17. Alamro, H., et al., *Exploiting machine learning models to identify novel Alzheimer's disease biomarkers and potential targets*. Sci Rep, 2023. **13**(1): p. 4979.

18. Odusami, M., et al., *Machine learning with multimodal neuroimaging data to classify stages of Alzheimer's disease: A systematic review and meta-analysis.* Cognitive Neurodynamics, 2023. 18(3): p. 775–794.
19. Dobson, R., and G. Giovannoni, *Multiple sclerosis - a review.* Eur J Neurol, 2019. **26**(1): p. 27–40.
20. The National Multiple Sclerosis Society. *National Multiple Sclerosis Society.* March 12, 2024 [cited 2024 March 20, 2024]; Available from: https://www.nationalmssociety.org.
21. Tatli, S., et al., *Transfer-transfer model with MSNet: An automated accurate multiple sclerosis and myelitis detection system.* Expert Syst Appl, 2024. **236**: p. 121314.
22. Ponce de Leon-Sanchez, E.R., et al., *A deep learning approach for predicting multiple sclerosis.* Micromachines, 2023. **14**(4): p. 749.
23. Margolis, R.L., and C.A. Ross, *Diagnosis of Huntington disease.* Clin Chem, 2003. **49**(10): p. 1726–32.
24. Walker, F.O., *Huntington's disease.* Lancet, 2007. **369**(9557): p. 218–28.
25. Ross, C.A., and S.J. Tabrizi, *Huntington's disease: From molecular pathogenesis to clinical treatment.* Lancet Neurol, 2011. **10**(1): p. 83–98.
26. Gusella, J.F., et al., *A polymorphic DNA marker genetically linked to Huntington's disease.* Nature, 1983. **306**(5940): p. 234–8.
27. Vaughan, R.A., and J.D. Foster, *Mechanisms of dopamine transporter regulation in normal and disease states.* Trends Pharmacol Sci, 2013. **34**(9): p. 489–96.
28. Angus, J.A., et al., *The acetylcholine paradox: A constrictor of human small coronary arteries even in the presence of endothelium.* Clin Exp Pharmacol Physiol, 1991. **18**(1): p. 33–6.
29. Berman, J.A., D.A. Talmage, and L.W. Role, *Cholinergic circuits and signaling in the pathophysiology of schizophrenia.* Int Rev Neurobiol, 2007. **78**: p. 193–223.
30. Grimm, M.O., et al., *Neprilysin and abeta clearance: Impact of the APP intracellular domain in NEP regulation and implications in Alzheimer's disease.* Front Aging Neurosci, 2013. **5**: p. 98.
31. Bhat, S., et al., *Parkinson's disease: Cause factors, measurable indicators, and early diagnosis.* Comput Biol Med, 2018. **102**: p. 234–41.
32. Coelho, B.F.O., et al., *Parkinson's disease effective biomarkers based on hjorth features improved by machine learning.* Expert Syst Appl, 2023. **212**: p. 118772.
33. Junaid, M., et al., *Explainable machine learning models based on multimodal time-series data for the early detection of Parkinson's disease.* Comput Methods Programs Biomed, 2023. **234**: p. 107495.
34. Transverse Myelitis Consortium Working Group. *Proposed diagnostic criteria and nosology of acute transverse myelitis.* Neurology, 2002. **59**(4): p. 499–505.
35. Riveros, C., et al., *A transcription factor map as revealed by a genome-wide gene expression analysis of whole-blood mRNA transcriptome in multiple sclerosis.* PLoS One, 2010. **5**(12): p. e14176.
36. Artur, M., *Review the performance of the Bernoulli naïve Bayes classifier in intrusion detection systems using recursive feature elimination with cross-validated selection of the Best number of features.* Procedia Comput Sci, 2021. **190**: p. 564–70.

Index

Note: Page numbers in **bold** and *italics* refer to tables and figures, respectively.

I

J

K

L

M

N

O

P

Q

R

S

T